The Mechanics of Solids

Photograph ©2000 by Doug Baker

The Mechanics of Solids
History and Evolution

A Festschrift in Honor of Arnold D. Kerr

Edited by Michael H. Santare
and
Michael J. Chajes

DELAWARE

Newark: University of Delaware Press

Associated University Presses
2010 Eastpark Boulevard
Cranbury, NJ 08512

The paper used in this publication meets the requirements of the American National Standard for Permanence of Paper for Printed Library materials Z39.48-1984.

PRINTED IN THE UNITED STATES OF AMERICA

Contents

Symposium Papers

Preface

The *Arnold D. Kerr Engineering Mechanics Symposium*, held on April 29 and 30, 2004, at the University of Delaware, focused on the state of the art of this field through invited lectures by distinguished experts in Mechanics.

For many years, Arnold talked about his vision for an Engineering Mechanics Colloquium. His idea was to invite several speakers a year to come to campus, and deliver a lecture, or short series of lectures, on a topic in Mechanics. The talks would not focus on the speakers' current research, this is what seminar series are for. Instead, the speakers would be asked to survey the evolution of a topic in Mechanics of Solids. They would present this survey, along with their own perspectives on the subject. Then, there would be ample time set aside for discussion.

We have always agreed that this was an excellent idea. After all, this kind of intellectual exchange is what the university is all about. However, other university business was always a higher priority and the colloquium idea was repeatedly put on the back burner.

When Arnold announced his retirement, we decided it would have to be now or never. This was the colloquium Arnold had long envisioned … compressed into a day and a half.

We were excited by the idea of bringing together many of the pioneers in this field to share their wealth of accumulated knowledge with others working in the area of engineering mechanics. We decided that the most effective way to do this was to have Professor Kerr compile a list of potential topics and speakers. As expected, he chose major figures in the field: our final program included seven members of the National Academy of Engineering, two of whom are also members of the National Academy of Sciences.

While the topic of engineering mechanics has been covered in other meetings, the *Kerr Symposium* was unique in featuring only invited talks by recognized leaders in the field, whose contributions span decades of research, teaching, and practice. The *Kerr Symposium* was also different from other conferences and symposia on engineering mechanics in featuring survey-type reviews of the field rather than covering specific problems.

Representing academia, government, and industry, our internationally known speakers addressed topics ranging from contact mechanics, nondestructive evaluation of structures, ice mechanics, stability of structures, engineering of railway tracks and concrete pavements, sandwich structures, biomechanics and biomaterials, fracture and dislocation mechanics and carbon nanotechnology and applied mathematics. Their talks are compiled in these proceedings, which we hope will enhance your understanding and enjoyment of engineering mechanics and prove valuable in your pursuit of further knowledge on this broad and fundamental topic.

In closing, we would like to thank the many people who helped to make the *Kerr Symposium* a resounding success: the faculty organizers and session chairs; the speakers, who were willing to travel here to share their scientific and engineering wisdom; the very capable staff of the Department of Civil and Environmental Engineering and the University's Laird Conference Center; Eric Kaler, Dean of the College of Engineering; and Tsu-Wei Chou, Chair of the

Department of Mechanical Engineering, without whose support the symposium would not have been possible; and, last but not least, Professor Kerr himself who was very helpful with advice and encouragement. A pioneer in the field of engineering mechanics, he collaborated with, taught, and learned from most of the symposium speakers during his long career.

We also greatly appreciate the financial support provided by the College of Engineering, the Department of Civil and Environmental Engineering, and the Department of Mechanical Engineering.

Michael H. Santare and Michael J. Chajes
Proceedings Co-editors

Biography of Arnold Kerr

Arnold D. Kerr earned his Dipl. Ing. degree in Civil Engineering in 1952 at the Technical University of Munich. He continued his education at Northwestern University, in Evanston, Illinois, where he completed his MS degree in Mechanics in 1956 and his PhD in Theoretical and Applied Mechanics in 1958. He spent his postdoctoral year, from 1958 to 1959, at Courant's Institute of Mathematical Sciences, New York University, doing research in the non-linear theory of elasticity and stability theory.

Subsequently, he joined the faculty of the Department of Aeronautics and Astronautics of New York University. He was Assistant Professor from 1959, Associate Professor from 1961, and professor from 1965 to 1973. He received tenure in 1965.

At NYU, he planned and built the Laboratory for Engineering Mechanics and was appointed its Director. In the spring of 1973, when plans were made to merge the New York University School of Engineering and Science with the Polytechnic Institute, he decided not to join the merged school and accepted an invitation to go to Princeton University as a Visiting Professor.

At Princeton, in addition to teaching engineering mechanics courses, he developed a research program in railway track engineering. Dr. Kerr organized the *International Symposium on Railroad Track Mechanics and Technology* in 1975 at Princeton and was editor of the proceedings, published by Pergamon Press in 1978. He was co-organizer (with A. L. Kornhauser) of the *Symposium on Productivity in U.S. Railroads* in 1978 and was co-editor of the proceedings, published by Pergamon Press in 1980.

In September 1978, Arnold Kerr became Professor of Civil Engineering at the University of Delaware, where he taught courses in engineering mechanics and railway track engineering and also built a strong graduate program in these areas.

In cooperation with his wife, Berta, he founded the Institute for Railway Engineering in Wilmington, Delaware, in 1978. The Institute has been offering short courses on railroad engineering to practicing engineers who work for freight and transit railroads, engineering consulting companies, federal and state DOTs, and the U.S. armed forces.

Dr. Kerr is the author of over 100 papers in journals and conference proceedings, coauthor (with V. A. Squire, R. J. Hosking, and P. J. Langhorne) of *Moving Loads on Ice Plates* (Kluwer Academic, 1996), and author of *Fundamentals of Railway Track Engineering* (Simmons-Boardman, 2003). He has served as a reviewer for various journals in engineering mechanics, ice engineering, railway engineering, and applied mathematics.

During his career, Prof. Kerr advised 16 PhD students and some 30 MS students. Many of his former students and advisees have gone on to distinguished careers in academia and industry, including several who are members of the National Academy of Engineering.

Publications of Arnold Kerr

Books

Fundamentals of Railway Engineering, by A. D. Kerr (in preparation).

Moving Loads on Ice Plates, by V. A. Squire, R. J. Hosking, A. D. Kerr, and P. J. Langhorne, Kluwer Academic Publishers, London/Dordrecht/Boston, 1996.

Productivity in U. S. Railroads, Proceedings of a Symposium held at Princeton University, A. D. Kerr and A. L. Kornhauser, Eds., Pergamon Press, 1980.

Railroad Track Mechanics and Technology, Proceedings of a Symposium held at Princeton University, A. D. Kerr, Ed., Pergamon Press, 1978.

Papers

(with M. Andrews) "Generalized Analysis of Railroad Tracks," submitted for publication.

(with A. Babinski) "Rail Creep According to the Johnson Hypothesis," submitted for publication.

(with G. Grissom) "Analysis of Lateral Track Buckling Using the New Frame-Type Equations," International Journal of Mechanical Sciences, Vol. 48, 2006.

"On the Determination of Rail Support Modulus k," *International Journal of Solids and Structures*, Vol. 37, 2000 (reprinted in AREMA Proceedings, 2002).

(with L. Bathurst) "A Method for Upgrading the Performance of Track Transitions for High-Speed Service, FRA Report, September 2001.

(with L. Bathurst) "Pads Ease Track Transitions, *Railway Track and Structures*, August 2000.

(with J. E. Cox) "Analysis and Tests of Rail Joints Subjected to Vertical Wheel Loads," *International Journal of Mechanical Sciences*, Vol. 41, 1999.

(with L. Bathurst) "An Improved Analysis for the Determination of Required Ballast Depth,*" Proceedings of the American Railway Engineering and Maintenance-of-Way Association*, Annual Conference on Track and Structures, 916–47, 1999.

(with L. Bathurst) "Zur Berechnung Querkraftbelasteter Phähle in Kohösiven Boden" (A Method for the Analysis of Laterally Loaded Piles in a Soil with Cohesion), *Bautechnik*, Vol. 76, Heft 12, 1999.

(with D.D. Pollard) "Toward More Realistic Formulations of the Analysis of Laccoliths," *Journal of Structural Geology*, Vol. 20, No. 12, 1998.

"The Assessment of Concrete Pavement Blowups," *ASCE Journal of Transportation Engineering*, No. 3/4, 1997.

(with A. Babinski) "Einfluss des Axialwiderstandes auf die Stabilität eines Kontinuierlich Gelagerten Stabes" (The Effect of Axial Resistance on the Buckling of Continuously Supported Beams, in German), *Bautechnik*, 1997.

(with A. Babinski) "Rail Travel (Creep) Caused by Moving Wheel Loads," *Proceedings of the American Railway Engineering Association*, Vol. 98, 1997.

(with M. Dalaei) "Natural Vibration Analysis of Clamped Rectangular Orthotropic Plates," *Journal of Sound and Vibration*, Vol. 189, No. 3, 1996. Erratum, Vol. 192, No. 5, 1996.

(with N. E. Soicher) "On a Peculiar Set of Problems in Linear Structural Mechanics," *International Journal of Solids and Structures*, Vol. 33, No. 6, 1996.

"Analysis of Plates that Rest on Expansive Subgrades," *Bautechnik*, Vol. 73, Heft 7, 1996.

"Bearing Capacity of Floating Ice Covers Subjected to Static, Moving, and Oscillatory Loads," *Applied Mechanics Reviews*, Vol. 49, No. 11, 1996.

(with M. Dalaei) "Analysis of Clamped Rectangular Orthotropic Plates Subjected to a Uniform Lateral Load," *International Journal of Mechanical Sciences*, Vol. 37, No. 5, 1995.

"Blowup of Concrete Pavement Adjoining a Rigid Structure," *International Journal of Non-Linear Mechanics*, Vol. 29, No. 3, 1994.

"On Peculiar Problems in Structural Mechanics" *Proceedings of the R. H. Scanlan Symposium*, The Johns Hopkins University Press, 1994.

"The Evolution of Foundation Models and Analyses for Concrete Pavements," *Proceedings of the Third International Workshop on the Design and Evaluation of Concrete Pavements*, Krumbach, Austria, 1994.

(with S. S. Kwak) "The Clamped Semi-Infinite Plate on a Winkler Base Subjected to a Vertical Force," *Quarterly Journal of Mechanics and Applied Mathematics*. Vol. 46, Part 3, 1993.

(with S. S. Kwak) "The Semi-Infinite Plate on a Winkler Base, Free Along the Edge, and Subjected to a Vertical Force," *Archive of Applied Mechanics* (formerly *Ingenieur Archiv*), Vol. 63, 1993.

(with F. D. Haynes and C. R. Martinson) "Effect of Oscillatory Loads on the Bearing Capacity of Floating Ice Sheets," *Cold Regions Science and Technology*. Vol. 21, 1993.

"Mathematical Modeling of Airport Pavements," in *Proceedings of ASCE Specialty Conference, Airport Pavement Innovation—Theory to Practice*, J. W. Hall, Jr., Ed., 1993.

(with B. E. Moroney) "Track Transition Problems and Remedies," *Proceedings of the American Railway Engineering Association*, Vol. 94, 1993.

"Berechnung Seitlich Belasteter Pfähle in Kohäsionslosen Boden," (On the Analysis of Laterally Loaded Piles in a Cohesionless Soil. In German.) *Die Bautechnik*, Vol. 69, Heft 2, 1992.

(with A. W. Eberhardt) "The Stress Analysis of Railroad Tracks with a Nonlinear Base Response," *Rail International*, No. 3, 1992.

"Upgrading Engineering Education," (for railroads), *Progressive Railroading*, No. 4, 1992.

(with D. W. Coffin) "Beams on a Two-Dimensional Pasternak Base Subjected to Loads that Cause Lift-Off," *International Journal of Solids and Structures*, Vol. 28, No. 4, 1991.

(with D. W. Coffin) "On Membrane and Plate Problems for which the Linear Theories are Not Admissible," *Journal of Applied Mechanics*, Vol. 57, No. 1, 1990.

(with M. A. El-Sibaie) "Green's Functions for Continously Supported Plates," *Zeitschrift für Angewandte Mathematik und Physik* (ZAMP), Vol. 40, 1989.

"Tests and Analyses of Footings on a Sand Base," *Soils and Foundations*, Vol. 29, No. 3, 1989.

"Additional Comments on Buckling Analyses of Embedded Layers," *Tectonophysics*, Vol. 169, 1989.

(with F. D. Haynes) "On the Determination of the Average Young's Modulus for a Floating Ice Cover," *Cold Regions Science and Technology*, Vol. 15, 1988.

"On the Buckling of Slender Piles," *Soils and Foundations*, Vol. 28, No. 2, 1988.

"Stability of a Water Tower," *Ingenieur-Archiv*, Vol. 58, 1988.

(with M. A. El-Sibaie) "Validation of New Equations for Dynamic Analyses of Tall Frame-Type Structures," *International Journal of Earthquake Engineering and Structural Dynamics*, Vol. 15, 1987.

(with F. D. Haynes) "The Effect of Oscillatory Loads on the Bearing Capacity of Floating Ice Covers," *Cold Regions Science and Technology*, Vol. 13, 1987.

(with M. L. Accorsi) "Numerical Validation of the New Track Equations for Static Problems," *International Journal of Mechanical Sciences*, Vol. 29, No. 1, 1987.

(with M. G. Donley) "Thermal Buckling of Curved Railroad Tracks," *International Journal of Non-Linear Mechanics*, Vol. 22, 1987.

(with M. A. El-Sibaie) "On the New Equations for the Lateral Dynamics of a Rail-Tie Structure," *Journal of Dynamic Systems, Measurement, and Control*, ASME, Vol. 109, 1987.

"A Few Concluding Remarks," *Mechanical Engineering*, ASME, 1987, p. 2.

"On the Vertical Modulus in the Standard Railway Track Analyses," *Rail International*, November, 1987.

(with R. B. Pipes) "Why We Need Hands-On Engineering Education," *Technology Review*, MIT, October 1987.

"Response of Floating Ice Beams and Plates with Partial Flooding," *Ice Technology, Proceedings of the First International Conference*, MIT, Cambridge, Mass., Springer-Verlag: Berlin, New York, 1986.

(with A. M. Zarembski) "On the New Equations for the Cross-Tie Track Response in the Lateral Plane," *Rail International*, No. 6, 1986.

(with G. E. Frankenstein) "Ice Cover Research—Present State and Future Needs," *Cold Regions Engineering, Proceedings of the Fourth International Conference*, W. L. Ryan, Ed., ASCE, 1986.

"The Real Crisis in Engineering Education," *Proceedings of the American Railway Engineering Association*, Vol. 87, Bulletin 706, 1986. Reprinted in *Indiana Civil Engineer*, ASCE, Summer 1986.

"A Question of Balance," Editorial on Engineering Education in *Mechanical Engineering*, ASME, July 1986.

"On the Buckling Analyses of Embedded Layers," *Tectonophysics*, Vol. 128, 1986.

(with D. A. Stafford) "Ice Forces on a Circular Pier Due to Water Level Changes," *Proceedings of the Eighth IAHR Symposium on Ice*, Vol. III, Iowa City, IA, 1986.

(with H. W. Shenton III) "Railroad Track Analyses and Determination of Parameters," *Journal of Engineering Mechanics*, ASCE, Vol. 112, No. 11, 1986.

"Analysis of Floating Ice Covers with Partial Flooding," *Cold Regions Science and Technology*, Vol. 10, 1985.

(with W. A. Dallis Jr.) "Blowup of Concrete Pavements," *Journal of Transportation Engineering*, ASCE, Vol. 111, No. 1, 1985.

(with M. L. Accorsi) "Generalization of the Equations for Frame-Type Structures: A Variational Approach," *Acta Mechanica*, Vol. 56, 1985.

"On the Determination of Foundation Model Parameters," *Journal of Geotechnical Engineering*, ASCE, Vol. 111, No. 11, 1985.

(with H. W. Shenton III) "On the Reduced Area Method for Calculating the Vertical Track Modulus," *Proceedings of the American Railway Engineering Association*, Vol. 86, 1985.

"Analysis of Piles Frozen-In to an Ice Cover and Subjected to Forces That Cause Pile Bending," *Proceedings of the IAHR Symposium on Ice*, Vol. II, Hamburg, Germany, 1984.

(with P. J. Shade) "Analysis of Concrete Pavement Blowups," *Acta Mechanica*, Vol. 52, 1984.

"On the Formal Development of Elastic Foundation Models," *Ingenieur Archiv*, Vol. 54, 1984.

"Mechanics of Ice Cover Breakthrough," *Proceedings of the Ice Penetration Technology Workshop*, W. F. Weeks, Ed., Cold Regions Research and Engineering Laboratory, Hanover, New Hampshire, 1984.

"The Critical Velocities of a Load Moving on a Floating Ice Plate That Is Subjected to In-Plane Forces," *Cold Regions Science and Technology*, Vol. 6, 1983.

"On the Method for Determining the Track Modulus Using a Locomotive or Car on Multi-Axle Trucks," *Proceedings of the American Railway Engineering Association*, Vol. 84, 1983.

(with P. J. Shade) "Analysis of Concrete Pavement Blowups," *Proceedings, ASCE, Engineering Mechanics, Recent Advances in Engineering Mechanics and Their Impact on Civil Engineering Practice*, Vol. II, 1983.

(with P. J. Shade) "An Analytical Approach to Concrete Pavement Blowups," *TRB Transportation Research Record*, 1983.

(with S. B. Bassler) "Effect of Rail Lift-Off on the Analysis of Railroad Tracks," *Rail International*, Nr. 10, 1982.

"Continuously Supported Beams and Plates Subjected to Moving Loads—A Survey," *Solid Mechanics Archives*, Vol. 6, Issue 4, December, 1981.

(with A. M. Zarembski) "The Response Equations for a Cross-Tie Track," *Acta Mechanica*, Vol. 40, 1981.

"Remarks to the Buckling Analyses of Floating Ice Sheets," *Proceedings* of the IAHR International Symposium on Ice, Quebec, Canada, 1981.

"On the Buckling Force of Floating Ice Plates," *Physics and Mechanics of Ice*, Per Tryde, Ed., Springer Verlag: Berlin, Heidelberg, New York, 1980.

"An Improved Analysis for Thermal Track Buckling," *International Journal of Non-Linear Mechanics*, Vol. 15, 1980.

"Improved Stress Analysis for Cross-Tie Tracks," *Engineering Mechanics*, EM4, ASCE, August 1979.

"On the Unbonded Contact Between Elastic and Elastic-Rigid Media," *Acta Mechanica*, Vol. 33, 1979.

"On Thermal Buckling of Straight Railroad Tracks and the Effect of Track Length on its Response," *Rail International*, No. 9, 1979.

"A New Iterative Scheme for the Solution of Partial Differential Equations Based on the Principle of Least Squares," SIAM *Journal on Applied Mathematics*, Vol. 34, 1978.

"On the Determination of Horizontal Forces a Floating Ice Plate Exerts on a Structure," *Journal of Glaciology*, Vol. 20, 1978.

"Analysis of Thermal Track Buckling in the Lateral Plane," *Acta Mechanica*, Vol. 30, 1978.

(with Y. El-Aini) "Determination of Admissible Temperature Increases to Prevent Vertical Track Buckling," *Journal of Applied Mechanics*, Vol. 45, 1978.

"Thermal Buckling of Straight Tracks: Fundamentals, Analyses, and Preventive Measures," *Proceedings of the American Railway Engineering Association*, Vol. 80, 1978.

"Lateral Buckling of Railroad Tracks Due to Constrained Thermal Expansions - A Critical Survey," *Railroad Track Mechanics and Technology*, Proceedings of a Symposium, A. D. Kerr, Ed., Pergamon Press, 1978.

"An Indirect Method for Evaluating Certain Infinite Integrals," *Zeitschrift fur Angewandte Mathematik und Physik* (ZAMP), Vol. 29, 1978.

"Forces an Ice Cover Exerts on Rows or Clusters of Piles Due to a Change of the Water Level," *Proceedings of the Fourth IAHR International Symposium on Ice Problems*, Vol. I, Luleå, Sweden, 1978.

"On the Adjacent Equilibrium Method in the Stability Theory of Conservative Elastic Continua," *International Journal of Non-Linear Mechanics*, Vol. 12, 1977.

"Problems and Needs in Track Structure Design and Analysis," Track Systems and Other Related Railroad Topics, *TRB Transportation Research Record* 653, National Academy of Sciences, Washington, D.C., 1977.

"On the Derivation of Well Posed Boundary Value Problems in Structural Mechanics," *International Journal of Solids and Structures*, Vol. 12, No. 1, 1976.

"The Effect of Lateral Resistance on Track Buckling Analyses," *Rail International*, No. 1, 1976.

"The Bearing Capacity of Floating Ice Plates Subjected to Static or Quasi-Static Loads," *Journal of Glaciology*, Vol. 17, No. 76, 1976.

"On the Dynamic Response of a Prestressed Beam," *Journal of Sound and Vibration*, Vol. 49, No. 4, 1976.

"On the Stress Analysis of Rails and Ties," *Proceedings of the American Railway Engineering Association*, Vol. 78, October 1976.

"Principles and Criteria for the Design of a Railroad Track Test Facility," *Proceedings of the American Railway Engineering Association*, Vol. 77, 1975.

"Ice Forces on Structures Due to Change of Water Level," *Proceedings of the IAHR Third International Symposium*, Ice Problems, Hanover, New Hampshire, November 1975.

(with R. Brantman) "On the Non-Existence of Adjacent Equilibrium States at Bifurcation Points," *Acta Mechanica*, Vol. 23, No. 1–2, 1975.

"On the Stability of the Railroad Track in the Vertical Plane," *Rail International* (Journal of the International Railway Congress Association, Brussels, Belgium), No. 2, 1974.

"The Stress and Stability Analyses of Railroad Tracks," *Journal of Applied Mechanics*, Vol. 41, No. 4, December 1974.

(with L. El-Bayoumy) "On the Stability of a Shallow Arch," *Acta Mechanica*, Vol. 18, 1973.

"A Model Study for Vertical Track Buckling," *High Speed Ground Transportation Journal*, Vol. 7, No. 3, 1973.

(with W. T. Palmer) "The Deformation and Stresses in Floating Ice Plates," *Acta Mechanica*, Vol. 15, 1972.

"The Continuously Supported Rail Subjected to an Axial Force and a Moving Load," *International Journal of Mechanical Sciences*, Vol. 14, 1972.

(with L. El-Bayoumy) "On the Nonunique Equilibrium States of a Shallow Arch Subjected to a Uniform Lateral Load," *Quarterly of Applied Mathematics*, October 1970.

(with H. Alexander) "Recommendations for a Revision of the Balloon Specifications MIL-P-4640 ANUSAFE," *Proceedings of the Sixth AFCRL Scientific Balloon Symposium*, L. A. Grass, Ed., 1970.

"Buckling of Continuously Supported Beams," *Engineering Mechanics*, ASCE, February 1969.

"On the Strength of High Altitude Balloons," *Facilities for Atmospheric Research*, No. 9, June 1969.

"An Extended Kantorovich Method for the Solution of Eigenvalue Problems," *International Journal of Solids and Structures*, Vol. 5, 1969.

(with M. T. Soifer) "The Linearization of the Prebuckling State and Its Effect on the Determined Instability Loads," *Journal of Applied Mechanics*, December 1969.

"Balloon Strength in the Troposphere as Affected by Creep at Launch," *Proceedings of the Fifth AFCRL Scientific Balloon Symposium*, L. A. Grass, Ed., 1968.

"An Extension of the Kantorovich Method," *Quarterly of Applied Mathematics*, July 1968.

(with H. Alexander) "An Application of the Extended Kantorovich Method to the Stress Analysis of a Clamped Rectangular Plate," *Acta Mechanica*, Vol. 6, 1968.

(with S. Tang) "The Instability of a Rectangular Elastic Solid," *Acta Mechanica*, Vol. IV/1, 1967.

(with R. S. Becker) "The Stress Analysis of Plates Sealing a Compressible Liquid," *International Journal of Mechanical Sciences*, Vol. 9, 1967.

"Experimental Study of Balloon Material Failures," Proceedings of the Fourth AFCRL Scientific Balloon Symposium, 1967.

"On Plates Sealing an Incompressible Liquid," *International Journal of Mechanical Sciences*, Vol. 8, 1966.

(with T. v. Kárrmán) "Instability of Spherical Shells Subjected to External Pressure," *Collected Works of Theodore von Kármán* 1952–1963, von Kármán Institute for Fluid Dynamics, Rhode-St-Genese, Belgium (1975), 292–311. Reprinted paper from "Topics in Applied Mechanics," Elsevier, 1965.

(with T. U. Myint) "The Stability of Core-Filled Long Cylinders Subjected to Uniform Outside Pressure," *International Journal of Mechanical Sciences*, Vol. 7, 1965.

"A Study of a New Foundation Model," *Acta Mechanica*, Vol. 1, No. 2, 1965.

"Bending of Circular Plates Sealing an Incompressible Liquid," *Journal of Applied Mechanics*, September 1965.

(with T. von Kármán) "Instability of Spherical Shells Subjected to External Pressure," *Topics in Applied Mechanics*, Elsevier Publishing Company, 1965.

"On Creep Failure of Balloons," *Proceedings of the AFCRL Scientific Balloon Workshop*, A. O. Korn, Ed., 1965.

(with S. Tang) "The Effect of Lateral Hydrostatic Pressure on the Instability of Elastic Solids, Particularly Beams and Plates," *Journal of Applied Mechanics*, 1965.

"Elastic and Viscoelastic Foundation Models," *Journal of Applied Mechanics*, Vol. 31, September 1964.

"Elastic Plates on a Liquid Foundation," *Engineering Mechanics*, ASCE, June 1963.

"On the Instability of Circular Plates," *Journal of Aerospace Sciences*, April 1962.

"On the Instability of Elastic Solids," *Proceedings*, 4th U.S. National Congress of Applied Mechanics, 1962.

"Viscoelastic Winkler Foundation with Shear Interactions," *Engineering Mechanics*, ASCE, June 1961.

"Uniformly Stretched Plates Subjected to Concentrated Transverse Forces," *Quarterly Journal of Mechanics and Applied Mathematics*, Oxford, November 1960.

"A Study of the Effect of the Capillary Zone on the Flow Through Homogeneous Earth Dams," *Geotechnique*, London, June 1959.

Abstracts of Lectures in Conference Proceedings

"Analyses of Floating Ice Covers Subjected to Moving Loads," 1st International Conference on Industrial and Applied Mathematics, Invited lecture, MS 37, Paris, France, 1987.

"Effect of Partial Flooding on the Response of Floating Plates," Proceedings of the 10th U. S. National Congress of Applied Mechanics, Austin, TX, 1986.

"Thermal Buckling of Concrete Pavements," Proceedings of the 20th Annual Meeting, Society of Engineering Science, 1983.

"Analysis of Thermal Track Buckling," Proceedings of the 14th International Conference on Theoretical and Applied Mechanics, Delft, Holland, 1976.

"An Extension of the Kantorovich Method," Proceedings of the 5th National Congress of Applied Mechanics, 1966.

Symposium Schedule

1:00–1:10   **Welcoming Remarks**
Eric W. Kaler
Dean, College of Engineering, University of Delaware

Michael J. Chajes
Chair, Department of Civil & Environmental Engineering
University of Delaware

Tsu-Wei Chou
Chair, Department of Mechanical Engineering, University of Delaware

1:10–1:25   **Symposium Overview**
Michael H. Santare
Associate Professor, University of Delaware

SESSION 1:   **MECHANICS AND ITS APPLICATIONS I**
Session Chair, Harry W. Shenton III
Associate Professor, University of Delaware

1:30–2:00   **The Evolution of Contact Mechanics**
Leon M. Keer
Chaired Professor, Northwestern University, and Member, National Academy of Engineering

2:10–2:40   **Crack Detection by Laser-Based Techniques**
Jan D. Achenbach
Chaired Professor, Northwestern University, and Member, National Academy of Engineering and National Academy of Sciences

2:50–3:20   **Engineering Mechanics of Structures That Talk Back: Biomechanics and Biomaterials**
Harold Alexander
President, Orthogen Corporation

SESSION 2:   **STABILITY OF STRUCTURES**
Session Chair, Michael D. Greenberg
Professor, University of Delaware

3:40–4:10   **The Evolution of Buckling Modes in Continuously Supported Structures: From Eigenfunctions to Actual Buckled Shapes**
Alan Needleman
Chaired Professor, Brown University, and Member, National Academy of Engineering

4:20–4:50   **History and Lessons of a Paradox Associated with Follower Forces**
Isaac E. Elishakoff
Professor, Florida Atlantic University

5:00–5:30   **Compression-Induced Ductile Failure of Brittle Materials and Brittle Failure of Ductile Materials: Experiments and Modeling**
Sia Nemat-Nasser
Chaired Professor, University of California–San Diego, and Member, National Academy of Engineering

6:45–9:00   **Dinner and Remarks**
Charles H. Thornton
Member, National Academy of Engineering, Chairman and Managing Principal, the Thornton-Tomasetti Group, Inc.
Harry Armen
President, ASME

SESSION 3:   **RAILWAY TRACK ENGINEERING AND MECHANICS**
Session Chair, Robert McCown
Director, Federal Railroad Administration

9:00–9:30   **Engineering of Railway Tracks**
Allan M. Zarembski
President, Zeta-Tech Associates

9:40–10:10   **Track-Train Interaction: Recent Advances**
Magdy El-Sibaie
Chief, Track Research Division, Federal Railroad Administration

SESSION 4:   **MECHANICS OF CONTINUOUSLY SUPPORTED PLATES**
Session Chair, Jaqueline Richter–Menge
U.S. Army Cold Regions Research and Engineering Laboratory

10:40–11:10   **Design Ice Forces and Fracture Scaling**
John P. Dempsey
Professor, Clarkson University

11:20–11:50   **Fracture Analysis and Size Effects in Failure of Sea Ice**
Zdeněk P. Bažant
Chaired Professor, Northwestern University and Member, National Academy of Engineering and National Academy of Sciences

12:00–12:30   **Follow That White Concrete Road: The Evolution of Concrete Pavement Analysis**
Anastasios M. Ioannides
Associate Professor, University of Cincinnati

SESSION 5: **MECHANICS AND ITS APPLICATIONS II**
Session Chair, Michael J. Chajes
Professor, University of Delaware

1:30–2:00 **History of Sandwich Structures and Their Analyses**
Jack R. Vinson
Chaired Professor, University of Delaware

2:10–2:40 **The Evolution of the Concept of Green's Function and Its Use in the Mechanics of Solids**
Ivar Stakgold
Professor Emeritus, University of Delaware

2:50–3:20 **Dislocations as Green's Functions in Plane Elasticity**
John Dundurs
Professor Emeritus, Northwestern University
Michael H. Santare
Associate Professor, University of Delaware

3:30–4:00 **On the Mechanical Behavior of Composite Nanotubes in Hexagonal Arrays**
R. Byron Pipes
Chaired Professor, University of Akron, and Member, National Academy of Engineering

4:00 **Concluding Remarks**
Michael J. Chajes and Arnold D. Kerr

Symposium Remarks by Michael J. Chajes, Chair, Department of Civil and Environmental Engineering, University of Delaware

Arnold Kerr was born Aronek Kierszkowski in Poland on March 9, 1928. His family was "traditionally Jewish but not fanatically religious." They maintained a kosher home, and his parents spoke to each other in Yiddish but to their children in Polish. Arnold's father ran a fur business, both import and export.

The family lived in Suwalki, a town that was itself isolated and one where the Jews were isolated further. Arnold recalls that like the rest of Poland, Suwalki was a very anti-Semitic place.

Arnold Kerr as a young child

Arnold Kerr with his parents

In 1939, when the war broke out in Europe, Kerr and his mother and brothers became refugees, fleeing to Wilno, a border town, to escape the Germans and the Russians. Unfortunately, his father went to Warsaw with three loads of furs, believing that he could still conduct business as usual. Arnold never saw him again, and he learned later that his father had been shot at a place near Warsaw where concentration camp guards were trained. As Arnold says, his father marched "straight into the mouth of the dragon."

Although Arnold and his mother and brothers thought they were safe in Wilno, by 1941, the town was occupied by Germans. Degrading pronouncements about acceptable behavior were issued daily, and a gradual thinning of the Jewish population began, with groups of people marched out of town each day and murdered. The Jews that remained in Wilno, like Kerr and his family, were packed into a medieval ghetto, where they were often hungry and forced to scrounge for food.

In 1943, the Germans liquidated the ghetto of Wilno, sending the men and older boys to labor camps. As Arnold and two of his brothers marched toward the ghetto gate, they passed their mother and youngest brother. That was the last time they saw each other. Arnold learned later that his mother and two of his brothers had been gassed at Auschwitz.

In 1944, Arnold ended up alone in a little-known concentration camp in a tiny village called Stutthof at the southern end of the Baltic Sea. His oldest brother—and last remaining relative—had been shot in Estonia. He was depressed, a dangerous state to be in when he needed to have all of his wits about him just to survive. But Arnold attributes his survival to a biblical experience—he heard the voice of an aunt calling his name. Although he never saw her or heard her voice again, the incident brought him out of his depression and back to life.

But he admits that his survival was also a matter of luck. There were no mass killings at Stutthof, but Arnold had one very close call. After his arrival there, all of the teenagers—516 of them—were collected to be sent to another camp, allegedly to be reunited with their relatives. The cattle cars brought to take them away could carry only 500 people. Arnold was one of only 16 randomly picked to stay. The other 500 were all taken to Auschwitz and gassed.

Arnold Kerr as a young man

In the bitter winter of 1945, Arnold was condemned, along with about 250,000 others, to the infamous death marches. Although he was dressed in thin ragged pajamas, Arnold managed to get a pair of boots in trade for some bread. He and others also found sugar and sausages on a derailed train that they were sent to put back on track. He filled the boots with sugar and rationed it to himself at the rate of two teaspoonfuls a day. At the end of the war, he still had some.

The death march brought Arnold and his group to a place called Rieben in West Prussia. They slept on church floors and were given little to eat by their Ukrainian guards but were not otherwise seriously abused. By March of 1945, Arnold sensed that things were beginning to change, but liberation came gradually rather than all of a sudden.

The day before his seventeenth birthday, the camp commandant, Miesel, gathered up the inmates and took them West. However, the march was more or less voluntary, and Arnold and his friends decided to not to go. Again, luck was with him. Although he could have been killed for his choice, he and the rest of the group were left unharmed and were fortunate enough to find food in a warehouse that the SS and the Ukrainian guards had left behind. They began to eat, one bite at a time to avoid death from sudden overeating after starvation.

They soon left the camp at the urging of the Russians who had come in after Miesel left. With the war over, things had turned around to the extent that the Germans were now afraid of the Jewish survivors. Arnold and his friends were actually fed and offered the use of a bathhouse by Germans.

Arnold intended to return home to Poland, but mistaken for liberated Russian POWs, he and the rest of the group he was traveling with were arrested by the Soviet NKVD. Amid the confusion, Arnold escaped and made his way to Berlin.

Although Arnold never saw his father again after his father's fateful decision to go to Warsaw in 1939, Arnold discovered that he had been left a valuable legacy. Only two weeks before the war had broken out, Arnold's father had been in London and from his transactions there had left

$75,000 in England. Although it was in sterling and had been severely devalued, Arnold received about $15,000 after the war was over and used the money to get an education.

He graduated from the Technical University of Munich in 1952 with a degree in Civil Engineering. After graduation, Arnold stayed at the University for another year, doing research at the Soil Mechanics Laboratory. He arrived in the United States on December 29, 1954, and began a new life when he created a new name for himself: Arnold D. Kerr. The "D," he points out, does not stand for anything. When challenged on this by an immigration officer, Arnold pointed out that if Harry S. Truman could do it, so could he, even if he wasn't President of the United States.

Upon arriving in the U.S., Arnold worked for about a year at Hazelet & Erdal, Consulting Engineers in Chicago, as a bridge design engineer. He continued his education at Northwestern University, in Evanston, Illinois, where he completed his MS degree in Mechanics in 1956 and his PhD in Theoretical and Applied Mechanics in 1958. He spent his postdoctoral year, 1958 to 1959, at Courant's Institute of Mathematical Sciences, New York University, doing research in the non-linear theory of elasticity and stability theory.

Subsequently he joined the faculty of the Department of Aeronautics and Astronautics of New York University, where he stayed until 1973. At NYU, he planned and built the Laboratory for Engineering Mechanics in the Department of Aeronautics and Astronautics and was appointed its Director. In the spring of 1973, when plans were made to merge the New York University School of Engineering and Science with the Polytechnic Institute, he decided not to join the merged school and accepted an invitation to go to Princeton University as a Visiting Professor.

In September 1978, Arnold Kerr became Professor of Civil Engineering at the University of Delaware. He quickly established premier programs in railroad engineering and engineering mechanics and built a strong graduate program in structural mechanics and structural engineering.

Arnold Kerr with Pioneering Researcher Theodore von Karman

Dr. Kerr is the author of over 100 papers in journals and conference proceedings and a book, Fundamentals of Railway Engineering, published by Simmons-Boardman in 2003. He has served as a reviewer for various journals in engineering mechanics, ice engineering, railway engineering, and applied mathematics. He is credited with organizing several international symposia and with establishing the Institute for Railroad Engineering in Wilmington, Delaware. During his career at UD, Prof. Kerr advised 16 PhD students and some 30 MS students.

***Arnold Kerr with His Wife Berta and
Children Orin and Regina***

Many of his former students and advisees have gone on to distinguished careers in academia and industry, including several who are now members of the National Academy of Engineering and have joined us for this symposium. One of them, Charles Thornton, will share his thoughts and memories of Arnold with us next.

You may have noticed that I've said little about Arnold's personal life after he came to the United States. I'm leaving that story for his daughter Regina to tell as she narrates the slide show that we've put together with photos of Arnold's life from 1928 to the present. The slide show will follow Thornton's remarks.

Symposium Remarks by Charles Thornton, Former Chairman and Managing Principal, the Thornton-Tomasetti Group Inc.

Introduction of Charles Thornton by Dr. Michael J. Chajes

Dr. Charles Thornton, a former student of Professor Kerr, was chairman and managing principal of the Thornton-Tomasetti Group Inc. headquartered in New York City. An expert in the area of building collapse and structural failure analysis, Mr. Thornton has participated in engineering team investigations of the causes of the collapse of not only the World Trade Centers in New York City, but also the Hartford Coliseum roof and the New York State Freeway Schoharie Creek Bridge. He also contributed to the federal government's assessment of the bombing of the Alfred Murrah Building in Oklahoma City. A graduate of Manhattan College with a bachelor of science degree in Civil Engineering, Dr. Thornton received his masters and PhD degrees in Civil Engineering from NYU. He is a registered professional engineer in twenty-six states. During his nearly forty years with the Thornton-Tomasetti Group, he was involved in the design and construction of projects all over the world, including the world's tallest buildings, the Petronas Twin Towers of Kuala Lumpur in Malaysia, as well as the New York City Hospital, Chicago's Comiskey Park, the fifty-story Americas Towers in New York, the sixty-five-story One Liberty Place building in Philadelphia, and the fifty-story Chifley Tower in Sydney, Australia. Many of these projects have set industry standards for innovative thinking and creativity. In 1997, Dr. Thornton was elected to the National Academy of Engineering on the basis of his significant contributions to the design of major structures worldwide. He is also the past recipient of the National Institute of Building Sciences honor award, which he received for his efforts in developing a nationwide program to attract high school students to the building industry. In 2003, he was a Franklin Institute Nobel Laureate; about 50% of Nobel laureates are Nobel Prize Winners. His partner, Richard Tomasetti, a member of the National Academy of Engineering, was also one of Professor Kerr's students at NYU.

Remarks by Dr. Charles Thornton

Thank you, it is a pleasure to be here this evening. When they closed NYU's engineering school, I lost track of Professor Arnold Kerr. Four years ago, I moved to the Eastern Shore of Maryland. Shortly thereafter, I was asked to serve on the External Advisory Council of the Civil and Environment Engineering Department here at Delaware. Then, two years ago, I was asked to come to campus to give a presentation on the World Trade Center collapse. It was on that visit that I reconnected with the greatest professor I ever had, Dr. Kerr.

Let me give you a little background on my graduate education at NYU. I grew up in the Heights, up by the Hall of Fame in the West Bronx. NYU was a great institution. It had great interdisciplinary activities. It had outstanding programs in civil, structural, environmental, aeronautical, mechanical engineering, etc. It also had the Courant Institute and the downtown campus. Dr. James Michael, chairman of the Civil Engineering Department at the time I was enrolled, once told me: "Charlie, you can go anywhere you want." And in 1962, I went over to the aeronautical program and took the Theory of Elastic Stability from Dr. Kerr, and here are my notes! Dr. Kerr also taught me the Theory of Plates and Shells in 1963, and again, here are the notes. I don't use them much anymore (perhaps I should donate them), but these are the notes that during the first half of my career, I used all of the time. Dr. Kerr taught me two things: He said you don't need a slide rule or a calculator when you take my course, you need to know how to formulate the problem, how to state the boundary conditions. And then he said, "Although I love math, hire a mathematician to solve the equations." He made me think about what engineering mechanics, solid mechanics, plates and shells, and structural engineering was all about. No one had ever done that before. Arnold, I love you.

Today, we heard Dr. Harry Alexander talk about a tooth implant, brilliant! In Kuala Lumpur we put down seventy teeth under each of the world's tallest buildings. They were one meter by two meters by one hundred forty meters deep, the world's deepest foundation. My dream, and the dreams of many others, were made possible by Dr. Kerr. In Kuala Lumpur, we did exactly the same thing as Dr. Alexander did with the tooth—we looked at how the loads got from the building into the pile and then how they got into the soil mass. We conducted nonlinear finite element modeling, which I could not have done if I didn't have the background I have. We also cast 800 by 700 foot floor plates below grade in Kuala Lumpur, 560,000 square feet of concrete with no expansion joints and no control joints. We conducted nonlinear creep and shrinkage finite element analyses. Half our staff in the New York office of Thornton-Tomasetti were either from Manhattan College or NYU, and a lot of them had Dr. Arnold Kerr. It's sad that the engineering school was sold off. I personally pledge, if Dr. Alexander will help me, to aggregate the Heights alumni who had Dr. Kerr as a teacher (and some of them are here) to help form an endowment fund here at UD. I'll make a contribution personally. We need to continue the legacy of Dr. Kerr.

Symposium Photographs

Shown with Arnold Kerr (center) are former students (from left) Jackie Richter-Menge, Steve Brauer, Michael Accorsi, Shirish Patel, Todd Euston, Martin Andreas, Joe Palese, Allen Zarembski, Luke Bathurst, Greg Grissom, and Harry (Tripp) Shenton.

Posing with Kerr are speakers and symposium chairs (from left), Allen Zarembski, Magdy El-Sibaie, Jackie Richter-Menge, Harry (Tripp) Shenton, Zdeněk Bažant, John Dempsey, Leon Keer, (Kerr), Alan Needleman, Anastasios Ionnaides, Isaac Elishakoff, Mike Santare, Byron Pipes, and Michael Chajes.

The Evolution of Contact Mechanics

Leon M. Keer, Northwestern University

Abstract

Using the Hertz theory as a starting point for the development of contact mechanics, a survey of the evolution of this field is given in light of the improvements that were made in such areas as surface characterization, materials development, power requirements and other technological advances. The approach is to focus on how improvements in computer hardware and software have led analysis from the arena of a closed-form solution of a single smooth contact (Hertz theory) to the semi-analytical analysis of rough contact, in which there may occur a large number of contacts due to the presence of many asperities. To show the evolution of contact mechanics best, this paper will limits itself to the following four issues: Hertz contact, Non-Hertzian contact—some geometrical issues, friction and its relation to Hertz Contact, and finally, purely numerical issues.

To illustrate how the development of high-speed computers enabled the solution of relatively complex problems, two specific examples are given. The first is the case of rough contact, which is solved by a combination of fast Fourier transform, multi-level summation, and conjugate gradient optimization. The second example is elastic-plastic indentation and calculation of residual stresses. Both examples rely of algorithms that increase the computational speed significantly. It is anticipated that future evolution of contact mechanics will involve calculations at even smaller length scales, which will depend upon the ever-increasing speed of computation and the corresponding ability to investigate materials at these scales.

Introduction

Contact mechanics began with the Hertz theory (1882 a,b) for the contact between two dissimilar ellipsoids. The geometry was approximated as the contact of two half spaces whose mismatch was considered as two paraboloids with the geometrical properties of the two ellipsoids at the point of contact. The contact was assumed to occur in a region so small that all lengths associated with the contact areas were much smaller than the ellipsoidal geometry. Thus the problem was solved as a kinematic problem of the contact between two half spaces. Abstracting ideas from the electrostatic problem of the charge on an electrified disk, Hertz was able to develop a closed form solution, which proved of great importance to the technical application of the contact of bearings.

This paper intends to use the Hertz theory as a starting point of contact mechanics and then look at the evolution of the field as improvement were made in such areas as surface characterization, materials development, power requirements and other advances in technology. The approach is to consider how improvements in computer hardware and software have led from the closed-form solution of a single contact to the analysis of rough contact, in which there may occur a large number of contacts due to the presence of many asperities. To show the evolution of contact mechanics best, this paper will limit itself to the following four issues:

Hertz Contact
Non-Hertzian Contact – Geometrical Issues
Friction and Relation to Hertz Contact
Numerical Issues

Although there are many more issues to consider in this field, this paper will restrict itself to the above four aspects of contact mechanics. Thus the evolution will stress advances of a mathematical nature, beginning with closed form solutions, such as those of Hertz, Mindlin, Cattaneo, Galin, Mossakovsky and others. The solutions from this group are all in closed form. Then there is the class of non-Hertzian contact solutions which use semi-analytical methods. These generally require a numerical solution, since they involve boundaries, such as edges, layers, wedges, and others. The numerical scheme of choice is usually the integral transform: Fourier transforms for planar surfaces and layers, Mellin transforms, for two-dimensional wedge problems, Hankel transforms for axially symmetric problems involving planar surfaces and layers, and Kantorovich-Lebedev transform for three-dimensional wedges. As such, the problems usually require the solution of integral equations to determine functions that determine important physical parameters, such as surface and subsurface stresses. A special section is devoted to friction because, although non-Hertzian, they also offer a closed form solution. There are also many excellent books available that survey contact mechanics, among them the following: Galin (1953), Gladwell (1980), Johnson (1985), and Goryacheva (1998).

Hertz Theory (1882)

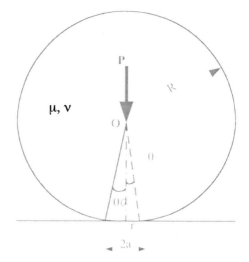

Fig. 1. Sphere indenting a rigid surface

Hertz noted the analogy between frictionless elastic contact and the electrostatic potential. He assumed that in the contact between two dissimilar bodies the area of contact would be elliptical and small compared with the dimensions of the two contacting bodies. The deflections of the surfaces would depend on the elastic constants of the two bodies and the radii of curvatures at the contact, which would be assumed to be large compared with the dimensions of the contact. The contact is depicted in Fig. 1 The elastic sphere indents a rigid surface under a load P. The radius of the contact is a, the radius of the sphere is R, and the angle between the normal to the surface and edge of contact is θ_0. Here θ is the angle, $0 < \theta < \theta_0$.

Geometrically this means, $a = R\sin\theta \approx R\theta$, with an elastic deflection in the contact as

$$u_z = \delta - (R - R\cos\theta)$$
$$= \delta - R[1 - (1 - \frac{1}{2}\theta^2 + ...)] \qquad (1)$$

$$u_z \approx \delta - R\frac{\theta^2}{2} = \delta - \frac{r^2}{2R} \qquad (2)$$

Here, δ is the relative approach, or the distance the center of the sphere moves from its unloaded position. In the case that the contact pressure is given as

$$p(r) = p_0(a^2 - r^2)^{1/2}/a \qquad (3)$$

then the load caused by the pressure is

$$P = \int_0^a 2\pi p_0(1 - r^2/a^2)^{1/2} r\,dr = \frac{2}{3}p_0\pi a^2 \qquad (4)$$

where p_0 is the peak stress, r is the radius and a is the radius of contact. For this load the displacement in the contact region can be found from the elasticity solution for a distributed load on a half space (see, e.g., Timoshenko and Goodier, 1970) as

$$u_z = \frac{1-v}{8\mu}\frac{\pi p_0}{a}(2a^2 - r^2) \qquad (5)$$

where μ, v are the shear modulus and Poisson's ratio. Using the boundary conditions yields

$$u_z = \frac{1-v}{8\mu}\frac{\pi p_0}{a}(2a^2 - r^2) = \delta - \frac{r^2}{2R}$$
$$\delta = \frac{1-v}{4\mu}\pi p_0 a \quad ; \quad \frac{1-v}{8\mu}\frac{\pi p_0}{a} = \frac{1}{2R} \qquad (6)$$

If two bodies with moduli E_1, v_1, and E_2, v_2, and respective radii of curvature R_1, R_2 are in contact, then the relations between the important parameters become (Johnson, 1985).

$$a = \left(\frac{3PR}{4E^*}\right)^{1/3}, \quad \delta = \frac{a^2}{R} = \left(\frac{9P^2}{16RE^{*2}}\right)^{1/3},$$

$$p_0 = \frac{3P}{2\pi a^2} = \left(\frac{6PE^{*2}}{\pi^3 R^2}\right)^{1/3}, \quad \frac{1}{E^*} = \frac{1-v_1^2}{E_1} + \frac{1-v_2^2}{E_2} \text{ and } \frac{1}{R} = \frac{1}{R_1} + \frac{1}{R_2} \qquad (7)$$

The relationship between load P and deflection δ corresponds to the stiffness for two nonlinear springs in series. It is important to note that all solutions are given in a closed form.

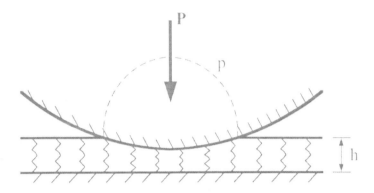

Fig. 2. Winkler foundation

Foundation models In some cases it is convenient to represent a foundation through use of models, such as representing a foundation with springs. The advantage of the assumption is that results can be usually obtained in a closed form. Considerable work has been done through foundation models developed by Winkler, Pasternak, Filoninko-Borodich. Excellent summaries and new results are found in papers by Kerr (1984, 1985), where the application is to layered structures. Further information is found in Johnson (1985). However, such models are appropriate to layers, where the analytical solutions do not lend themselves to a closed form, as will be discussed later.

For instance, a Winkler-type foundation is not an appropriate model for a half space, as is seen by the following example given in Johnson (1985) and shown in Fig. 2. For spherical contact, the elastic displacement is

$$u_z = \begin{cases} \delta - z(x,y) & \delta > z \\ 0 & \delta \le z \end{cases} \qquad (8)$$

Here $z(x,y) = h = Ax^2 + By^2 = \dfrac{x^2}{2R'} + \dfrac{y^2}{2R''}$. Assuming a Winkler foundation requires that the pressure is proportional to the deflection, i.e.

$$p(x,y) = \frac{K}{h} u_z(x,y) \qquad (9)$$

where K is the foundation modulus. Using the boundary condition, $u_z = \delta - \dfrac{x^2}{2R'} - \dfrac{y^2}{2R''}$, a pressure distribution can be easily obtained as

$$p(x,y) = \frac{K\delta}{h}\left[1 - \left(\frac{x}{a}\right)^2 - \left(\frac{y}{b}\right)^2\right] \qquad (10)$$

The load is given as $P = \int_A p(x,y)dA = K\pi ab\delta / 2h$. The pressure is thus seen to be parabolic, not elliptic. Moreover, the deflection at r=a differs from Hertz theory. The values are close to Hertz, though they require adjustment, as pointed out by Johnson. However, for a thin layer on a half space, this approach gives a good approximate result.

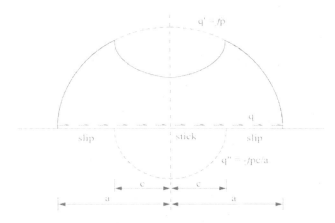

Fig. 3. Sliding with stick and slip (Johnson, 1985)

Special case: Method of Cattaneo (1938) and Mindlin (1949) Sometimes it is possible to use Hertz type of results to obtain a more complicated solution. Such is the case of two normally loaded identical elliptical bodies, which experience a subsequent tangential load. This problem is also detailed in Johnson (1985) and his notation is used here. The argument made by Cattaneo and Mindlin was that the displacement u_x for a tangentially applied Hertzian load in an elliptical region is constant. However, it is also noted that the displacement u_y in this region may not be. Their argument leads to the situation shown in the figure for the case of spheres: the contact stresses will be assumed to consist of two Hertzian distributions of assumed tangential loading.

$$q'(r) = \mu p_0 (1 - \frac{r^2}{a^2})^{1/2}$$

$$q''(r) = -\frac{c}{a} \mu p_0 (1 - \frac{r^2}{a^2})^{1/2}$$

(11)

The displacements caused by these tangential stresses are found to be

$$u_x' = \frac{\pi \mu p_0}{32 Ga} \left[4(2-v)a^2 + (4-v)x^2 + (4-3v)y^2 \right] \quad u_y' = \frac{\pi \mu p_0}{32 Ga}(2vxy)$$

$$u_x'' = -\frac{c}{a} \frac{\pi \mu p_0}{32 Gc} \left[4(2-v)c^2 + (4-v)x^2 + (4-3v)y^2 \right] \quad u_y'' = -\frac{c}{a} \frac{\pi \mu p_0}{32 Gc}(2vxy)$$

(12)

The resultant displacements in $r < c$

$$u_x = u_x' + u_y'' = \frac{\pi\mu p_0}{8Ga}(2-v)(a^2-c^2), \; u_y = 0 \tag{13}$$

and the condition for no-slip gives

$$\delta_x = \frac{3\mu P(2-v)(a^2-c^2)}{16Ga^3} \tag{14}$$

hence

$$Q_x = \int_0^a 2\pi q'' r dr - \int_0^c 2\pi q'' r dr = \mu P\left(1 - \frac{c^3}{a^3}\right) \tag{15}$$

or

$$\frac{c}{a} = \left(1 - \frac{Q_x}{\mu P}\right)^{1/3} \tag{16}$$

This solution is clearly approximate, since the slip in the region $c < r < a$ will not be in the direction of the applied friction force. However, the deviation, which depends on Poisson's ratio is small so that the theory can be applied to many practical situations.

Non-Hertzian Contact – Geometrical Issues

Problems involving corners, edges, and layers   It was seen that Hertzian problems were ones that had specific boundary values that lead to closed-form elasticity solutions. Even the friction case of Cattaneo and Mindlin was shown to result in a closed form. In many cases of corners and edges a closed form solution could also be obtained. However, in cases involving layers there is no possibility of obtaining a closed-form solution, except by asymptotic methods, which require a small parameter. Some examples follow:

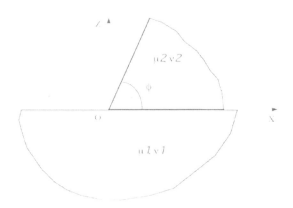

Fig. 4. Wedge-type contact

i) Wedge indentation [Dundurs and Lee (1972)]

A general approach to this problem (as shown in Fig. 4) assumes a power singularity at origin and uses boundary conditions to find the appropriate potential function. For this case the power singularity is determined by solution of an eigenvalue problem, where the eigenvalue problem was found to depend on the following Dundurs parameters:

$$\alpha = \left\{\frac{(1-v_1)/\mu_1 - (1-v_2)/\mu_2}{(1-v_1)/\mu_1 + (1-v_2)/\mu_2}\right\}$$

$$\beta = \frac{1}{2}\left\{\frac{(1-2v_1)/\mu_1 - (1-2v_2)/\mu_2}{(1-v_1)/\mu_1 + (1-v_2)/\mu_2}\right\}$$

(17)

Solutions using this approach were solved by Knien (1926), Williams (1957), and Rice and Sih (1965). The method was developed mainly for solution of fracture mechanics problems; however, the Dundurs and Lee application leads to important solutions of contact for wedges for many different types of bonding between the wedge and substrate.

A more general contact problem for an elastic wedge was solved by Hanson and Keer (1991). As Mellin transforms are appropriate for a two-dimensional wedge, the Kontorovich-Lebedev transform must be used for the three-dimensional wedge (Ufliand, 1965). The analysis of Hanson and Keer, among other results, found that for wedges in contact with a spherical indenter, an obtuse wedge angle would produce a singularity at the edge of the wedge; however, an acute wedge angle would not produce a singularity. The reason is that bending for the acute angle will cause the wedge to deform away from the sphere as the load is applied. This feature is illustrated in the Fig. 5, which shows the stress distribution of an obtuse wedge (126°) and an acute wedge (60°) as the indenter is moved from a position at the edge.

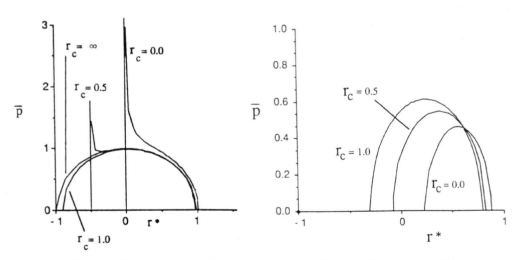

Fig. 5a. Wedge angle 126°   Fig. 5b. Wedge angle 60°

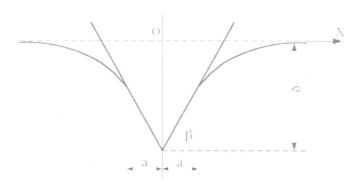

Fig. 6. Geometry and coordinate system for conical indent

ii)   Conical Indentation

Problems involving conical indentation have been solved by integration of the Boussinesq solution and by integral transforms. A recent historical survey of problems relating to the conical indenter has been given by Borodich and Keer (2004), who also look at the effects of friction on the lead-deflection relations. Although the paper focuses on problems related to nanoindentation, a useful and complete literature study is provided. It should be noted that problems for an indenter of monomial shape can be solved in a closed form, whether friction is present or not. For the non-adhesive case, it is found that

$$P = \frac{4a\mu}{(1-v)}\left(\delta - \frac{\pi a}{4}\tan\beta\right), \quad \delta = \frac{\pi a}{2}\tan\beta \tag{18}$$

Thus, the average load is seen to be

$$\frac{P}{\pi a^2} = P_m = \frac{\mu}{1-v}\tan\beta \tag{19}$$

where

$$\frac{\tau_{zz}}{P_m} = -\cosh^{-1}\left(\frac{a}{r}\right) \quad 0 < r < a$$

This solution gives a logarithmic singularity at r=0, and is clearly non-Hertzian. However, as in the Hertz theory this solution is in a closed form. More will be said later about such problems with friction.

iii)   Problems Involving Layers – Integral transforms

Problems that involve an elastic layer must be solved through use of integral transform theory. For many cases involving single or multiple layers, their solution is obtained by reducing the problem to a Fredholm integral equation of the second kind. Such problems involve the use of Papkovich Neuber potentials or Galerkin vector potential, where the potential functions that occur with these equations can be written as integral transforms. There have been two approaches to solving such problems: the utilization of double Fourier series or integrals for rectangular domains and techniques using Hankel transforms for circular domains. Some of the early work associated these problems have been done by Chen (1971), Chen and Engle (1972), Vorovich and Ustinov (1959), Aleksandrov and Vorovich (1960). There have also been solutions performed by asymptotic methods, most notably by Aleksandrov (1968). These solutions are for an elastic layer on a rigid or multi-layered substrate under normal loading by an indenter. The solution methods used to solve such problems were semi-analytical. Their solutions were therefore limited in the sense that problems involving real roughness, where multiple contacts were involved, would carry a high time penalty.

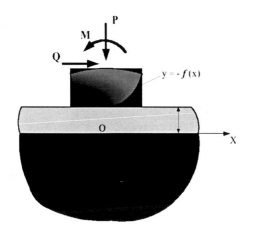

Fig. 7a. 2D indentation of a layer in half-plane

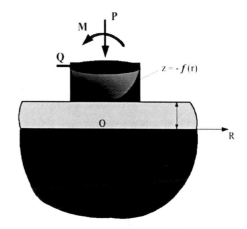

Fig. 7b. 3D indentation of a layer in a half space

The method of solution using Hankel transforms for normal loading without friction suggests boundary conditions of the following type:

$$\tau_{zr} = \tau_{z\theta} = 0, \qquad z = 0,\ 0 < r < \infty$$

$$u_z = \sum w_n(r)\cos(n\theta), \quad z = 0,\ 0 < r < a$$

$$\tau_{zz} = 0, \qquad z = 0,\ a < r < \infty$$

$$u_r = u_\theta = u_z = 0, \qquad z = h,\ 0 < r < \infty$$

(20)

where $w_n(r)$ is a prescribed indenter shape, a is the radius of contact and h the layer thickness. Geometries for 2D and 3D layer problems are shown in Fig. 7. The last equation expresses the fact that the layer is fixed at z=h. This is essentially the approach to the axially symmetric problem solved by Hayes et al. (1972). The problem was reduced to a Fredholm integral equation of the second kind, which is solved by asymptotic methods, noting that the ratio a/h is a small parameter. This method is essentially a perturbation around the half space solution. For the more general case above, the method of Papkovich-Neuber potentials can be used, as given in Green and Zerna (1954). The potentials are written in the form of inverse Hankel transforms of order n and in a form suitable for satisfying the boundary conditions on z=0,h.

If instead the problem is formulated to use double Fourier series, then the two-dimensional Fourier integral is used instead of Hankel transforms, where the boundary conditions for a rectangular load replace the trigonometric expansion. In this case the problem has stresses prescribed on z=0, and the rectangles are used to define a grid. However, this method lends itself to the numerical methods discussed later. A better method utilizes the closed form solution for a rectangular distribution of stress on a half space (Love, 1927).

The case of tangential loading of an indenter on an elastic layer fixed to a rigid substrate was solved by Goodman and Keer (1975). This problem represents the layer analog to the Cattaneo-Mindlin problem for a half space mentioned earlier. Assuming that the displacements can be put into the form

$$u_r = \sum_{n=0}^{\infty} u_r^{(n)}(r,z)\cos n\theta, \quad u_\theta = \sum_{n=0}^{\infty} u_\theta^{(n)}(r,z)\sin n\theta, \quad u_z = \sum_{n=0}^{\infty} u_z^{(n)}(r,z)\cos n\theta \qquad (21)$$

$$\tau_{zr} = \sum_{n=0}^{\infty} \tau_{zr}^{(n)}(r,z)\cos n\theta, \quad \tau_{z\theta} = \sum_{n=0}^{\infty} \tau_{z\theta}^{(n)}(r,z)\cos n\theta \qquad (22)$$

then the solutions can be represented as a Fourier-Bessel series. For the layer case it is convenient to write the boundary conditions in terms of the coefficients in the Fourier-Bessel series as follows

On z=0:

$$\tau_{zz}(r,z) = 0, \quad 0 < r < \infty$$
$$u_r^{(n)} + u_\theta^{(n)} = \Delta_{n+1}, \quad n = 0,1,......, \ 0 < r < a$$
$$u_r^{(n)} - u_\theta^{(n)} = \Delta_{n-1}, \quad n = 1,2,......, \ 0 < r < a \qquad (23)$$
$$\tau_{zr}^{(n)} + \tau_{z\theta}^{(n)} = 0, \quad n = 0,1,......, \ 0 < r < a$$
$$\tau_{zr}^{(n)} - \tau_{z\theta}^{(n)} = 0, \quad n = 0,1,......, \ 0 < r < a$$

On z=h:

$$u_r^{(n)} + u_\theta^{(n)} = \Delta_{n+1}, \quad n = 0,1,......, \ 0 < r < a$$
$$u_r^{(n)} - u_\theta^{(n)} = \Delta_{n-1}, \quad n = 1,2,......, \ 0 < r < a \qquad (24)$$

These boundary conditions will lead to coupled pairs of dual integral equations, which in turn lead to coupled Fredholm integral equations of the second kind. Owing to the computer available at the time, asymptotic procedures for a/h<1 were applied to obtain a valid solution. Using the computers available today, an asymptotic solution for small a/h is no longer a restriction. However, the case for real roughness will carry a severe time penalty for both half space and layer problems in this case.

Friction

The role of friction is an important one in contact mechanics. Most half space problems of friction can usually be reduced to a Hilbert problem as in Mossakovskii (1949), who considered the mixed problem for a half space with a circular line separating the boundary conditions. He obtained most of the physical quantities in a closed form. Goodman (1962) obtained a closed form solution to the Hertz problem with non-slip adhesion by using an approximation that ignored the vertical displacements caused by tangential shear. He showed that the approximation did not produce significant error. Spence (1975) solved the problem exactly using self-similarity methods, thus showing that the Hertz problem for non-slip adhesion can also be solved using a Weiner-Hopf technique. It is interesting that although the problems mentioned here are not technically Hertzian, they still result in a closed form. Borodich and Keer (2004) solve the problem for a monomial indenter with specific application made to the practical nanoindentation application. A brief review is presented next.

If the equation for the surface of a rigid indenter of arbitrary shape is pressed into an elastic half space and has the form

$$x_3 = -f(r), \quad f(0) = 0 \tag{25}$$

then Galin (1953) has shown that for frictionless contact the relation between load P and indenter shape f is given as

$$P = \frac{2E}{1-v^2}\int_0^a r_1 \Delta f(r_1)\sqrt{a^2 - r_1^2}\,dr_1, \quad h = \int_0^a \rho_1 \Delta f(r_1)\operatorname{arctanh}(\sqrt{1 - r_1^2/a^2})dr_1 \tag{26}$$

where $\Delta = \frac{\partial^2}{\partial x_1^2} + \frac{\partial^2}{\partial x_2^2} = \frac{\partial^2}{\partial r^2} + \frac{1}{r}\frac{\partial}{\partial r}$, $\operatorname{arctanh} v = \int_0^v \frac{dt}{1-t^2} = \frac{1}{2}\ln\frac{1+v}{1-v}$, a is the radius of contact and

h is the depth of penetration. Later, Sneddon (1965) showed that the relations could be put into the form

$$P = \frac{4\mu a}{1-v}\int_0^1 \frac{\xi^2 w'(\xi)d\xi}{\sqrt{1-\xi^2}}, \quad h = \int_0^1 \frac{w'(\xi)d\xi}{\sqrt{1-\xi^2}} \tag{27}$$

Now if the indenter has the profile

$$x_3 = -f(r), \quad f(0) = 0 \quad f(\rho) = w(r/a) \tag{28}$$

where the shape is of a monomial indenter, then Galin (1946, 1953) shows the relations for

$$f(\rho) = B_d r^d \tag{29}$$

leads to

$$P = \frac{E}{1-v^2} B_d \frac{d^2}{d+1} 2^{d-1} \frac{\left[\Gamma(d/2)\right]^2}{\Gamma(d)} a^{d+1}, \quad h = B_d d 2^{d-2} \frac{\left[\Gamma(d/2)\right]^2}{\Gamma(d)} a^d \tag{30}$$

This result can be shown to give a relation between force and deflection as

$$P = \frac{E}{1-v^2} \left[B_d^{-\frac{1}{d}} 2^{2/d} d^{\frac{d-1}{d}} \frac{1}{d+1} \left[\Gamma(d/2)\right]^{-\frac{2}{d}} \left[\Gamma(d)\right]^{\frac{1}{d}} \right] h^{\frac{d+1}{d}} \tag{31}$$

When there is adhesive contact (no-slip case) Mossakovskii (1954) presented first solution. Developing the Mossakovskii approach, Borodich and Keer (2004) obtained a formula for the relation between load and deflection as

$$P(a) = \frac{16\mu(1-v)\ln(3-4v)}{\pi(1-2v)\sqrt{3-4v}} \mathbf{I} \int_0^a h'(t) t \, dt, \quad \mathbf{I} = \int_0^a \frac{\chi(x,a)}{\sqrt{a^2-x^2}} dx \tag{32}$$

and $\chi(x,a) = \cos\left(\beta \ln \frac{a-x}{a+x}\right)$. where $\beta = \frac{1}{2\pi} \ln(3-4v)$

It is found that the integral can be evaluated as $\mathbf{I} = \frac{\pi}{4} \frac{\sqrt{3-4v}}{1-v}$. For punches of monomial shape

$$f(r) = B_d r^d \tag{33}$$

the relation between contact force P and contact radius a and between the displacement h and a becomes

$$P = \frac{E\ln(3-4v)}{(1+v)(1-2v)} B_d \frac{d}{d+1} 2^{d-1} \frac{\left[\Gamma(d/2)\right]^2}{\Gamma(d)} \frac{1}{\mathbf{I}*(d)} a^{d+1},$$

$$h = B_d 2^{d-2} \frac{\left[\Gamma(d/2)\right]^2}{\Gamma(d)} \frac{1}{\mathbf{I}*(d)} a^d. \tag{34}$$

From these relations a relationship between P and h can be shown to be

$$P = \frac{E\ln(3-4v)}{(1+v)(1-2v)} \frac{d}{d+1} \left[\frac{4\mathbf{I}*(d)}{B_d} \frac{\Gamma(d)}{\left[\Gamma(d/2)\right]^2} \right]^{1/d} h^{\frac{d+1}{d}} \tag{35}$$

where

$$\mathbf{I}*(d) = \int_0^1 t^{d-1} \cos\left(\beta \ln \frac{1-t}{1+t}\right) dt$$

It is thus seen that the relationships given above can be especially used in the technologically important area of nanoindentation. It is also noteworthy that the area of frictional contact leads to a closed form solution. However, again if layers or wedges are involved, then numerical schemes must be employed.

Numerical Issues

For many cases of current interest only a numerical procedure can be used. Even for a half space with a known closed-form Green's function the solution for contact between two rough surfaces may involve numerous contacts. This case is therefore as challenging as it is important for industries that have to assign roughness values for its machine components. As was seen earlier, the presence of a layer or multi-layered contacts involve numerical computations, but at a complexity that could be termed semi-analytical. Such problems can often be solved using numerical techniques that are relative close to the analysis.

When the problem considers a large number of contacts whose parameters are given in a statistical sense or from actual industrial data, then a host of numerical techniques are available to avoid severe time penalties. The author notes that the approach taken here is to investigate two specific examples as case studies: rough dry contact and elastic-plastic contact. This limits the paper in the sense that finite element and other computational methods will not be discussed here. Mention should be made of two areas: variational methods as developed for example by Kalker (1977) and adhesion mechanics as addressed by Johnson, Kendall and Roberts (1971) and by Derjaguin, Muller and Toporov (1975) for the different physical cases polymeric materials and metals. A good review of adhesion mechanics is found in a book by Maugis (1999).

Rough Dry Contact A dry contact problem is one that can be modeled without including the lubricant action, although in practice it can also be applied, with some accuracy, to boundary-, mixed-, and even full-lubricated contacts. When modeling rough contacts, it is important to take into account the complex and largely random topography of rough surfaces. The only existing approach that accounts for the random, multi-scale nature of real surfaces and asperity interactions is direct numerical solution of contact problems using surface topography samples collected from real rough surfaces. Such analyses were performed by Lai and Cheng (1985), Webster and Sayles (1986), Seabra and Berthe (1987), Ren and Lee (1993), and many others, using the numerical methods developed earlier by Kalker and van Randen (1972), Kubo et al. (1981), and Francis (1983).

Rough contact problems of practical interest often need to be solved on large computational grids (10^4–10^6 nodes), necessitating the use of particularly fast numerical methods. One such method is based on the multi-level multi-summation (MLMS) technique (Brandt and Lubrecht, 1990), also known as multi-level matrix multiplication, multi-level multi-integration, and fast integration. The MLMS technique, combined with a multigrid iteration scheme, was first applied to dry rough contacts by Lubrecht and Ioannides (1991). Polonsky and Keer (1999) combined an alternative MLMS algorithm with an iteration scheme based on the conjugate gradient (CG) method (Hestenes, 1980), thus obtaining a fast contact solver that converges for arbitrary rough contacts.

Another common method, fast Fourier Transform, was used to solve contact problems in 2D by Ju and Farris (1996). A fast method for the numerical analysis of 3D rough contacts was developed by Nogi and Kato (1997) by combining the fast Fourier transform (FFT) technique with a CG-based iteration scheme. This method is ideally suited for analyzing the contacts of surfaces that can be adequately described by periodic functions. Non-periodic contacts were efficiently treated by Polonsky and Keer (2000), who employed periodicity error correction, and by Liu et al. (2000), using the Discrete-Convolution FFT scheme (DC-FFT).

Solution of the contact problem When the size of the contact region is small compared to the size of the bodies in contact, and surface roughness features are small compared to the size of the contact region, contact between two rough surfaces, one or both of which may have an elastically dissimilar coating, can be modeled by the following system of equations (Fig. 8):

$$h(x, y) = w(x, y) + h_0 + v(x, y)$$
$$h(x, y) = 0 \quad (p_{co} > p > 0)$$
$$h(x, y) \geq 0 \quad (p = 0) \tag{36}$$
$$h(x,y) \leq 0 \quad (p = p_{co})$$
$$p_{co} \geq p \geq 0$$

$$v(x, y) = \iint_{\Omega_c} K(x - x', y - y') p(x', y') dx' dy' \tag{37}$$

$$\iint_{\Omega_c} p(x, y) dx dy = W \tag{38}$$

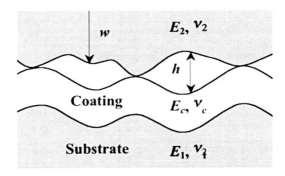

Fig. 8. Rough contact between layer and half space

where h – the gap between the two surfaces, w – the composite surface shape before deformation, h_0 – the normal (rigid body) approach, v – the composite elastic deformation of the surfaces, p – the contact pressure, W – the total normal force acting between the two bodies. The kernel $K(x, y)$ stands for the surface deflection produced by a concentrated normal contact load of unit magnitude acting at the origin, Fig. 9. For homogeneous solids, the kernel $K(x, y)$ is given by the Boussinesq formula:

$$K(x,y) = \frac{1-\nu}{2\pi\mu}\left[(x-x')^2 + (y-y')^2\right]^{1/2} = \frac{1-\nu}{2\pi\mu r} \tag{39}$$

where E is the Young's modulus, and ν is the Poisson's ratio. The actual contact region, Ω_c, is not known *a priori* and is determined in the course of the solution, so as to satisfy Eqs. (36). In rough contact, the contact region is usually discontinuous.

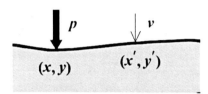

Fig. 9. Influence function geometry

Eqs. (36-39) reflect the elastic contact problem, except for the use of the cut-off pressure p_{co}, which is introduced to account approximately for the possibility of plastic yielding at highly stressed contact locations. This cutoff value can be picked to correspond to the least hardness value of the two contacting surfaces.

The system of equations and inequalities (36-38) is discretized over an equally spaced rectangular grid and solved for pressure p. The iterative solution scheme is based on the conjugate gradient method, described in Polonsky and Keer (1999). The solution consists of determining the normal approach, the real area of contact, and the pressure distribution that would satisfy the geometric constraints (36) and the load balance (38).

The multi-summation introduced by the discrete version of Eq. (37) presents the most computational challenge:

$$v_{ij} = \sum_{k=1}^{n_x}\sum_{l=1}^{n_y} K_{i-k,j-l}p_{kl} = K_{i-k,j-l} \otimes p_{kl} \tag{40}$$

This type of summation is known as *discrete convolution*. Its calculation is made much faster through the use of the multi-level multi-summation (MLMS) technique or the fast Fourier transform (FFT), either of which can reduce the computational complexity from $O(N^2)$ to $O(N \cdot \log_2 N)$. The latter technique is particularly useful when modeling theoretically infinite (i.e. relatively large) contacts, which can be represented as a periodic array of contacts. The FFT technique is also used for modeling layered contacts, for which the kernel K_{mn} in Eq. (40) can be found analytically only in the Fourier transform domain (Chen, 1971).

In rolling and/or sliding contacts, the contact stress always has a tangential component. The tangential contact stress may arise from dry friction between the contacting solids or from lubricant traction. In the event of gross sliding, the tangential contact stress q can be calculated using the Coulomb friction law:

$$q(x, y) = \mu p(x, y) \tag{41}$$

Here μ is the (dynamic) average friction coefficient, or the average traction coefficient.

Calculation of subsurface stresses After the contact pressure p and tangential contact stress q distributions have been determined, the corresponding subsurface stress distribution can be calculated as follows:

$$
\begin{aligned}
\sigma_{mn}(x,y,z) = & \iint_{\Omega_c} K_S^{mn}(x-x',y-y',z)p(x',y')dx'dy' \\
& + \iint_{\Omega_c} K_T^{mn}(x-x',y-y',z)q(x',y')dx'dy'
\end{aligned}
\tag{42}
$$

where K_S^{mn} and K_T^{mn} are the stress influence functions for the normal and tangential contact loading, respectively. These functions play a role similar to that of the displacement influence function K in Eq. (37): they describe stresses generated by unit contact pressures and tangential contact stresses. For homogeneous solids, closed-form analytical expressions for the six components of K_S^{mn} and K_T^{mn} are available (Kalker, 1986), but for layered solids, closed-form expressions are available only for their Fourier transforms (Chen, 1971). The subsurface stresses at given depths are computed similarly to the surface deflections, using MLMS or FFT techniques.

Example Contact of two rough steel rollers, one of which is coated with a hard coating. The contact is modeled as a line problem, i.e. infinite along the y axis and finite along the x axis. The radii of the rollers are 1 in., and the load is 1.5 kN. Digitized rough surface samples used in the calculation were scanned from real machined surfaces using an optical profilometer.

Figures 10-12 show a rough surface sample used in the calculation, the calculated pressure distribution in the contact region, and an octahedral shear stress distribution in a vertical cross-section. The octahedral shear stress is defined as

$$
\sigma_{oct} = \frac{1}{3}\left[(\sigma_x - \sigma_y)^2 + (\sigma_y - \sigma_z)^2 + (\sigma_z - \sigma_x)^2 + 6(\tau_{xy}^2 + \tau_{yz}^2 + \tau_{zx}^2)\right]^{1/2}
\tag{43}
$$

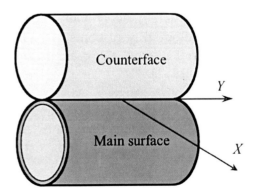

Fig. 10. Contact between two

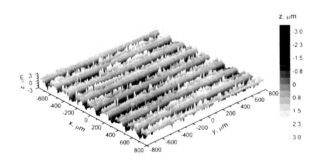

Fig. 11a. Rough surface sample used in the dry contact simulation

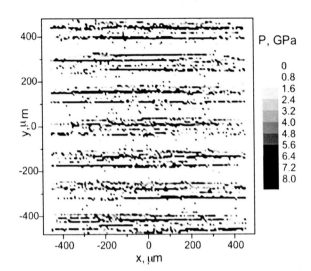

Fig. 11b. Contact pressure distribution

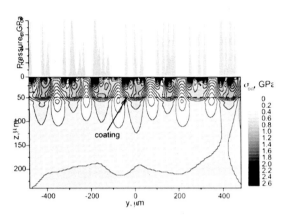

Fig. 12. Octahedral shear stress distribution in the vertical plane along the centerline

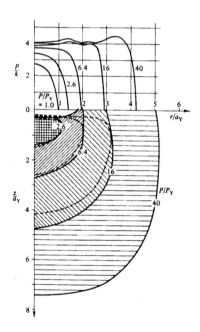

Fig. 13. Development of the plastic zone (Johnson, 1985)

Numerical Simulation for Elastic-Plastic Contact Problem The theoretical approach for elastic contact is well developed; however, it cannot be used in non-elastic conditions. Elastic-plastic contact is certainly more complicated than pure elastic contact.

It is found that the development of the plastic strain with the loading process can be divided into three ranges, Fig. 13: I Purely elastic, II Elastic-plastic, and III Perfectly plastic. Suppose the loading is applied gradually from zero to a high value. In range I, there is no plastic deformation, and the stress fields in the contact body increase with the loading until reaching the yield value. In range II, the yield point is exceeded, and the plastic zone appears. But the plastic zone is very small, which does not reach the surface. Thus there is a plastic core surrounded by the elastic region, and plastic strain has the same order of magnitude as the surrounding elastic strain. The plastic zone will expand with the increase of the load. Finally the plastic zone breaks out to the free surface and the displaced material is free to escape by plastic flow to the side of indenter, which includes range III. There are not any satisfactory methods to calculate elastic-plastic contact.

The three ranges respond to different stress-strain relationships as follows:
    I   Purely elastic: (1)
    II Elastic-plastic: (3) or (4)
    III Fully plastic: (2)

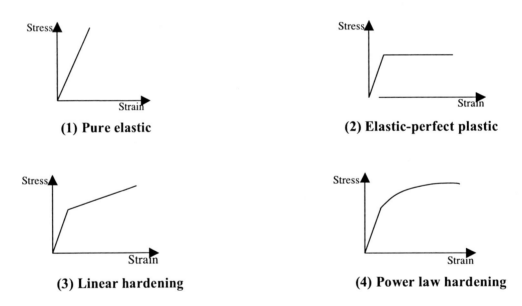

Fig. 14. Stress-strain relation for different load ranges

Basic formulas for elastic-plastic contact The method to follow is that developed by Jacq et al. (2002). For elastic-plastic contact, the normal deflection is induced not only by contact pressure, but also depends on plastic strain.

$$u_{k(total)} = u_{k(pressure)} + u_{k(r)}$$ (44)

where

$$u_{k(pressure)} = \int_{\Gamma_c} u_i^*(M, p^*(A)) p_i(M) d\Gamma \qquad k=1,2,3$$

$$u_{k(r)} = 2\mu \int_{\Omega_p} \varepsilon_{ij}^p(M) \varepsilon_{ij}^*(M, p^*(A)) d\Omega \qquad k=1,2,3$$

Due to Hooke's law

$$\acute{o}_{total} = \acute{o}_{pressure} + \acute{o}_r \qquad\qquad\qquad (45)$$

Based on the eigenstrain theory (see, e.g., Mura (1993), Y.P. Chiu (1977) was able to calculate the stress field for an eigenstrain of a cuboidal shape. Using this result, the residual stress σ_r is expressed as

$$\sigma_{ij}^r(M) = \sum_{n=1}^{N} A_{ijkl}(M,n)\varepsilon_{kl}^p(n) \qquad\qquad\qquad (46)$$

Resolving Process for Elastic-Plastic Contact For elastic-plastic contact, the total displacement is the sum of residual displacement u_r and the displacement produced by the contact pressure $u_{pressure}$. The displacement can be calculated using an iteration method. First pure elastic contact is calculated, giving the pressure distribution p, stress field $\acute{o}_{pressure}$, and residual displacement $u_{pressure}$. The displacement due to plastic strain u_r is then calculated. Since u_r changes the shape of the surface, elastic contact must be calculated again based on the new geometry of the surface. The calculation is repeated until the solution convergence. The resolving process is shown as Fig. 15.

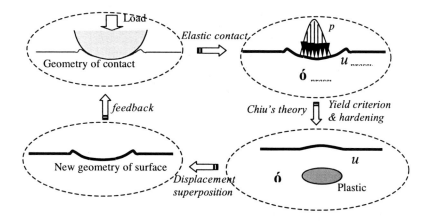

Fig. 15. Computational scheme

Methods to increase computational speed The computational speed of the calculation can be greatly enhanced by application of the following techniques:

1. Applying Fast Fourier Transforms (FFT) to increase the speed of computing elastic contact. In the calculation of residual stress, multi-summation must be operated in a 3D domain, which costs much more time than in just elastic contact. Thus, FFT technique is often preferred.
2. Calculating the influence coefficients **A** in advance, outside the loading loop
3. Using stress domain to calculate residual stress

Example-Nanoindentation of a steel A three-dimensional numerical simulation is performed for nanoindentation test using the code we developed. The punch is assumed as a spherical rigid body with the radium of $105\,\mu m$. The indentation object is a half space of steel. The smooth surfaces are supposed for the punch and the half space. The mechanical properties of the steel material are shown in the following Table. Where E is Young's modulus, υ is Poisson's ratio, and B, C, n are constants of Swift's law, which is used to describe the hardening behavior of the material.

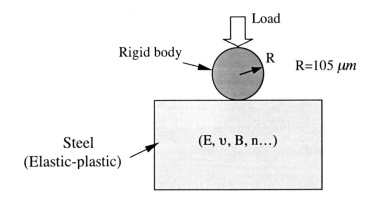

Fig. 16. Description of numerical example

Table 1 Material properties

R (μm)	E (GPa)	υ	B (MPa)	n	C
105	210	0.3	1240	30	0.085

Results Analysis The calculation results demonstrate that there is no plastic deformation before loading. Plastic strain first appears when the load increases to approximately 20 mN. The plastic deformation occurs below the contact region on the surface. Then the plastic region expands gradually with the increase of load. Since this plastic region does not touch the surface there is a plastic core surrounded by the elastic region. The second range, elastic-plastic, is proved to exist in the indentation process. The plastic core looks like a half-ellipsoid. The upper surface of this half-ellipsoid is a plane when the load is less than 100 mN and will eventually reach the surface if the load is increased enough.

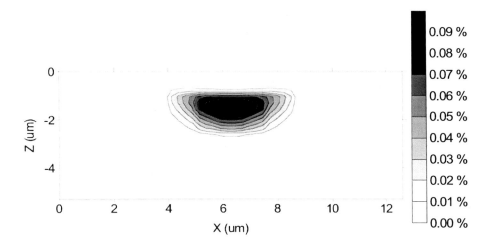

Fig. 17a. Equivalent strain, load=70 mN

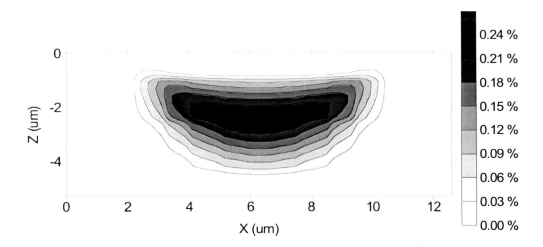

Fig. 17b. Equivalent strain, load=190 mN

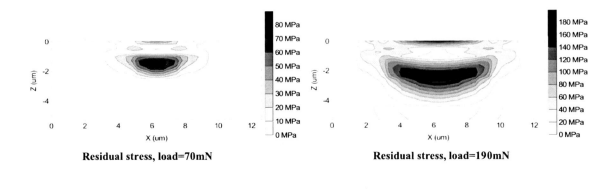

Residual stress, load=70mN   Residual stress, load=190mN

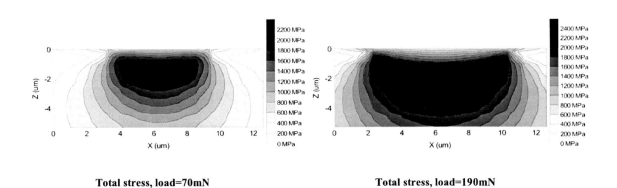

Total stress, load=70mN   Total stress, load=190mN

Fig. 18. Residual stress and total stress for two loads.

The distribution of contact pressure along x is shown below. The maximum value is 4 GPa when the load equals 190 mN. Unlike the results of elastic contact, the distribution of pressure becomes flat around the center of the contact for elastic-plastic contact. The size of the nonzero pressure region is identical to the size of the contact region.

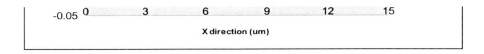

Fig. 19a. Initial contact surfaces, no load

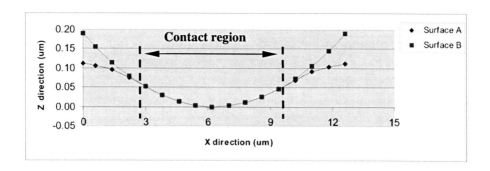

Fig. 19b. Deformed contact surfaces, load=190 mN.

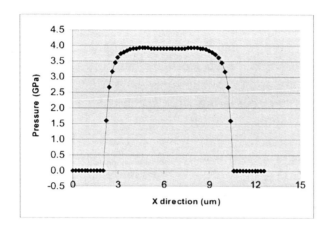

Fig. 19c. Surface deformation and pressure distribution for load of 190 mN

The curves of dimensionless strain versus mean pressure obtained from different sources are shown below. The curve calculated by the present code fits the test and FEM results very well. Test results are from Tabor (1951), and FEM results are from Hardy et al. (1971).

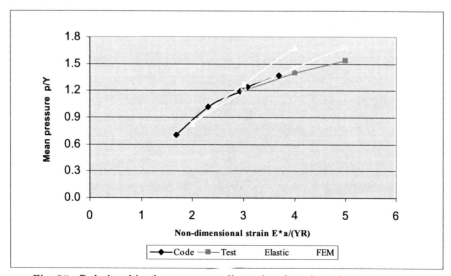

Fig. 20. Relationships between non-dimensional strain and mean pressure.

Conclusions

This survey shows an evolution of contact mechanics through a one hundred and twenty year period. From the types of examples given it is clear that solutions have undergone a change from purely analytical and closed form to methods that are mostly computational. This evolution appears to have been brought about for two reasons: the evolution of high speed computing machinery and the need for industry to have available solutions at every smaller length scales. When Hertz theory was the major theory used, the need to investigate the effects of surface roughness was not so immediate. However, as steels became better, a more complete knowledge of the surface became important and thus an understanding of the effects of surface roughness was necessary. Also the materials aspects of surface modification and development of hard protective coatings made it necessary to look more carefully at layers. However, rough coated layers can not be analyzed without a mostly computational agenda.

In addition to the development of analytical and computational tools, the materials scientists were developing steels whose quality was so high that deep crack nucleation, as is the example of Hertz, became not as important as the study of surface phenomena. Thus the study of engineered surfaces came into being along with the need for analysis with high-speed computers and advanced computational tools.

While it is difficult to predict the future in this field, it is clear that computation, along with selected analysis will be the driving force. As the computational machinery improves, it is likely that system performance at the large scale can be coupled with nucleation of flaws at the small scale, providing a more accurate and refined picture.

Acknowledgments

The author would like to express his appreciation for the support from the Center for Surface Engineering and Tribology at Northwestern University.

References

Aleksandrov, V.M., 1968. Asymptotic methods in contact problems of the theory of elasticity. *J. Appl. Math. Mech.*, **32**, 672-683.

Aleksandrov, V.M. and Vorovich, I.I., 1960. The action of a die on an elastic layer of finite thickness. *Prik. Mat. I Mekh.*, **24**, 323-333.

Borodich, F.M., and Keer, L.M., 2004. Contact problems and dept-sensing nanoindentation for frictionless and frictional boundary conditions. *Int. J. Solids and Structures*, **41**, 2479-2499.

Brandt, A., and Lubrecht, A.A., 1990. Multilevel matrix multiplication and fast solution of integral equations. *Journal of Computational Physics*, **90**, 348–370.

Brebbia, L.C., 1980. *The Boundary Element Method for Engineers*. London: Pentech Press.

Cattaneo, C., 1938. Sul contatto di due corpi elasticii. *Acc. Dei Lincei*. Rend Ser., 6, **27**, Pt. I, 342-348, Pt. II, 434-436, Pt. III, 474-478.

Chen, W.T., 1971. Computation of stresses and displacements in a layered elastic medium. *International Journal of Engineering Science*, **9**, 775–800.

Chen, W.T., and Engle, P.A., 1972. Impact and contact stress analysis in multi-layered medium. *Int. J. Solid and Structures,* **14**, 1257-1281.

Chiu, Y.P., 1977. Stress-field due to initial strains in a cuboid surrounded by an infinite elastic space. *ASME J. of Appl. Mech.,* **44**, 587-590.

Chiu, Y.P., 1978. Stress-field and surface deformation in a half space with a cuboidal zone in which initial strains are uniform. *ASME J. of Appl. Mech.,* **45**, 302-306.

Chiu, Y.P., and Hartnett, M.J., 1983. A numerical solution for layered solid contact problems with application to bearings. *J. Lubrication Tech.,* **105**, 585-590.

Derjaguin, V.M., Muller, V.M., and Toporov, Yu.P., 1975. Effect of contact deformation on adhesion of particles. *J. Colloid Interface Sci.,* **53**, 314-326.

Dundurs, J., and Lee, M.S., 1972. Stress concentrations at a sharp edge in contact problems. *J. Elas.,* **52**, 109-112.

Francis, H.A., 1983. The accuracy of plane strain models for the elastic contact of three-dimensional rough surfaces. *Wear,* **85**, 239–256.

Galin, L.A., 1946. Spatial contact problems of the theory of elasticity for punches of circular shape in planar projection. *J. Appl. Math. Mech. (PMM),* **10**, 425-448 (in Russian),

Galin, L.A., 1953. *Contact problems in the theory of elasticity.* Gostekhizdat, Moscow-Leningrad (English trans. Galin, L.A. (1961) I.N. Sneddon, ed., *Contact Problems in the Theory of Elasticity.* North Carolina State College, Dept. of Math. and Engineer Research, NSF Grant No. G16447, 1961),

Gladwell, G.M.L., 1980. *Contact problems in the classical theory of elasticity.* Sijthoff & Noordhoff.

Goodman, L.M., 1962. Contact stress analysis of normally loaded rough spheres. *ASME J. Appl. Mech.,* **29**, 515-522.

Goodman, , and Keer, L.M., 1975. Influence of an elastic layer on the tangential compliance of bodies in contact. *The mechanics of contact between deformable bodies* (Ed. A.D. de Pater and J.J. Kalker), 127-151.

Goryacheva, I.G., 1998. *Contact mechanics in tribology.* Kluwer Academic Publishers.

Green, A.E., and Zerna, W., 1954. *Theoretical Elasticity.* Oxford University Press, 169-170.

Hanson, M.T., and Keer, L.M., 1991. Analysis of edge effects on rail-wheel contact. *Wear,* **144**, 39-55.

Hardy, C., Baronet, C. N., and Tordion, G. V., 1971. Elastoplastic indentation of a half-space by a rigid sphere. *Journal of Numerical Methods in Engineering,* **3**, 451.

Hayes, W.C., Keer, L.M., Herrmann, G., and Mockros, L.F., 1972. A mathematical analysis for indentation tests of articular cartilage. *J. Biomechanics,* 541-551.

Hertz, H., 1882a. Über die Berührung fester elastischer Körper. J. reine angewandte Mathematik., 92, 156–171. (English transl. Hertz, H. (1896) On the contact of elastic solids. In: Miscellaneous Papers by H. Hertz. Eds. D.E. Jones and G.A. Schott, Macmillan, London, 146–162).

Hertz, H., 1882b. Über die Berührung fester elastischer Körper und über die Harte. Verhandlungen des Vereins zur Beförderung des Gewerbefleißes Berlin, Nov, 1882. (English transl. Hertz, H. (1896) On the contact of elastic solids and on hardness. In: Miscellaneous Papers by H. Hertz. Eds. D.E. Jones and G.A. Schott, Macmillan, London, 163–183).

Hestenes, M.R., 1980. *Conjugate Direction Methods in Optimization.* Springer-Verlag, New York, Chaps. 2, 3.

Jacq, C., Nelias, D., Lormand, G., and Girodin, D., 2002. Development of a three-dimensional semi-analytical elastic-plastic contact code. *ASME J. Tribol.*, **124**, 653-667.

Johnson, K.L., 1985. *Contact mechanics*. Cambridge University Press, 452.

Johnson, K.L., Kendall, K., and Roberts, A.D., 1971. Surface energy and the contact of elastic solids. *Proc. Roy. Soc.,* **A323**, 301-313.

Ju, Y., and Farris, T.N., 1996. Spectral analysis of two-dimensional contact problems. *ASME J. Tribol.*, **118**, 320-328.

Kalker, J.J., 1977. Variational principles in contact elastostatics. *J. Inst. Math. & Appl.*, **20**, 199-219.

Kalker, J.J., 1986. Numerical calculation of the elastic field in a half-space. *Comm. Appl. Numer. Methods*, **2**, 401–410.

Kalker, J.J. and van Randen, Y.A., 1972. A minimum principle for frictionless elastic contact with application to non Hertzian problems. *Journal of Engineering Mathematics*, **6**, 193–206.

Kerr, A.D., 1984. On the formal development of elastic-foundation models. *Ing. Arch.*, **54**, 455-464

Kerr, A.D., 1985. On the determination of foundation model parameters. *J. Geotech. Engng.*, ASCE, **111**, 1334-1340.

Knien, M., 1926. Zur theorie des druekversuchs. *ZAMM*, **6**, 414-416,

Kubo, A., Okamoto, T., and Kurokawa, N., 1981. Contact stress between rollers with surface irregularity. *ASME Journal of Mechanical Design*, **103**, 492–498.

Lai, W.T., and Cheng, H.S., 1985. Computer simulation of elastic rough contacts. *ASLE Transactions*, **28**, 172–180.

Liu, S., Wang, Q., and, Liu, G., 2000. A versatile method of discrete convolution and FFT (DC-FFT) for contact analyses. *Wear*, **243**, 101-110.

Love, A.E.H., 1927. *A treatise on the mathematical theory of elasticity*, 4th ed. Cambridge University Press.

Lubrecht, A.A., and Ioannides, E., 1991. A fast solution of the dry contact problem and the associated subsurface stress field, using multilevel techniques. *ASME J. Tribol.*, **113**, 128–133.

Maugis, D., 1999. *Contact, adhesion and rupture of elastic solids*. Springer.

Mayeur, C., 1995. Modélisation du contact rugueux élastoplastique. PhD Thesis, INSA Lyon, France.

Mindlin, R.D., 1949. Compliance of elastic bodies in contact. *ASME J. Appl. Mech.*, **16**, 259-268.

Mossakovskii, V.I., 1949. The fundamental mixed problem of the theory of elasticity for a half-space with a circular line separating he boundary conditions. *J. Appl. Math. Mech.* (PMM), **18**, 187-196.

Mura, T., 1993. *Micromechanics of Defects in Solids*, 2nd ed. Kluwer.

Nogi, T., and Kato, T., 1997. Influence on a hard surface layer on the limit of elastic contact—part 1: analysis using a real surface model. *ASME J. Tribol.*, **119**, 493-500.

Polonsky, I.A., and Keer, L.M., 1999. A numerical method for solving rough contact problems based on the multi-level multi-summation and conjugate gradient techniques. *Wear*, **231**, 206-219.

Polonsky, I.A., and Keer, L.M., 2000. A fast and accurate method for numerical analysis of layered elastic contacts. *ASME J. Tribol.*, **122**, 30-35.

Ren, N., and Lee, S.C., 1993. Contact simulation of three-dimensional rough surfaces using moving grid method. *ASME J. Tribol.*, **115**, 597–601.

Rice, J.R., and Sih, G.C., 1965. Plane problems of cracks in dissimilar media. *ASME J. Appl. Mech.*, **32**, 418-423.

Seabra, J., and Berthe, D., 1987. Influence of surface waviness and roughness on the normal pressure distribution in the Hertzian contact. *ASME J. Tribol.*, **109**, 462–470.

Sneddon, I.N., 1965. The relation between load and penetration in the axisymmetric Boussinesq problem for a punch of arbitrary profile. *Int. J. Eng. Sci.*, **3**, 47-57.

Spence, D.A., 1975. Self similar solutions to adhesive contact problems with incremental loading. *Proc. R. Soc. London*, **A305**, 55-80.

Tabor, D., 1951. *Hardness of Metals*. Oxford University Press.

Timoshenko, S.P., and Goodier, J.N., 1970. *Theory of elasticity*, 3rd ed. McGraw Book Company.

Ufliand, I.S., 1965. *Survey of articles on the applications of integral transforms in the theory of elasticity*. English Translation, Sneddon, I.N. (Ed.) Raleigh, North Carolina State University, Department of Mathematics.

Vorovich, I.I., and Ustinov, I.A., 1959. Pressure of a die on an elastic layer of finite thickness. *Prik. Mat. I Mekh.*, **23**, 445-455.

Webster, M.N., and Sayles, R.S., 1986. A numerical model for the elastic frictionless contact of real rough surfaces. *ASME J. Tribol.*, **108**, 314–320.

Williams, M.L., 1957. On the stress distribution at the base of a stationary crack. *ASME J. Appl. Mech.*, **24**, 104-114

Crack Detection by Laser-Based Techniques

Jan D. Achenbach, Northwestern University

Abstract

The generation of ultrasound by heat deposition due to laser irradiation has several advantages over conventional generation by piezoelectric transducers. This paper explores the applicability of laser-based ultrasonics to the detection of surface-breaking cracks in rail heads. A simplified approach is presented to determine the ultrasonic surface wave motion generated by laser irradiation of a linearly elastic solid. It combines three novel ideas. The first is a way to directly determine the mechanical loading equivalent to laser irradiation by considering local thermoelastic effects in the area that is heated by the laser. The second idea is the use of a new formulation for surface waves which considers a surface wave as a carrier wave that satisfies Helmholtz' equation in surface coordinates and that supports depth-dependent motions that are the same for any kind of carrier wave. The third idea is to directly determine the surface wave motion generated by the equivalent mechanical loading by applying the elastodynamic reciprocity relation for the generated surface wave motion and an appropriately selected "virtual" wave motion. The analysis based on the reciprocity relations is carried out for the time-harmonic case, but Fourier superposition is used to obtain the surface wave pulses corresponding to pulsed laser irradiation. For a homogeneous isotropic solid the details have been worked out for the surface wave pulses generated by laser irradiation of arbitrary time dependence at a point, an infinitely long line, and a line of finite length.

Introduction

Railway track engineering requires attention to a broad variety of problems in engineering mechanics. Arnold Kerr has used his considerable skills in the analytical and computational methods of engineering mechanics to provide solutions to many significant problems in that field. These include his work on the stability against buckling of rails and his studies of the load-carrying capability of railroad tracks.

A different class of problems that has not been so fortunate as to receive Arnold's attention concerns the formation of cracks in rail heads. The appearance of cracks presents an important safety problem. To a large extent, the appearance of cracks is a fact of life, and the next best thing then is to have reliable methods for crack detection in order that timely repair or replacement can be implemented.

Indeed, as noted by Kenderian et al. (2002),

> maintenance of railroad rails is one of the greatest problems facing the transportation industry today. In one four month period in 1998, a major railroad company experienced 10 derailments due to broken rails at an expense of over $1.3 million. In its *NTIAC Newsletter*, in September of 2000, the Texas Research Institute estimated that every 90 minutes a derailment, an accident or other rail related incident takes place in the United States (Texas Research Institute, 2000).

A variety of testing techniques have been used since the introduction of railways, but none of them is satisfactory for the detection of the many possible defects. The earliest of these methods, visual testing, is obviously too slow and incapable of detecting internal discontinuities in the rail. Furthermore, many discontinuities on the surface of the rail are missed by the visual testing method because of surface coverings of dirt, grease or other foreign matter.

In this paper, we will consider ultrasonic testing, which is the most common method used today for the testing of railroad tracks. Again quoting Kenderian et al. (2000),

> the most current method uses piezoelectric ultrasonic transducers in rolling rubber wheels filled with water or oil, in constant contact with the railroad track. This ultrasonic method can detect both surface and internal horizontal cracks (parallel to rail top surface), but cannot detect transverse cracking, vertical cracking or cracks inclined to the top surface beyond a certain limit. Both the magnetic induction and ultrasonic methods interrogate only the rail top surface because of obstacles regularly appearing along the sides of the rail, which fasten the rails and the ties together. In addition, these methods are currently limited to testing speeds of no more than 16 to 24 km/h (10 to 15 mi/h). Federal Railroad Administration rules require that any indication considered suspect by the test equipment on the test car are hand verified immediately. This leads to a stop/start mode, which effectively reduces the overall test speed in any given workday.

In recent years it has become apparent that faster and more reliable testing methods need to be available in order to prevent property damage and life threatening injury. With that purpose in mind, we consider here a non-contact method, namely, a fully laser-based system for detection of surface-breaking cracks. In such a system, ultrasound is generated by laser-pulse irradiation of the rail surface. The generated ultrasound interacts with the crack, and the interaction, which contains information that can be used to characterize the crack, is detected by a laser interferometer. In this paper, particular attention is devoted to the so-called laser-scanning-source technique, (Kromine et al., 2000).

We should also mention related hybrid techniques, based on laser ultrasound generation and air coupled ultrasound detection. These hybrid systems have been proposed for the detection of internal and surface rail defects (Lanza di Scalea, 2000; Kenderian et al., 2002).

Another noncontact technique uses air coupled transducers for both ultrasound generation and detection. Air coupled transduction overcomes the problems associated with the low repetition rate of common pulsed lasers, typically 20 Hz, and allows for a more controllable excitation signal through analog circuitry. This comes at the expense of a poor penetration efficiency into the test material due to the large acoustic impedance mismatch with the air. The mismatch can be exploited by a method that measures Doppler shifts in the airborne reflections from the rail surface for the detection of surface breaking cracks, (Wooh, 2000).

Before we proceed to a detailed discussion of laser-based ultrasonics, and its use for crack detection in railroad tracks, we will as a matter of general interest, briefly review some aspects of the history and the current state of quantitative nondestructive evaluation (QNDE).

Quantitative Nondestructive Evaluation

A retrospective look at the field of nondestructive testing as it existed some thirty years ago shows a field that was much smaller than it is now, but one that was already concerned with problems that are still being considered today: detection and characterization of cracks and cavities, disbonds and corrosion, and the nondestructive determination of material properties, among others. In those days the attention was primarily directed towards suitable approaches for specific configurations and problems. In more recent years the emphasis has shifted from the treatment of specific problems to general approaches. Also, experimental ingenuity, improved hardware and better signal processing techniques have produced significant progress. In addition, an important extension of the field has been provided by the use of analytical techniques for the development of measurement models. Such models are essential. Measured data generally cannot be understood in a quantitative way without a measurement model.

A model's principal purpose is to predict, from first principles, the measurement system's response to material properties and anomalies in a given material or structure, (e.g. cracks, voids, distributed damage, corrosion, deviations in material properties from specification, and others). For the ultrasonic case, a measurement model should include the configuration of probe and component being inspected, as well as a description of the generation, propagation and reception of the interrogating ultrasonic signals. Detailed modeling of the ultrasonic interactions, which generate the measurement system's response function, should also be included.

The availability of a measurement model has many benefits. First, numerical results, based on a reliable model, are very helpful in the design and optimization of efficient testing configurations. Second, a good model is indispensable in the interpretation of experimental data and in the recognition of characteristic signal features. The relative ease of parametrical studies, based on a measurement model, facilitates an assessment of probability of detection of anomalies. A measurement model is a virtual requirement for the development of an inverse technique based on quantitative data. Finally, a measurement model whose accuracy has been tested by comparison with experimental data provides a practical way of generating a training set for a neural network or a knowledge base for an expert system.

A more detailed review can be found in a paper by Achenbach (2000).

Laser-Based Ultrasonics

The irradiation of the surface of a solid by pulsed laser light generates wave motion in the solid material. Since the dominant frequencies of the generated wave motion are generally above 20,000 Hz, the waves are not audible to the human ear, and they are therefore termed ultrasonic waves. There are generally two mechanisms for such wave generation, depending on the energy density deposited by the laser pulse. At high energy density, a thin surface layer of the solid material melts, followed by an ablation process whereby particles fly off the surface, thus giving rise to forces which generate the ultrasonic waves. At low energy density, the surface material does not melt, but it expands at a high rate and wave motion is generated due to thermoelastic processes. As opposed to generation in the ablation range, laser generation of ultrasound in the thermoelastic range does not damage the surface of the material. For applications in nondestructive evaluation, ultrasound generated by laser irradiation in the thermoelastic regime is of interest and will be dealt with in this paper.

The generation of ultrasound by laser irradiation provides a number of advantages over conventional generation by piezoelectric transducers. These are: higher spatial resolution, non-contact generation and detection of ultrasonic waves, use of fiber optics, narrow-band and broad-band generation, absolute measurements, and ability to operate on curved and rough surfaces and at hard-to-access locations. On the receiving side, surface ultrasonic waves can be detected by using piezoelectric (PZT) or EMAT transducers, or optical interferometers in a completely laser-based system. Ultrasound generated by laser irradiation contains a large component of surface wave motion, and is therefore particularly useful for the detection of surface-breaking cracks. A Scanning Laser Source technique (SLS) has been proposed by Kromine et al. (2000) for this purpose. A corresponding analysis has been presented by Arias and Achenbach (2003b).

Since White (1963) first demonstrated the generation of high frequency acoustic pulses by laser irradiation of a metal surface, considerable progress has been made in developing theoretical models to explain and provide fruitful interpretation of experimental data. Scruby et al. (1980) assumed that, in the thermoelastic regime, the laser-heated region in the metal sample acts as an expanding point volume at the surface, which then was postulated to be equivalent to a set of two mutually orthogonal force dipoles. Based on intuitive arguments, these authors related the strength of the dipoles to the heat input and certain physical properties of the material. In this manner, the thermoelastic area source was reduced to a purely mechanical point-source acting on the surface of the sample. This point-source representation neglects optical absorption of the laser energy into the bulk material and thermal diffusion from the heat source. Furthermore, it does not take account of the finite lateral dimensions of the source. Rose (1984) gave a rigorous mathematical basis for the point-source representation on an elastic half-space, which he called a Surface Center of Expansion (SCOE), starting from a general representation theorem for volume sources. Although a formal solution for the double (Hankel-Laplace) transformed solution was given, its inverse could only be determined in a convenient analytical form for special configurations.

The SCOE model predicts the major features of the waveform and agrees with experiments particularly well for highly focused and short laser pulses. However, it fails to predict the so-called precursor in the ultrasonic epicentral waveform. The precursor is a small, but relatively sharp initial spike observed in metals at the longitudinal wave arrival. Doyle (1986) proved that the presence of the precursor in metals is due to subsurface sources equivalent to those arising from thermal diffusion. Although the focus was on metallic materials, these results showed that the precursor is present whenever subsurface sources exist. In metals, the subsurface sources arise mainly from thermal diffusion, since the optical absorption depth is very small compared to the thermal diffusion length.

The early work discussed above, suggested that a complete theory based on the treatment of the thermoelastic problem was necessary in order to provide understanding of the characteristics of the generated waveforms and assess the approximations introduced in the formulation of previously proposed models. Based on previous work by McDonald (1989), Spicer (1991) used the generalized theory of thermoelasticity to formulate a realistic model for the circular laser source, which accounted for both thermal diffusion and the finite spatial and temporal shape of the laser pulse.

All the works cited to this point deal with the modeling for a circular spot of laser illumination. One major problem associated with laser ultrasonics is poor signal-to-noise ratio. By focusing the laser beam into a line rather than a circular spot, the signal-to-noise ratio can be improved, since more energy can be injected into the surface while keeping the energy density low enough to avoid ablation. In addition, the generated surface waves have almost plane wavefronts parallel to the line-source, except near the ends of the line, which is advantageous for surface crack detection and sizing and for material characterization. Therefore, line-sources are used in inspection techniques such as the Scanning Laser Source technique for detection of surface-breaking cracks (Kromine et al., 2000).

Although the laser line-source offers several advantages, it has received considerably less attention than the circularly symmetric source. Three-dimensional representations for a line of finite width and length can be derived by superposition of surface centers of expansion. In some particular situations, when the effects of thermal diffusion and optical penetration can be neglected and interest is directed only to specific features of the generated field, such as surface wave displacements, for instance (Doyle and Scala, 1996), the superposition can be performed readily by analytical integration of the formal expressions put forth by Rose (1984). However, in more general cases, when the finite size of the laser source and the effects of thermal diffusion and optical penetration are accounted for, no analytical solution is available in the physical domain. Then, the superposition has to be performed numerically, resulting in a considerable computational effort. For these more general cases, a two-dimensional approach in which the line-source is considered to be infinitely long becomes highly advantageous. Bernstein and Spicer (2000) formulated a two-dimensional representation for an infinitely long and thin line-source. Their model results in a line of force dipoles acting normal to the line of laser illumination. Thus, they considered neither thermal diffusion nor optical penetration nor the finite width of the laser line-source.

A recent paper by Arias and Achenbach (2003a) considers a two-dimensional model for the line-source based on a unified treatment of the corresponding thermoelastic problem in plane strain, rather than integration of available results for the point-source. As opposed to Bernstein and Spicer (2000), this model takes account of the finite width of the source, the shape of the pulse and the subsurface sources arising from thermal diffusion and optical penetration. The thermoelastic problem in an isotropic half-space is solved analytically in the Fourier-Laplace transform domain. The doubly transformed solution is inverted numerically to produce theoretical waveforms. This approach alleviates much of the computational effort required by the superposition of point-source solutions.

Governing Equations

The thermoelastic fields are governed by the coupled equations of thermoelasticity. Based on the hyperbolic generalized theory of thermoelasticity, the governing equations for an isotropic solid are

$$k\nabla^2 T = \rho c_V \tau \ddot{T} + \rho c_V \dot{T} + T_0\beta\nabla\cdot\ddot{u} - q \tag{1}$$

$$\mu\nabla^2 u + (\lambda + \mu)\nabla(\nabla\cdot u) = \rho\ddot{u} + \beta\nabla T, \tag{2}$$

where T_0 is the ambient temperature, u is the displacement vector, λ and μ are Lamé's elastic constants, τ is the material relaxation time, β is the thermoacoustic coupling constant, $\beta=(3\lambda+2\mu)\alpha$, and α is the coefficient of linear thermal expansion. Also, k is the thermal conductivity, and κ is the thermal diffusivity, $\kappa=k/\rho c_V$, ρ and c_V being the mass density and the specific heat at constant deformation. In the thermoelastic regime the heat produced by mechanical deformation, given by the term $T_0\beta\nabla\cdot\ddot{u}$ can be neglected. With this approximation, Eq. (1) reduces to

$$\nabla^2 T - \frac{1}{\kappa}\dot{T} - \frac{1}{c^2}\ddot{T} = -\frac{q}{k}, \qquad c^2 = \frac{k}{\rho c_V \tau}. \tag{3}$$

An appropriate mathematical expression for the source term, q, will be given later. Eq. (3) is hyperbolic because of the presence of the term $\ddot{T}/c^2$. On the other hand, its counterpart in the classical theory, i.e., Eq. (3), without the term $\ddot{T}/c^2$, is parabolic. In the parabolic description of the heat flow, an infinite heat propagation speed is predicted, while the hyperbolic description introduces a finite propagation speed c, which is not known.

Both the classical and hyperbolic heat equations have been used to model thermoelastic laser generated ultrasound. As an accurate determination of the temperature field is vital to accurate predictions of laser generated ultrasonic waves, the question of which equation should be used arises naturally, and has been addressed in the literature by Sanderson et al. (1997).

In order to provide an answer to the question of which equation should be used for modeling purposes, the temperature field generated by line illumination of a half-space has been determined based on the two heat equations. Unlike the hyperbolic solution, the classical solution shows no distinct wavefront, and temperature increase starts at the initial time, as expected. However, the differences in the predicted temperature between the two theories are small and only apparent for very small time scales (on the order of hundred picoseconds). In the case of laser generation of ultrasound for NDE applications, we are typically interested in time scales on the order of microseconds. These time scales are large enough for the solutions given by both theories to be numerically undistinguishable. Consequently, the selection of the theory for the time scales of interest can be done for convenience, with no practical effect on the calculated results. Likewise, the choice of a specific value for the heat propagation speed in the hyperbolic equation does not effect the results. From the practical point of view, the choice of a value for the heat propagation speed equal to the speed of the longitudinal waves in the hyperbolic formulation, presents some numerical advantages in that it simplifies the inversion of the transforms.

Equivalent Dipole Loading

Several physical processes may take place when a solid surface is illuminated by a laser beam depending on the incident power (Scruby and Drain, 1990). Here, only low incident powers will be considered, since high powers produce damage on the material surface, rendering the technique unsuitable for nondestructive testing. At low incident powers, the laser source induces heating, the generation of thermal waves by heat conduction, and the generation of elastic waves (ultrasound). The geometry is shown in Fig.1.

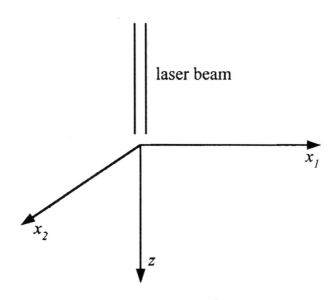

Fig. 1. Geometry of laser-irradiated body.

For application in NDE, generation of elastic waves is required in the ultrasonic frequency range and with reasonable amplitudes. This can be achieved without damage of the material surface only with short-pulsed lasers. The majority of published work has employed Q-switched laser pulses of duration of 10-40 ns. A suitable expression for the heat deposition in the solid along an infinitely long line is

$$q = E(1 - R_i)h(z)f(r)g(t), \tag{4}$$

where

$$h(z) = \gamma e^{-\gamma z}, \quad f(r) = \frac{2}{\pi}\frac{1}{R_G^2}e^{-2r^2/R_G^2}, \quad \text{and} \quad g(t) = \frac{8t^3}{v^4}e^{-2t^2/v^2}. \tag{5a, b, c}$$

In these expressions, E is the energy of the laser pulse, R_i is the surface reflectivity, R_G is the Gaussian beam radius, v is a laser pulse rise time parameter and γ is the extinction coefficient. For both the temporal and the spatial profile, the functional dependence of q has been constructed so that in the limits $v \to 0$, $R_G \to 0$ and $\gamma \to \infty$, an equivalent concentrated point source of the form

$$q = E(1 - R_i)\delta(z)\frac{\delta(r)}{2\pi r}\delta(t) \tag{6}$$

is obtained.

In the next step it is assumed that the energy deposition is so fast that it is not affected by wave propagation and heat conduction effects. The solution of Eq. (3) then reduces, and the increment of temperature, ΔT, becomes

$$\Delta T = \frac{E}{\rho c_V}(1 - R_i)h(z)f(r)\int_0^t g(s)\,ds. \tag{7}$$

To proceed further, we do a thought experiment on the mechanism of ultrasound generation by laser irradiation. Suppose the heat energy of the laser beam is deposited in a surface element of the illuminated body, and let us hypothetically remove this element from the body. The removed element has radius r_0 and thickness d. The element, which is now free on all sides, is free to expand due to the temperature rise ΔT. The strains due to thermal expansion are

$$\varepsilon_r = \varepsilon_\theta = \alpha\overline{\Delta T}, \tag{8}$$

where $\overline{\Delta T}$ is the averaged temperature increment,

$$\overline{\Delta T} = \frac{E}{\rho c_V}(1 - R_i)\frac{H(d, r_0)}{\pi r_0^2 d}\int_0^t g(s)\,ds, \tag{9}$$

and

$$H(d, r_0) = 2\pi\int_0^d h(z)\,dz\int_0^{r_0} f(r)r\,dr \tag{10}$$

The removal of the element from the body has created a cavity. To put the expanded element back into the cavity, it has to be compressed isothermally to its original radius. The compressive radial stresses are provided by the material surrounding the element, and their reactions on the walls of the cavity generate the ultrasound in the irradiated body.

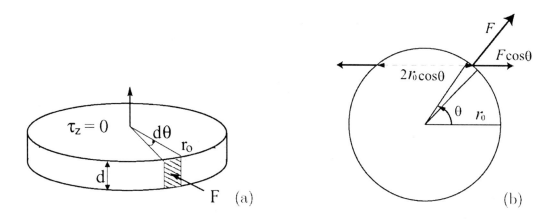

Fig. 2. (a) Elementary surface disk, (b) schematic of forces acting on the surrounding material.

Thus, to place the element back into the half-space, the surface element is subjected to an imposed deformation in its plane of the same magnitude as Eq. (8), but opposite sign. Let us call this imposed state of strain $\bar{\varepsilon}_r = \bar{\varepsilon}_\theta = -\alpha\overline{\Delta T} = \varepsilon$. The corresponding normal stress in the z direction is

$$\tau_z = (\lambda + 2\mu)\varepsilon_z + 2\lambda\varepsilon = 0, \tag{11}$$

and thus

$$\varepsilon_z = -\frac{2\lambda}{\lambda + 2\mu}\varepsilon. \tag{12}$$

Hence

$$\tau_r = 2(\lambda + \mu)\varepsilon + \lambda\varepsilon_{33} = \frac{2\mu(3\lambda + 2\mu)}{\lambda + 2\mu}\varepsilon$$

$$= -\frac{2\mu}{\lambda + 2\mu}(3\lambda + 2\mu)\alpha\overline{\Delta T} \equiv \sigma,$$

$$\tau_\theta = \sigma, \quad \tau_{r\theta} = 0, \tag{13}$$

where the relevant expression for $\overline{\Delta T}$ is given by Eq. (9). The reaction of the state of stress given in Eq. (13) on the surface of the hole, shown in Fig. 2(b), is equivalent to that produced by two orthogonal dipoles of magnitude D. As illustrated in Fig. 2(a), the force acting on an elementary sector of the circular disc is directed along its normal and its magnitude is

$$F = \sigma dS = \sigma dr_0 d\theta, \tag{14}$$

where d is the thickness of the element. Consider now two orthogonal directions defined by $\theta = 0$ and $\theta = \pi / 2$ in the plane of the element. As shown in Fig. 2(b), the component along one of these directions ($\theta = 0$) of the force acting on the surrounding material is given by $F \cos \theta$. The elementary dipole is obtained multiplying by the separation of the corresponding elementary forces, i.e., $2r_0 \cos \theta$, as

$$D = 2r_0 F \cos^2 \theta \ , \tag{15}$$

and hence, integration along the half-circumference of the hole yields

$$D = \int_{-\frac{\pi}{2}}^{\frac{\pi}{2}} 2r_0^2 \sigma d \cos^2 \theta d\theta = \pi r_0^2 \sigma d \ . \tag{16}$$

Thus, finally

$$D = \frac{2\mu}{\lambda + 2\mu} (3\lambda + 2\mu)\alpha \frac{E}{\rho c_V} (1 - R_i) H(d, r_0). \tag{17}$$

The same result is obtained for the dipole in the $\theta = \pi / 2$ direction.

The dipole has dimension force-length. Note that Eq. (17) coincides with the expression given by Scruby and Drain (1990) except for the factor $2\mu/(\lambda + 2\mu)$. For materials such as aluminum for which $\lambda \doteq 2\mu$, this factor is approximately 0.5. The simple derivation presented above shows that, contrary to the conclusion by Scruby and Drain (1990), the free surface does in fact reduce the strength of the surviving dipoles by a factor of $2\mu/(\lambda + 2\mu)$. In a recent work Royer (2001) reached the same conclusion by comparing a model for an infinitely long and thin line source based on a mixed matrix formulation to the line-source representation obtained by superposition of point-sources.

Based on the same arguments, the line-source can be modeled as a line of dipoles acting on the surface perpendicularly to the axis of the line.

Surface Waves Generated by Laser Irradiation

This Section follows earlier work by the author (Achenbach, 2003a). As discussed in the previous section, the wave motion generated by laser irradiation of a surface can be approximated by the waves generated by an equivalent mechanical loading. The analysis of the generated wave motion then becomes a purely isothermal problem of elastodynamics. The application of a surface disturbance generates a set of body waves and a surface wave. Due to the different rates of geometrical decay, only the surface wave needs to be considered sufficiently far from the area of application of a surface disturbance.

To directly calculate the surface wave, we use the elastodynamic reciprocity theorem, see Achenbach (2003a). For a region V with boundary S, and for two time-harmonic states defined by superscripts A and B, this theorem may be written as

$$\int_V (f_j^A u_j^B - f_j^B u_j^A)\, dV = \int_S (\tau_{ij}^B u_j^A - \tau_{ij}^A u_j^B)\, n_i\, dS, \tag{18}$$

where n is the outward normal.

From Eq. (13) we find that the relevant harmonic component of the traction on the surface may be written as

$$t_r = -\frac{D}{\pi r_0^2}\, \delta(r - r_0). \tag{19}$$

Following earlier work by the author (Achenbach, 2003a), the surface wave that is generated by this condition can be obtained by an application of the reciprocity theorem over a cylindrical body, $z \geq 0$, $0 \leq r \leq b$, $0 \leq \theta \leq 2\pi$, and bounded by $z=0$ and $r=b$. For State A, we select surface waves that propagate away from the circular area of application of the axially symmetric surface tractions. These surface waves may be written in the general form given by

$$u_r = -A V^R(z)\, H_1^{(1)}(kr) \tag{20}$$

$$u_z = A W^R(z)\, H_0^{(1)}(kr), \tag{21}$$

where $A(\omega)$ is an unknown amplitude factor. The relevant corresponding stresses are given by

$$\tau_{rr} = A\,[T_{rr}(z) H_0^{(1)}(kr) + \overline{T}_{rr}(z)\frac{1}{kr}H_1^{(1)}(kr)] \tag{22}$$

$$\tau_{rz} = -A T_{rz}(z)\, H_1^{(1)}(kr), \tag{23}$$

where, for the isotropic case $T_{rr}(z)$, $\overline{T}_{rr}(z)$ and $T_{rz}(z)$ can be found in Achenbach (2003a), Eqs. (56)-(57).

For State B, the virtual wave, we select a combination of outgoing and incoming surface waves so that the displacements are bounded at $r=0$:

$$u_r^B = -\frac{1}{2} B V^R(z)\,[H_1^{(1)}(k_R r) + H_1^{(2)}(k_R r)] \tag{24}$$

$$u_z^B = \frac{1}{2} B W^R(z)\,[H_0^{(1)}(k_R r) + H_0^{(2)}(k_R r)] \tag{25}$$

The corresponding stresses follow from Hooke's law; they are explicitly given elsewhere (Achenbach, 2003a). Since there are no internal body forces, the left hand side of Eq. (18) vanishes. We obtain

$$-2\frac{D}{r_0}BV^R(0)J_1(k_Rr_0) = b\int_0^{2\pi}\int_0^{\infty}\{[u_r^A\tau_{rr}^B - u_r^B\tau_{rr}^A] + [u_z^A\tau_{rz}^B - u_z^B\tau_{rz}^A]\}d\theta\,dz.\qquad(26)$$

It is very useful that the products of terms representing counter-propagating waves on the right-hand side of Eq. (26) cancel each other, and the remaining terms reduce to a very simple result

$$-2\frac{D}{r_0}V^R(0)J_1(k_Rr_0) = 4i\frac{\mu}{k_R}A(\omega)J,\qquad(27)$$

where

$$J = \int_0^{\infty}[T_{rr}^R(z)V^R(z) - T_{rz}^R(z)W^R(z)]\,dz\qquad(28)$$

This integral has a constant value that has been calculated for an isotropic solid (Achenbach, 2003b, p. 140). It is noted that $A(\omega)$ as solved from Eq. (2) would contain the length parameter r_0, which can be taken as R_G, the Gaussian beam radius, i.e., a parameter defining the irradiated area. A still simpler result is obtained by considering the Bessel functions $J_1(k_Rr_0)$ at very small values of k_Rr_0. Then we have $(1/r_0)J_1(k_Rr_0) \sim k_R/2$, and Eq. (27) yields

$$A(\omega) = -\frac{i}{2}C(-i\omega)^2,\qquad(29)$$

where $k_R=\omega/c_R$ has been used, and the constant C is

$$C = \frac{D}{2}\frac{1}{c_R^2}\frac{V^R(0)}{\mu J}.\qquad(30)$$

In the frequency domain, the vertical displacement at $z=0$ follows from Eq. (21), after restoring the time factor, as

$$\hat{u}_z(r,\omega) = -\frac{i}{2}CW^R(0)(-i\omega)\hat{g}(\omega)H_0^{(1)}(k_Rr)e^{-i\omega t}.\qquad(31)$$

This expression is the exponential Fourier transform of $u_z(r, t)$. Compatible with the definition of $\hat{g}(\omega)$, which is the Fourier transform of $g(t)$, i.e.,

$$\hat{g}(w) = \frac{1}{2\pi}\int_{-\infty}^{\infty}g(s)e^{i\omega s}ds,\qquad(32)$$

we write

$$u_z(r,t) = -\frac{i}{2} C W^R(0) \int_{-\infty}^{\infty} (-i\omega)\, \hat{g}(\omega) H_0^{(1)}(\omega \bar{r}) e^{-i\omega t} d\omega \,, \tag{33}$$

where $\bar{r} = r/c_R$. It is emphasized that C and $W_R(0)$ are constants that depend on the velocity of surface waves.

The integral in Eq. (33) can be simplified considerably. The final result has been derived elsewhere, Achenbach (2003a),

$$u_z(r,t) = -\frac{1}{\pi} \frac{D}{2} \frac{1}{c_R^2} \frac{V^R(0) W^R(0)}{\mu J} I \,, \tag{34}$$

where

$$I = \int_0^{t-r/c_R} \frac{\dot{g}(s)\, ds}{[(t-s)^2 - (r/c_R)^2]^{1/2}} \tag{35}$$

Based on the work of Rose (1984), Berthelot (1994) presented a further simplification of the point-focus pulse referred to as the half-order derivative of $g(t)$. This simplification can be obtained from Eq. (35) by noting that the term $[t - s + r/c_R]^{1/2}$ gets most of its contribution from the vicinity of the singularity of the integrand at $s=t-r/c_R$. Hence we can substitute this value for s in $[t - s + r/c_R]^{1/2}$. The result is given by Berthelot (1994).

Eq. (34) can be interpreted as the field generated by a source of time dependence $\dot{g}(t)$ applied at the origin as a spatial delta function, $\delta(x_1)\delta(x_2)$. In the time-space domain the field at position (t, x_1, x_2) has a domain of dependence defined by the cone

$$(t-s) - [(\bar{x}_1 - \bar{\zeta})^2 + (\bar{x}_2 - \bar{\xi})^2]^{1/2} \geq 0 \,, \quad t \geq s \geq 0 \,, \tag{36}$$

where $\bar{x}_1 = x_1/c_R$ and $\bar{x}_2 = x_2/c_R$. Now suppose a uniform distribution of sources is applied along the complete length of the x_2-axis. For the resulting two-dimensional problem, instead of Eq. (34), we then obtain an integration over the intersection of the cone with the tx_2-plane

$$I_1 = c_R \int_0^{t-\bar{x}_1} \dot{g}(s)\, ds \int_{\bar{x}_2 - [(t-s)^2 - \bar{x}_1^2]^{1/2}}^{\bar{x}_2 + [(t-s)^2 - \bar{x}_1^2]^{1/2}} \frac{d\bar{\xi}}{[(t-s)^2 - \bar{x}_1^2 - (\bar{x}_2 - \bar{\xi})^2]^{1/2}} \tag{37}$$

The second integral in Eq. (37) can be evaluated to yield π. It follows that Eq. (34) becomes

$$u_z(r,t) = -\frac{\overline{D}}{2} \frac{1}{c_R} \frac{V^R(0) W^R(0)}{\mu J} g\left(t \pm \frac{x_1}{c_R}\right) \tag{38}$$

Here we have placed a bar over the D to indicate that the $\overline{D}$ should now be interpreted as the dipole per unit length.

A comparison of Eqs. (34) and (38) shows that the surface wave pulses for line-focus and point-focus laser-irradiation are quite different in shape. For line-focus irradiation the pulse shape is proportional to the laser pulse, while for the point-focus case an initial pulse is followed by a smaller pulse with a somewhat extended tail.

Irradiation of a line segment $-a \leq x_2 \leq a$ is more interesting than either point- or infinite-line irradiation. An efficient way to obtain the ultrasonic field generated by illumination of a line segment is to use three building blocks: (1) "positive" irradiation over $-\infty < x_2 < \infty$; (2) "negative" irradiation over $a \leq x_2 < \infty$; and (3) "negative" irradiation over $-\infty < x_2 \leq -a$. The superposition of the three results yields "positive" irradiation over the strip $-a \leq x_2 \leq a$. The ultrasonic field for case (1) follows from Eqs. (34) and (37). For case (2) we will now evaluate the relevant superposition integral corresponding to Eq. (34) by integration over the appropriate area of dependence in the x_1, s-plane. We obtain for a point (x_1, x_2) where $x_2 < a$

$$I_2 = c_R \int_0^{t-[\bar{x}_1^2 + (\bar{a} - \bar{x}_2)^2]^{1/2}} \dot{g}(s)\,d\,s \int_{\bar{a}}^{\bar{x}_2 + [(t-s)^2 - \bar{x}_1^2]^{1/2}} \frac{d\,\bar{\xi}}{[(t-s)^2 - \bar{x}_1^2 - (\bar{x}_2 - \bar{\xi})^2]^{1/2}} \tag{39}$$

where $\bar{a} = a/c_R$ The integral can be evaluated further to yield

$$I_2 = \frac{\pi}{2} c_R g \left[t - \sqrt{\bar{x}_1^2 + (\bar{a} - \bar{x}_2)^2} \right]$$
$$+ c_R \int_0^{t-[\bar{x}_1^2 + (\bar{a} - \bar{x}_2)^2]^{1/2}} \dot{g}(s) \sin^{-1} \frac{\bar{x}_2 - \bar{a}}{[(t-s)^2 - \bar{x}_1^2]^{1/2}}\,d\,s \tag{40}$$

Similarly for the third problem we get

$$I_3 = \frac{\pi}{2} c_R g \left[t - \sqrt{\bar{x}_1^2 + (\bar{a} + \bar{x}_2)^2} \right]$$
$$- c_R \int_0^{t-[\bar{x}_1^2 + (\bar{a} + \bar{x}_2)^2]^{1/2}} \dot{g}(s) \sin^{-1} \frac{\bar{x}_2 + \bar{a}}{[(t-s)^2 - \bar{x}_1^2]^{1/2}}\,d\,s \tag{41}$$

The vertical displacement at (x_1, x_2), $|x_2| \leq a$ for line irradiation along the line segment $x_1 = 0$, $|x_2| \leq a$, may then be obtained by replacing I in Eq. (34) by $I_1 - I_2 - I_3$.

Scanning Laser Source Technique

Surface-breaking cracks in a structure can be ultrasonically detected using Rayleigh or Lamb waves. Conventional ultrasonic flaw detection methodologies require the generation of an ultrasonic wave packet that travels through a structure and interacts with existing flaws within the structure. Either reflected echoes or transmitted signals may be monitored in the pulse-echo or pitch-catch mode of operation. In the first method, the presence of an unexpected reflected signal is a possible indicator of a flaw, whereas in the second method, the transmitted signal may be significantly attenuated by the presence of an intervening flaw.

Pulse-echo and pitch-catch techniques can also be used with laser ultrasonics where a high-power pulsed laser is used to generate ultrasound thermoelastically, as discussed in a previous section. See also Scruby and Drain (1990). In both the pitch-catch and pulse-echo methods, the source is expected to generate a well-established ultrasonic wave, which then interacts with existing flaws. The limitations on the size of flaws that can be detected using pulse-echo or pitch-catch methods are determined by the ultrasonic reflectivity or transmittance of the flaws for the particular wavelength used, and by the sensitivity of the ultrasonic detector. The reflectivity and transmittance of Rayleigh waves as a function of defect geometry, and the change in frequency spectra of ultrasonic waves upon transmission or reflection by a defect, have been investigated. As might be expected, small flaws give rise to weak reflections and small changes in the amplitude of transmitted signals. These small variations are often too weak to be detected with existing laser detectors.

In this paper, we briefly discuss an alternate approach for ultrasonic detection of small surface-breaking cracks using laser-based techniques—the Scanning Laser Source (SLS) technique. This approach, discussed in detail by Kromine et al. (2000), does not monitor the interaction of a well-established ultrasonic surface wave with a flaw, but rather monitors the changes in the generated ultrasonic signal as the laser source passes over a defect. Changes in amplitude and frequency of the degenerated ultrasound are observed which result from the changed constraints under which the ultrasound is generated over uniform versus defective surface areas. These changes are quite readily detectable using existing laser detectors even for very small flaws. The geometry is shown in Fig. 3.

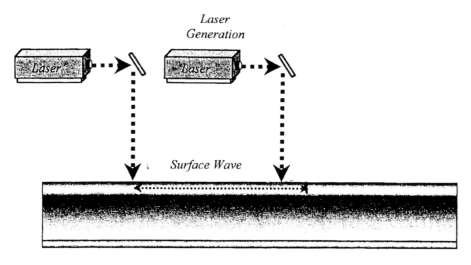

Fig. 3. Geometry for Scanning Laser Source (SLS) technique.

The SLS approach has been experimentally verified by testing of an aluminum specimen with a surface-breaking fatigue crack of 4 mm length and 50 μm width. A broad-band heterodyne interferometer with 1-15 Mhz bandwidth was used as an ultrasonic detector. The SLS was formed by focusing the irradiation of a pulsed Nd:YAG laser (pulse duration 10ns, energy 5 mJ) into a line of 5 mm length and 0.4 mm width. The generation of ultrasound was done in the thermoelastic regime. The distance between the source and detector was kept at 40 mm.

A plot of ultrasonic amplitude of the generated signal versus the scanning laser source as it was scanned over the crack is shown in Fig.4. A characteristic signature of the crack can be seen on this plot as a specific variation of the ultrasonic amplitude. The following aspects of this signature should be noted.

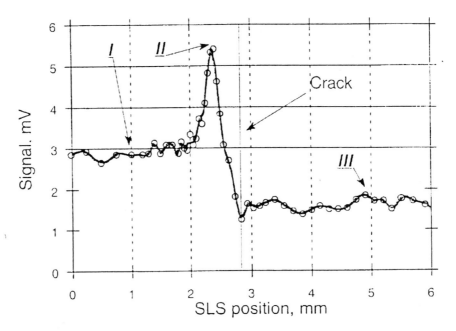

Fig. 4. Typical ultrasonic amplitude as the source is scanned over defect: (I) far ahead,
(II) close to, and (III) behind the defect.

(1) In the absence of a crack or when the source is far from reaching the crack, the amplitude of the generated ultrasonic *direct* signal is stable (see zone I in Fig. 4), and the signal is of sufficient amplitude above the noise floor to be unambiguously picked up by the laser detector, as shown in Fig. 5a. Note that a weak reflection from the crack is barely visible amidst the noise in Fig. 5a.

(2) As the source approaches the crack, the amplitude of the detected signal significantly increases (zone II in Fig. 4). This increase (from a level that was already sufficiently above the noise floor) is more readily detectable with a laser interferomener than any weak echoes from the crack (see Fig. 5b). We attribute this increase in signal amplitude to interference of the incident wave with the wave reflected by the crack.

(3) As the source is very close to, or right above the defect, a steep decay is observed, which is attributed to the changes in the conditions for generation of ultrasound, due to the presence of the crack.

(4) Subsequently, as the source moves behind the crack, the amplitude remains low due to attenuation of the signal generated by the crack (see zone III in Fig. 4). An example of this attenuated signal is shown in Fig. 5c. Note, however, that when the crack depth is smaller than the wavelength of the generated ultrasound, a significant portion of the ultrasound can pass underneath the crack when the source is behind it.

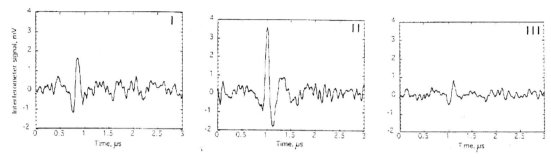

Fig. 5. Representative ultrasonic time-domain signals detected by the heterodyne interferometer at a fixed location, when the source is: (I) far ahead of, (II) close to, and (III) behind the defect.

Concluding Comments

This paper has advanced the contention that laser-based ultrasonics, e.g., the generation of ultrasound by laser illumination and its detection by the use of a laser interferometer, can be applied to the detection of surface-breaking cracks in rail heads. In particular, the paper has introduced a few ideas to simplify the calculation of the surface wave pulse that is generated by laser irradiation of a free surface. These simplifications provide an alternative method to the use of integral transform techniques. They offer a direct determination of the mechanical loading equivalent to laser irradiation, a more general formulation of free surface waves, and a direct way of determining the surface waves generated by the equivalent mechanical loading. Complete details were worked out for the case of a *homogeneous, isotropic elastic* solid, for which the surface wave pulses were determined for laser irradiation of arbitrary time dependence at a point, an infinitely long line and a line of finite length. In addition, a technique whereby the ultrasound generating beam of laser light was used to scan the surface, the scanning laser source (SLS) technique, was briefly discussed.

References

Achenbach, J.D., 2000. Quantitative nondestructive evaluation. *International Journal of Solids and Structures,* **37**, 13-27.

Achenbach, J.D., 2003a. Laser excitation of surface wave motion. *Journal of the Mechanics and Physics of Solids,* **51**, 1885-1902.

Achenbach, J.D., 2003b. *Reciprocity in Elastodynamics.* Cambridge University Press, Cambridge, UK.

Arias, I., and Achenbach, J.D., 2003a. Thermoelastic generation of ultrasound by laser line-source irradiation. *International Journal of Solids and Structures,* **40**, 6917-6835.

Arias, I., and Achenbach, J.D., 2003b. A model for the ultrasonic detection of surface-breaking cracks by the scanning laser source technique. *Wave Motion,* **39**, (1), 61-76.

Bernstein, J., and Spicer, J., 2000. Line source representation for laser-generated ultrasound in aluminum. *Journal of the Acoustical Society of America,* **107**, (3), 1352-1357.

Berthelot, Y.H., 1994. Half-order derivative formulation for the analysis of laser-generated Rayleigh waves. *Ultrasonics,* **32** (2), 153-154.

Doyle, P., 1986.   On epicentral waveforms for laser-generated ultrasound. *Journal of Physics D*, **19**, 1613-1623.

Doyle, P., and Scala, C., 1996. Near-field ultrasonic Rayleigh waves from a laser line source. *Ultrasonics*, **34**, 1-8.

Kenderian, S., Djordjevic, B.B., and Green, R.E. Jr., 2002.   Laser-based and air coupled ultrasound as noncontact and remote techniques for testing of railroad tracks. *Materials Evaluation*, **60**, (11), 65-70.

Kromine, A., Fomitchov, P., Krishnaswamy, S., and Achenbach, J.D., 2000. Laser ultrasonic detection of surface breaking discontinuities: Scanning Laser Source technique. *Materials Evaluation*, **58**, (2), 173-177.

Lanza di Scalea, F., 2000.   Advances in noncontact ultrasonic inspection of railroad tracks. *Experimental Techniques*. **24**, 23-26.

McDonald, F., 1989. Practical quantitative theory of photoacoustic pulse generation. *Applied Physics Letters*, **54**, (16), 1504-1506.

Rose, L., 1984. Point-source representation for laser-generated ultrasound. *Journal of the Acoustical Society of America,* **75**, (3), 723-732.

Royer, D., 2001. Mixed matrix formulation for the analysis of laser-generated acoustic waves by a thermoelastic line source. *Ultrasonics*, **39**, 345-354.

Sanderson, T., Ume, C., and Jarzynski, J., 1997. Laser generated ultrasound: a thermoelastic analysis of the source. *Ultrasonics*, **35**, 115-124.

Scruby, C., Dewhurst, R., Hutchins, D., and Palmer, S., 1980. Quantitative studies of thermally-generated elastic waves in laser irradiated metals. *Journal of Applied Physics*, **51**, 6210-6216.

Scruby, C., and Drain, L., 1990. *Laser Ultrasonics: Techniques and Applications.* Adam Hilger, New York.

Spicer, J., 1991. Laser ultrasonics in finite structures: comprehensive modelling with supporting experiment. PhD thesis, The John Hopkins University Press.

Texas Research Institute, 2000. Keeping railroads on track. *NTIAC Newsletter*, **25**, 1-6.

White, R., 1963. Generation of elastic waves by transient surface heating. *Journal of Applied Physics*, **34**, (12), 3559-3567.

Wook, S.C., 2000. Doppler-based airborne ultrasound for detecting surface discontinuities on a moving target. *Research in Nondestructive Evaluation*, **13**, 145-166.

Evolution of Buckling Modes in Continuously Supported Structures: From Eigenfunctions to Actual Buckled Shapes

Alan Needleman, Brown University

Abstract

The transition from a short wave length periodic buckling mode to the typically seen more localized final buckled shape is discussed. Attention is restricted to linear elastic material behavior in the structure. A simple model is used to illustrate the basic mechanisms of buckling localization and buckle propagation. Beams on nonlinear foundations illustrate further features of these processes. The specific example of thermal buckling of railroad tracks is considered.

Introduction

A widely observed phenomenon is that the final buckled configuration involves a localized deformation pattern whereas the critical buckling mode is associated with a periodic deformation pattern. Quite general considerations indicate that structures that are prone to such localization have the common property that the applied load-displacement curve attains a maximum (Tvergaard and Needleman, 1980) analogous to the necking localization that occurs in tensile bars (Considere, 1885). A related phenomenon is buckle propagation which, for example, occurs in pressurized cylinders (Kyriakides and Babcock, 1983; Dyau and Kyriakides, 1993). Here, we begin by illustrating the phenomena of buckling localization and buckle propagation in the context of a simple one dimensional model. Next, results for continuously supported beams are reviewed. We close with a consideration of the thermal buckling of railroad tracks.

One Dimensional Model

This section, which illustrates, in a simple context, many of the general issues involved in localization analyses is adopted from Tvergaard and Needleman (1980) and Needleman (1999). Here, attention is confined to those aspects related to structural buckling phenomena; issues related to finite deformation effects, tensile instabilities and material modeling are not discussed. These are discussed in Needleman (1999) and references cited therein. A long periodically buckled structure is modeled as a bar as sketched in Fig. 1.

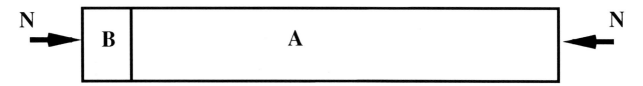

Fig. 1. One-dimensional model of a compressed structure.

With x the axial coordinate and t time, the displacement is denoted by $u(x,t)$ and the strain is given by

$$e = \frac{\partial u}{\partial x}$$
(1)

Balance of linear momentum requires

$$\frac{\partial N}{\partial x} = \rho \frac{\partial v}{\partial t}$$
(2)

where $N(x,t)$ is the axial force, ρ is the relevant density and $v(x,t)$ is the velocity, $v = \partial u/\partial t$.

Multiplying both sides of Eq. (2) by v and integrating with respect to x gives the identity

$$Nv \Big|_{x_1}^{x_2} = \int_x^{x_2} s\frac{\partial e}{\partial t}dx + \int_{x_1}^{x_2} \rho\left(\frac{\partial v}{\partial t}\right)vdx$$
(3)

where x_1 and x_2 are two points along the bar and

$$\frac{\partial e}{\partial t} = \frac{\partial v}{\partial x}$$
(4)

where here, and subsequently, a superposed dot is used to denote $\partial (\ )/\partial t$.

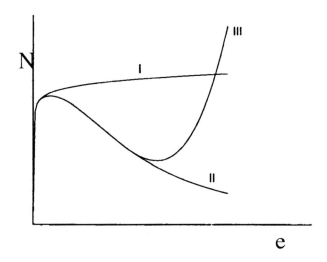

Fig. 2. Three stress-strain curves, N versus e giving rise to three types of behavior.

The circumstances considered are where strains remain small and nonlinearity arises from geometric effects so that the constitutive relation can be taken to be of the form

$$N(e) \tag{5}$$

where the dependence of N on e can be highly nonlinear in the post-buckling regime as illustrated for the three cases in Fig. 2.

Buckle Localization

One equilibrium solution to Eq. (2) corresponds to a state where N is independent of x. We consider the possibility of a non-uniform state; in particular, the possibility of two uniformly deformed regions denoted by A and B in Fig. 1. Equilibrium requires

$$N_A = N_B = N \tag{6}$$

Let the bar occupy the interval *[0,L]* and region B is *[0,L-B]* while region A is *[L-B,L]* so that L_A =$L - L_B$. Also, take the displacement at $x = 0$ to be zero, the displacement at $x = L_B$ to be U_1 and the displacement at $x = L$ to be U. Then, the strain-displacement relation is given by

$$e = \frac{U}{L} = \frac{U_1}{L} + \frac{U-U_1}{L} = \left(\frac{L_B}{L}\right)\left(\frac{U_1}{L_B}\right) + \left(\frac{L_A}{L}\right)\left(\frac{U-U_1}{L_A}\right) \tag{7}$$

with $\beta = L_B/L$ so that

$$e = (1-\beta)e_A + \beta e_B \tag{8}$$

Three forms of $N(e)$ are sketched in Fig. 2. For curve I, N is a monotonic function of e. For curves II and III, a maximum is attained where uniqueness is lost.

Equilibrium in the current state requires $N_A = N_B = N$ and hence $e_A = e_B = e$.

The change from the current state

$$\dot{N}_A = \dot{N}_B = \dot{N} \tag{9}$$

with

$$\dot{N} = \frac{dN}{de}\dot{e} = K\dot{e} \tag{10}$$

so that

$$K\left(\dot{e}_B - \dot{e}_A\right) = 0 \tag{11}$$

Hence $\dot{e}_B \neq \dot{e}_A$ if and only if $K = 0$; otherwise $\dot{e}_A = \dot{e}_B$ and only the uniform solution is available. Hence, localization is precluded for curve I in Fig. 2. At the peak, $K = 0$ (curves II and III in Fig. 2) and this is the point at which bifurcation from the homogeneous state first becomes possible. After the peak,

$$\dot{N}_A = K_A \dot{e}_A \qquad \dot{N}_B = K_B \dot{e}_B \tag{12}$$

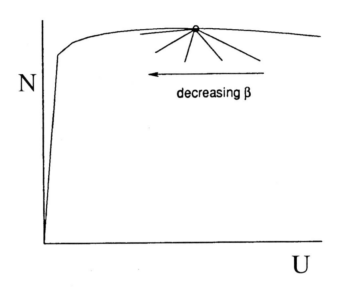

Fig. 3. Overall response as a function of $\beta = L_B / L$.

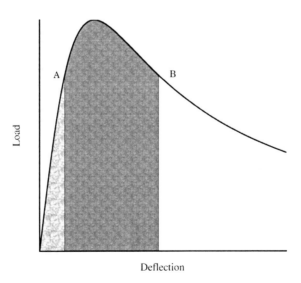

Fig. 4. The stored elastic energy in states A and B.

combining Eqs. (12), (9) and the rate form of Eq. (8) gives

$$\dot{e} = \frac{\dot{N}}{K_B}\left[\beta + (1-\beta)\frac{K_B}{K_A}\right]$$

(13)

For definiteness suppose that region B corresponds to the post-peak branch of *N(e)* and region A to the pre-peak branch (see Fig. 1). The post-bifurcation *N* versus *e* curve lies below the one for homogeneous straining and, the smaller β is, the more quickly *s* drops, as indicated in Fig. 3. The limiting case $\beta \to 0$ corresponds to a thin band in an otherwise uniform solid and gives the "most unstable" response.

The post-localization stiffness depends on the size scale of the localized region and there is nothing in the analysis to set the length scale, i.e., the value of β.

The stored elastic energy is

$$\Phi = (1-\beta)\Phi_A + \beta\Phi_B$$

(14)

with Φ being given by

$$\Phi = \int \frac{dN}{de}de$$

(15)

As seen from Fig. 4, the minimum energy is obtained in the limit $\beta \to 0$. However, β is undetermined by the analysis and the physical length scale that determines the localization width is set by factors outside this theoretical framework. For structural buckling, the size scale of the localization is set by geometry and is determined by analyzing more realistic structural models. Some types of structural buckling lead to a structural load-displacement response like that of curve II in Fig. 2 and this gives rise to the phenomenon of localization of buckling patterns, Tvergaard and Needleman (1980).

Buckle Propagation

Next, the propagation of a sharp wave front through the continuum is investigated. As shown in Fig. 5, the front is at *x* at time *t* and at *x* + δ*x* at time *t* + δ*t*. The speed of propagation of the front is *V* so that

$$\delta x = V \delta t$$

(16)

Attention is focused on the interval *[x, x+ δ x]* between times *t* and *t* + δ *t*. With *V* as shown in Fig. 5, the wave front travels from left to right and quantities to the right of the front are denoted by a subscript *A* and quantities to the left of the front by a subscript *B*. From Eqs. (2) and (16), balance of linear momentum on this interval gives

$$[N] = -\rho V [v] \tag{17}$$

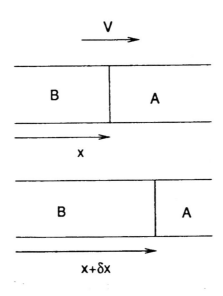

Fig. 5. A front propagating with velocity V.

where here, and subsequently,

$$[\ ] = (\ )_B - (\ )_A \tag{18}$$

The compatibility relation Eq. (4) together with Eq. (16) leads to

$$[v] = -V [e] \tag{19}$$

Similarly, the energy balance Eq. (3) for the material element δx over the time interval δt results in the energy relation

$$-[Nv] = V[\Phi] + V \frac{1}{2} \rho [v^2] \tag{20}$$

where

$$N\dot{e} = \dot{\Phi} \tag{21}$$

with Φ being given by Eq. (15).

Behavior like that in curve III in Fig. 2 arises in a variety of contexts including neck propagation in polymers and in the continuum theory of phase transitions, but attention here is confined to the context of structural buckling where such a response leads to the phenomenon of buckle propagation, see e.g. Kyriakides (1993). In Eq. (20) use

$$[Nv] = \frac{v_B + v_A}{2}[N] + \frac{N_B + N_A}{2}[v]$$

(22)

together with Eq. (17) and Eq. (19) to obtain

$$-[Nv] = \frac{v_B + v_A}{2}\rho V[v] + \frac{N_B + N_A}{2}V[e]$$

$$= V[\Phi] + V\frac{1}{2}\rho[v^2]$$

(23)

Since

$$\frac{v_B + v_A}{2}[v] = \frac{(v_B + v_A)(v_B - v_A)}{2} = \frac{v_B^2 - v_A^2}{2} = \frac{1}{2}[v^2]$$

the relation Eq. (23) reduces to

$$\frac{N_B + N_A}{2}[e] = [\Phi]$$

(24)

where a common factor V has been canceled from each term in Eq. (24), thus presuming $V \neq 0$.

For a quasi-statically propagating front, *[N]* ≈ *0* from Eq. (17) and $N_A = N_B = N_M$. Then, Eq. (24) simplifies to

$$N_M[e] = [\Phi]$$

(25)

The relation (25) is known as the Maxwell line construction in the literature on phase transitions and has the simple graphical interpretation that quasi-static propagation takes place at the value of *s* for which $R_1 = R_2$ as sketched in Fig. 6. The implication of this is the following; as the stress increases, the deformation remains uniform until the peak stress is attained. A non-uniform state then becomes possible. The stress then decreases as the front between the A and B states propagates quasi-statically. Thus, the propagation stress is less than the stress required for initiation. A number of propagating instability problems exhibit this general behavior, with one familiar example being the inflation of a party balloon, Chater and Hutchinson (1984).

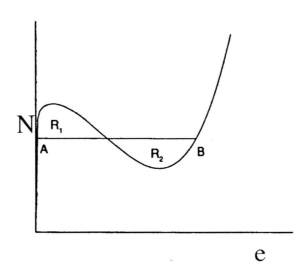

Fig. 6. Maxwell line construction for a slowly moving front.

Continuously Supported Beam-Column

A more realistic description of buckling mode evolution is obtained by considering a linear elastic beam-column continuously supported on a nonlinear foundation. The length scale associated with the localized collapse mode is an outcome of the analysis.

Foundation models are useful in representing a variety of nonlinear effects in an otherwise linear elastic structure. Indeed, a broad range of foundation models were considered in Kerr (1964) and as stated therein: (i) the interest is in the effect of the foundation on the structure and not in the stresses or displacements in the foundation; and (ii) the key is to find a relatively simple mathematical expression that represents the effect of the foundation.

Buckle localization

Moment equilibrium for a simply supported bar with imperfection $\overline{W}$ is expressed as

$$EI\frac{d^4W}{dX^4} + N\frac{d^2W}{dX^2} + F = -N\frac{d^2\overline{W}}{dX^2} \tag{26}$$

where E is the Young's modulus, I is the moment of inertia of the beam cross-section, $W(X)$ is the deflection in addition to the initial imperfection $\overline{W}(X)$, N is the axial force and F is the restoring force per unit length provided by the foundation.

For the bilinear foundation in Tvergaard and Needleman (1980)

$$F = \begin{cases} K_1 W & W < W_0 \\ K_2 (W - W_0) & W \geq W_0 \end{cases} \tag{27}$$

with $K_2/K_1 < 1$ for a softening foundation.

The critical value of N for buckling for a perfect column is

$$K_c = 2 (K_1 EI)^{1/2} \tag{28}$$

with the periodic bifurcation mode

$$W_c (X) = \sin \left(\frac{n \pi X}{L} \right) \tag{29}$$

where

$$n = \frac{L}{\pi} \left(\frac{K_1}{EI} \right)^{1/4} \tag{30}$$

In Tvergaard and Needleman (1980) analytical solutions to Eq. (26) were presented for the periodic buckling pattern for imperfect columns and for bifurcation away from the periodic deformation mode. Fig. 7 shows curves of normalized load versus deflection in the periodic mode for various imperfection amplitudes with n the normalized beam length, $\lambda = P/P_c$ and δ the imperfection magnitude in the periodic mode. Bifurcation from the periodic mode occurs beyond the maximum load, with the delay being smaller for longer columns (larger n) and greater for shorter columns (smaller n). The negative slope of the post-buckling response in the periodic mode decreases with increasing imperfection amplitude and so, for given n, does the delay between the maximum load and the bifurcation point. Fig. 8 shows the periodic mode, the mode associated with the bifurcation from the periodic mode and the superposition of this mode with the periodic mode showing that the bifurcation is indeed associated with a localization of the deformation pattern.

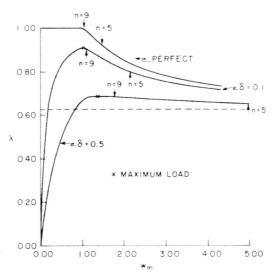

Fig. 7. Normalized load versus deflection for a column on a softening foundation. From Tvergaard and Needleman (1980).

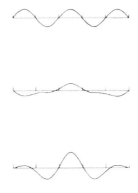

Fig. 8. The initial periodic bifurcation mode, the secondary bifurcation mode and an arbitrary linear combination of these illustrating the tendency to localization. From Tvergaard and Needleman (1980).

Numerical solution for an imperfection of the form

$$\overline{W} - W_0\left[\delta_1 \sin\frac{n\pi X}{L} - \delta_2 \sin\frac{(n-2)\pi X}{L}\right] \tag{31}$$

were presented in Needleman and Tvergaard (1982). As shown in Fig. 9 symmetry about the center of the beam was assumed. Fig. 10 shows the periodic mode that prevails at the maximum load and the localized mode that develops subsequently.

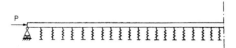

Fig. 9. Beam analyzed numerically with symmetry about the center assumed. From Needleman and Tvergaard (1982).

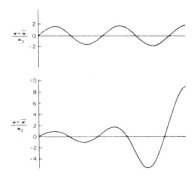

Fig. 10. Periodic mode that prevails at the maximum load and the localized mode that develops subsequently. From Needleman and Tvergaard (1982).

The mechanism of localization is a bifurcation subsequent to the maximum load. The final collapse mode bears no resemblance to the deformation pattern at the load maximum. The pre-localization deformation pattern determines the maximum load point.

Buckle Propagation

The phenomenon of buckle propagation occurs when bifurcation occurs in one part of a structure and then propagates. Chater and Hutchinson (1984) give two examples; inflation of a party balloon and collapse of a long pipe under external pressure. As seen in a previous section, what is needed for this phenomenon is a load-deflection curve like that for case 3 in Fig. 2. Kyriakides (1993) provides an overview of the phenomenon while the phenomenon of buckle propagation in cylinder shells is considered in detail in Dyau and Kyriakides. Here, some results from Chater, Hutchinson and Neale (1983) are discussed which show how the essence of the phenomenon is revealed by a model consisting of a continuously supported beam with an appropriate nonlinear description of the foundation restoring force.

The governing equation analyzed by Chater et al. (1983) is

$$EI \frac{d^4W}{dX^4} + F(W) = P \tag{32}$$

with P prescribed and with $F(W)$ given as shown in Fig. 11. The linear elastic beam is subject to a uniform lateral load P and the restoring force of the foundation per unit length $F(W)$ is as shown in Fig. 12. The spread of the zone of collapse from an initially weak region of the foundation was analyzed.

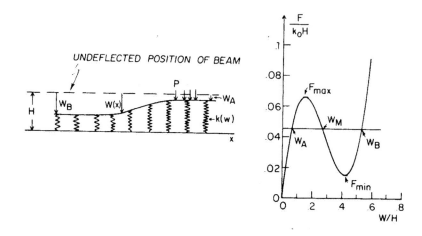

Fig. 11. Sketch of the beam and the spring force-deflection relation used in the analysis of Chater et al. (1983).

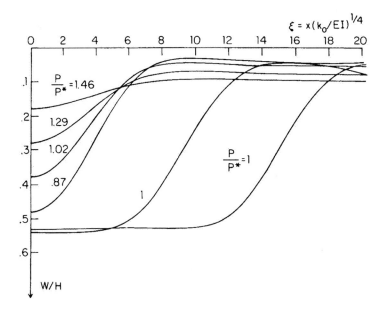

Fig. 12. Results from Chater et al. (1983) illustrating buckle propagation.

Fig. 12 shows numerical results from Chater, Hutchinson and Neale (1983). There is an initial imperfection that grows with increasing load. Eventually, however, a maximum value of P is attained and then the buckle expands under increasing load. At a value of P, given by the Maxwell construction, buckle propagation takes place quasi-statically at essentially constant load. Thus, the load required for initiation is greater than the load required for propagation. Dyau and Kyriakides (1993) carried out experiments on long cylindrical shells under external pressure where the collapse mode can be governed by this phenomenon.

Buckling of Railroad Tracks

Constrained thermal expansion of the rails can give rise to forces that induce buckling of railroad tracks on a hot day. Buckling can occur either out of the track plane or in the plane of the track. There is a large literature on thermal track buckling (see Kerr, 1978). Here, attention is focused on in-plane buckling. Two observations are worth noting: (i) tracks with noticeable imperfections undergo buckling with a smaller temperature rise than do tracks with smaller imperfections and (ii) the observed mode is a localized one.

The track can be modeled as a beam on a softening foundation that represents the ballast. The beam is constrained against thermal expansion and subject to a prescribed temperature increase ΔT. The beam is taken to be linear elastic so that

$$N = EA(e - \alpha \Delta T) \qquad M = EI\kappa \qquad (33)$$

with

$$e = \frac{du}{dx} + \frac{1}{2}\left(\frac{dw}{dx}\right)^2 - \frac{1}{2}\left(\frac{d\overline{w}}{dx}\right)^2 \qquad \kappa = \frac{d^2 w}{dx^2} - \frac{d^2 \overline{w}}{dx^2} \qquad (34)$$

The principle of virtual work can be written as

$$\int_0^L \left[N \delta e + M \delta \kappa + F \delta w + K \delta u\right] dx = 0 \qquad (35)$$

where F models the lateral resistance of the foundation and K the axial resistance. Some results from Tvergaard and Needleman (1981) are summarized in which F and K are taken to be nonlinear functions modeling the results of Birmann (1957) so that the track is modeled as a beam on a continuous foundation that softens with increasing track displacement. Issues concerning modeling of the properties of the continuous foundation are discussed in Kerr (2000).

For perfectly straight tracks, buckling into a periodic mode at ΔT_c with half wavelength a_c takes place at

$$\Delta T_c = \frac{2}{\alpha A}\sqrt{\frac{kI}{3}} \qquad a_c = \pi \left(\frac{EI}{k}\right)^{1/4} \qquad (36)$$

where k is the initial lateral spring stiffness. The critical bifurcation temperature is about one order of magnitude greater than what is observed for tracks with representative geometric imperfections. For the parameter values in Tvergaard and Needleman (1981), the critical bifurcation temperature for a perfect track is $\approx 600°C$.

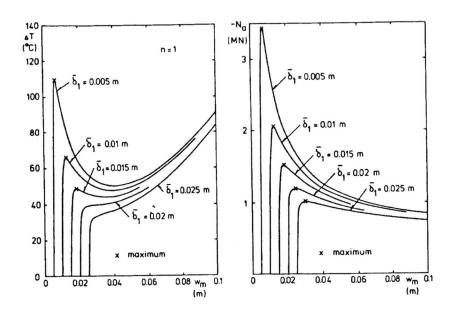

Fig. 13. Curves of temperature rise and axial force versus maximum deflection in the periodic buckling mode.
From Tvergaard and Needleman (1981).

The far post-buckling behavior of a railroad track in the localized mode has been analyzed by Kerr (1978). In Tvergaard and Needleman (1981), it is shown how a bifurcation into a localized mode serves as the mechanism for the transition from the periodic critical mode to the localized collapse mode.

In Tvergaard and Needleman (1981), the effects of initial geometric imperfections were analyzed. Fig. 13 shows curves of ΔT versus maximum lateral deflection and curves of axial force N versus maximum lateral deflection for periodic imperfections of the form $\overline{w} = \overline{\delta}_1 \sin \pi x / a$. In Fig. 13, the calculations are carried out requiring the deformation mode to remain periodic. For the smallest imperfection amplitudes a maximum temperature is reached. Hence, the branches where the temperature decays are unstable branches since temperature is the prescribed quantity. On the other hand, no temperature maximum occurs with a larger imperfection. However, in all cases a maximum compressive force is attained and attaining a maximum compressive force is what is required for a buckling localization bifurcation to occur.

Fig. 14 shows the response of a track with an initial imperfection of the form $\overline{w} = \left[\overline{\delta} + \overline{\delta}_2 \exp - \left((2x - L) / a \right)^2 \right] \sin \pi x / a$ while Fig. 15 shows the localized mode that develops.

The maximum values of ΔT are marked on Fig. 14. In Fig. 15, the small initial wave amplitude outside the region where localization has occurred has hardly grown (less than 1%) during the course of the deformation history.

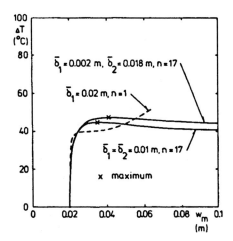

Fig. 14. Temperature rise versus maximum deflection allowing for buckling in the localized mode. From Tvergaard and Needleman (1981).

Fig. 15. Localized track mode. From Tvergaard and Needleman (1981).

Track buckling is found to be influenced by two bifurcation points; the first is associated with a periodic mode and the other gives the transition to the collapse mode after the axial force attains a maximum. Hence, collapse can occur with very little deflection in the periodic mode and the collapse mode can be highly localized. Also, a maximum force in the periodic mode does not necessarily imply a maximum temperature. Factors neglected in the analysis in Tvergaard and Needleman (1981) include lift-off in front of or behind a wheel, inhomogeneities in the track resistances, resistance against rotations of the rail relative to the cross ties.

Concluding Remarks

Structures that attain a maximum load in the post-buckling regime are susceptible to a mode transition involving localization or propagation. As a consequence, the final collapse mode of such a structure bears no resemblance to the critical buckling mode eigenfunction. Continuously supported structures, with appropriate foundation models, directly exhibit the mechanism of transition from eigenfunctions to actual buckled shapes.

References

Chater, E., and Hutchinson, J.W., 1984. On the propagation of bulges and buckles. *Journal of Applied Mechanics,* **51**, 269-277.

Chater, E., Hutchinson, J.W., and Neale, K.W., 1983. Buckle propagation on a beam on a nonlinear elastic foundation. In *Collapse: The Buckling of Structures in Theory and Practice*, (ed. by J.M.T. Thompson and G. W. Hunt), Cambridge University Press, 31-41.

Considére, A., 1885. L'Emploi du fer et de l'acier. *Annales des Ponts et Chaussées*, **9**, Ser. 6, 574-775.

Dyau, J.Y., and Kyriakides, S., 1993. On the propagation pressure of long cylindrical shells under external pressure. *International Journal of Mechanical Science,* **35**, 675-713.

Kerr, A.D., 1964. Elastic and viscoelastic foundation models. *Journal of Applied Mechanics,* **31**, 491-498.

Kerr, A.D., 1978. Analysis of thermal track buckling in the lateral plane. *Acta Mechanica,* **30**, 17-50.

Kerr, A.D., 2000. On the determination of the rail support modulus k. *International Journal of Solids and Structures,* **37**, 4335-4351.

Kyriakides, S., 1993. Propagating instabilities in structures. *Advances in Applied Mechanics,* **30**, 64-189.

Kyriakides, S., and Babcock, C.D., 1984. Buckle propagation phenomena in pipelines. *Collapse: The Buckling of Structures in Theory and Practice.* (ed. by J.M.T. Thompson and G.W. Hunt), Cambridge University Press, 75-91.

Needleman, A., 1999. Plastic strain localization in metals. *The Integration of Material, Process and Product Design* (ed. N. Zabaras et al.), A.A. Balkema, Rotterdam, 59-70.

Needleman, A., and Tvergaard, A., 1982. Aspects of plastic postbuckling behavior. Mechanics of Solids, The Rodney Hill 60th Anniversary Volume, (ed. by H.G. Hopkins and M.J. Sewell), Pergamon Press, 453-498.

Tvergaard, V., and Needleman, A., 1980. On the localization of buckling patterns. *Journal of Applied Mechanics*, **47**, 613-619.

Tvergaard, V., and Needleman, A., 1981. On localized thermal track buckling. *International Journal of the Mechanical Sciences*, **23**, 577-587.

Tvergaard, V., and Needleman, A., 1983. On the development of localized buckling patterns. *Collapse: The Buckling of Structures in Theory and Practice*, (ed. by J. M. T. Thompson and G. W. Hunt), Cambridge University Press, 1-17.

History and Lessons of a Paradox Associated with Follower Forces

Isaac E. Elishakoff, Florida Atlantic University

Abstract

This paper deals with the paradoxical result reported over three decades ago by Smith and Herrmann. They showed that the uniform column subjected to the follower force and placed on the uniform Winkler elastic foundation loses its stability at the same load as the column without an elastic foundation. This paper examines the above paradoxical result and outlines ways of its resolution.

Introduction

Smith and Herrmann (1972) published their study on the stability of Beck's column on a uniform elastic foundation (Fig. 1), described by the following equation:

$$EI\, d^4w/dx^4 + P\, d^2w/dx^2 + kw = 0 \qquad (1)$$

They showed that the buckling load of such a column does not differ from that of the column without foundation. They argued as follows:

"Although one may expect that increasing foundation stiffness would tend to make the columns less susceptible to flutter (*i.e.*, would increase the critical load), it can be argued on qualitative grounds that such is not the case: Flutter instability results from the fact that at sufficiently large loads the (non-conservative) applied force does positive work over each complete cycle. The free vibration mode shapes of the rod are the same with, or without, the supporting foundations (only the frequencies are affected), and therefore the critical load for flutter should also remain the same."

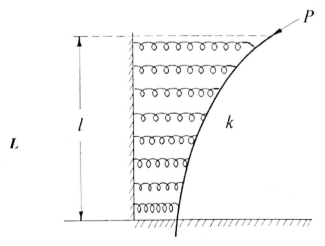

Fig. 1. Smith-Herrmann column is a Beck's column on an elastic foundation

In contrast the buckling load of the Euler's column on an elastic foundation is gradually increased by the introduction of the elastic foundation. For example, a pinned-pinned uniform column under constant-directional force, buckles at

$$P_{cr} = \pi^2 EI / L^2 \tag{2}$$

when no elastic foundation is present, and increases when the elastic foundation modulus is increased. If k satisfies the inequality

$$k \leq 4\pi^4 EI / L^4 \tag{3}$$

then the buckling load equals

$$P_{cr} = \pi^2 EI / L^2 + kL^2 / n^2 \tag{4}$$

and is reducible to Euler's familiar expression in Eq. (2) when k tends to zero. When k increases without bound the buckling load tends to the following expression

$$P_{cr} = 2\sqrt{kEI} \tag{5}$$

The buckling load of Euler's column can be arbitrarily increased by introducing a suitable elastic foundation. As far as the uniform Beck's column (i.e. the column subjected to the follower force) is concerned, its instability load cannot be increased by the uniform elastic foundation, however stiff.

At first, this result was not doubted by the researchers, including the present writer. This happened due to the following reason. It was shown that if either column or the elastic foundations were non-uniform, then the buckling load would depend on the elastic foundation (Celep, 1977, 1980). Sundararajan (1976) proved that only if the distribution of the mass density and of the elastic foundation are proportional (the premise that is automatically valid for the uniform case, considered by Smith and Herrmann (1972)) the buckling load would be unaffected. He proved the following theorem:

"The critical load of an undamped linearly elastic column subjected to either conservative or non-conservative stationary load does not decrease due to the introduction of a Winkler type elastic foundation having a modulus distribution geometrically similar to the mass distribution of the column."

Since in reality there cannot be purely uniform columns, or purely uniform elastic foundations, one may argue, the paradoxically appearing results of Smith and Herrmann would never materialize. The above finding of Sundararajan (1976) was generalized by Jacoby and Elishakoff (1986) who showed that if a uniform column with density with concentrated end mass M, supported by a uniform elastic foundation of stiffness k and the concentrated spring of stiffness K such that $\rho/k = M/K$, then the buckling load of such a (so-called Pflüger) column would not change; since uniformity is a particular case of non-uniformity, researchers, including this writer (naively) did not raise doubts on the results pertaining to uniform columns.

The following question was raised: How does a uniform column under a partial elastic foundation behave? Naturally, it was anticipated, following the conclusion made by Sundararajan (1976) and Jacoby and Elishakoff (1986), that only in the perfect case of proportionality of the mass distribution and the variation of elastic modulus the effect of the elastic foundation would be unfelt. The study by Elishakoff and Wang (1987) demonstrated that in the case of Beck's column under partial elastic foundation (Fig. 2) the critical load first increases with the increase of the extent a of the column supported by the elastic foundation, reaching a maximum at some a^*. Yet, once $a > a^*$, the critical load decreases and when a tends to the full length of the column, it reaches the critical load of the unstiffened Beck's column, thus recovering the result by Smith and Herrmann (1972). It must be noted here that in variance with the incorrect figure in the paper by Elishakoff and Wang (1987) the critical load of the column on partial elastic foundation may also take values below the column without an elastic foundation. At first glance, one may conclude that for the Beck's column under partial foundation some optimal length a_{opt} of the partial foundation exists, namely $a_{opt}=a^*$. On the other hand, the derived result of the decrease, when $a > a^*$, of the critical load seemed quite paradoxical.

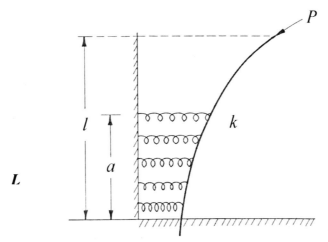

Fig. 2. A Beck's column on partial elastic foundation

Even more paradoxical results were reported by Vetter and Hauger (1976). They investigated Beck's column on the Winkler foundation with variable modulus

$$k(\xi) = k_0 \left[4(1-\gamma)(\xi^2 - \xi) + 1 \right] \qquad (6)$$

where $\xi = x/L$ is the non-dimensional axial coordinate. Numerical results showed that for $\gamma = 1$, *i.e.* constant foundation modulus $k = k_0$, the flutter load is independent of the magnitude of the modulus. Vetter and Hauger (1976) report:

"If $\gamma < 1$, however, then the critical load increases considerably, depending on the value of the parameter k_0. Since the case $\gamma < 1$ can be interpreted as a weakening of the foundation for $0 < x < L$, one arrives at the intuitively unexpected result that a weakening of the foundation may improve the stability of the column.

On the other hand, if $\gamma > 1$ the critical load decreases... The case $\gamma > 1$ can be interpreted as a strengthening of the foundation for $0 < x < L$ as compared to the foundation with constant modulus k_0. Hence, a strengthening of the foundation may have a destabilizing effect which is even increased with increasing values of k_0. This is again an unexpected result."

Lee and Yang (1994), using the transfer matrix method, reported a large variation of the critical flutter loads for a tapered column on an elastic foundation having a large variation in elastic modulus.

We know that paradoxical results may appear also for conservative problems (Parnes, 1977; Zaslavsky, 1979, 1981; Panovko and Sorokin, 1993). Still, a nagging question may arise: Why do these seemingly innocent columns under the "follower force" on elastic foundation generate so many paradoxical results? Some answers are appearing in the next section.

Criticism of Smith-Herrmann's Work

A strong criticism of the work by Smith and Herrmann (1972) is contained in the book by Panovko and Gubanova (1987). They write: " ... in 1972 a paper appeared by G. Herrmann (co-authored with T. Smith), in which the Beck's problem is considered with a complication in the form of continuous non-inertial elastic foundation, but the damping was set to be *absent* from the very start; it was found that the critical value of the load *does not depend* upon the coefficient of rigidity of the foundation.

As Voloshin and Gromov established in 1977, the result of Herrmann-Smith is *erroneous* (italics by the authors) and constituted direct consequence of too early a transfer to the ... ideal elastic model. I. I. Voloshin and V. G. Gromov considered the problem by Herrmann-Smith under the assumption that the material properties of the column are described by the standard linear viscoelastic model, and found that the critical force *depends* (italic by authors) on the coefficient of stiffness of the elastic foundation – even when the coefficients characterizing viscous properties vanish."

It must be noted that the papers by (Wahed, 1975; Anderson, 1975, 1976; Becker, Hauger, and Winzen, 1977; Kar, 1980; Celep, 1980; and Lee, 1996) addressed the issue of damping on the stability of columns with elastic foundation.

Anderson (1975, 1976) investigated the effect of internal damping, concentrated mass and rotary inertia on the stability of a column on a Winkler or a Wieghardt elastic foundation. He demonstrated that there was a dependence upon the moduli of elastic foundations. Wahed (1975) investigated Beck's column on elastic foundation taking into account the viscous damping. The solution was obtained by the Bubnov-Galerkin method. It was found that viscous damping increases the critical value of the load. The presence of the elastic foundation increased the stabilizing effect of damping. In the system without damping the result by Smith and Herrmann (1972) was recovered.

Becker, Hauger and Winzen (1977) studied the effect of external damping of the elastic support as well as internal damping of the column. Two distinct types of elastic foundations were included: a Winkler foundation and a rotary foundation, which causes restoring moments proportional to the slope of the column. The authors show that a small damping of the Winkler-type foundation "surprisingly does not change the critical load which, however, is considerably reduced by an inclusion of internal damping."

Sinha and Pawlowski (1984), studied a uniform cantilever column supported by a Maxwell type viscoelastic foundation and subjected to a constant tangential force. A Maxwell foundation is shown to produce a stabilizing effect. Moreover, an optimum combination of foundation parameters was shown to exist to yield the maximum flutter load. Increasing the foundation modulus beyond this optimum value leads to the decrease of the flutter load.

Kar (1980) studied the influence of a viscoelastic foundation on the stability of a linearly tapered cantilever beam of rectangular cross-section subjected to a follower end-load. The effect of internal damping in the beam is also included. A sufficiently small value of internal damping was shown to have a destabilizing effect for smaller values of the foundation stiffness. The author concludes that "Although an increase of the viscous damping parameter of the foundation raises the critical flutter load over a range of values of foundation elasticity, yet at very large values of foundation elasticity the value of critical flutter load for a lower value of viscous damping maybe higher than that calculated for a higher value of viscous damping."

Lee (1996) showed that the presence of a small amount of damping in a beam may reduce, increase, or may not affect the critical flutter load depending on the taper as well as the modulus of the elastic foundation.

Voloshin and Gromov (1977) conclude their study as follows: "… the elastic idealization of deformable systems in non-conservative stability problems becomes physically meaningless. Formal consideration of elastic systems leads Voloshin and Gromov (1976) to loss of solutions of the frequency equation and, as a consequence, to illegitimate results. We must apparently take the well-founded conclusions to be those that do not contradict the continuity principle when we go from real system to idealized systems." As Sedov (1980) notes: "During all time, in deep antiquity and in our days, scientists-thinkers always had the goal in their analyses to find the smallest number of basic hypotheses …"

The following question begs then to be asked: 'Does the concept of the undamped system under follower force entail a minimum number of parameters?" If indeed the damping must be included, then the system under follower force, but without damping, *i.e.* the elastic Beck's column would be an unrealistic system.

Koiter's Ideas on the Topic of Nonconservative Forces

In his paper, Koiter (1985) writes: "In the past decades much attention has been paid to stability problems of elastic structures under the action of non-conservative purely configuration-dependent loads, e.g., so-called follower forces whose directions follow directions of the structure."

Koiter (1980) reconsidered "the century-old Greenhill problem of buckling of a flexible shaft under torque loading conditions." The Greenhill's value for the critical torque W_{cr} for the simply supported shaft of flexural rigidity B and length L, equals $W_{cr} = \pi B / L$. The derivation was performed using a classical, static neutral equilibrium approach. Here Koiter (1980) makes the following comment: "... It appears impossible to achieve any non-conservative loading conditions in the laboratory by purely mechanical means."

He modified the Greenhill problem by considering a shaft supported by Cardan joints at its ends and the torque W transmitted by these joints. According to Koiter (1980) "the torque load transmitted by a Cardan joint is mechanically equivalent to the conservative quasi-tangential torque considered earlier by Ziegler. Analysis showed that the critical torque is roughly half the Greenhill value; moreover, it depended discontinuously on the relative position of Cardan joints in the undeformed and unloaded shaft."

One can visualize that even if the purely statically applied "follower forces" are non-existent and/or non-verifiable experimentally, they still may serve some *useful* purpose. For example, one may propose that they perhaps may serve as a (good) approximation of some *other* forces. The first candidate for such a realistic case is the pipe conveying fluid.

Is the Beck's Column a Good Model of a Fluid-Conveying Pipes?

This is a topic that attracted considerable attention in the literature. The reviews of (Païdoussis, 1987; and Païdoussis and Li, 1993) as well as the monographs by (Chen, 1987; Blevins, 1990; and Païdoussis, 1998) are recommended to the interested reader. The governing differential equation for the uniform pipe reads:

$$EI \frac{\partial^4 w}{\partial x^4} + \tilde{n} A v^2 \frac{\partial^2 w}{\partial x^2} + 2\tilde{n} A v \frac{\partial^2 w}{\partial x \partial t} + M \frac{\partial^2 w}{\partial t^2} = 0 \qquad (7)$$

where $M = m + \rho A$ is the mass per unit length of the pipe plus the fluid in the pipe. The first and the last terms in this equation are the usual stiffness and inertia terms.

According to Blevins (1990), "The second term from the left in Eq. (7) represents the force required to change the direction of the fluid to conform to the curvature of the pipe. The third term from the left represents the force required to rotate the fluid element as each point in the span rotates with angular velocity $\partial^2 w / \partial x \partial t$."

In the terminology of Païdoussis (1987) "… the various terms may be identified, sequentially, as the flexural restoring force, a "centrifugal" term, a Coriolis term, and the inertia term." The boundary conditions for a cantilever pipe clamped at $x = 0$ and free at $x = L$ are

$$\partial^2 w(L,t)/ \partial x^2 = \partial^3 w(L,t)/ \partial x^3 = 0 \tag{8}$$

One can observe a seeming similarity between Eqs. (1) and (7) without the term kw. If we note that the existing fluid is naturally tangential to the bent pipe one may think that in Eq. (7) one has a follower force system with an "added" Coriolis force.

Païdoussis (1987, p.165) writes: "… Herrmann and Nemat-Nasser (1967) …, elucidated the connection between this (pipe conveying fluid-I.E.) and Beck's classical problem of a column subjected to a tangential follower-type load at the free end (*i.e.*, a load remaining tangential to the oscillating free end)."

Indeed, Herrmann and Nemat-Nasser (1967) write, after deriving a governing equation for "a cantilevered bar of uniform cross-section with flexible pipes conveying fluid attached to it," "Equations… are analogous to those obtained by present authors Nemat-Nasser and Herrmann, (1966) for cantilevered bars subjected at the free end to follower forces except for the third term in the first equation which is due to Coriolis acceleration."

Crandall (1995) writes: "A pure follower force is difficult to realize physically, but can be approximated by the jet reaction of a fluid stream flowing through a hollow tube in the column and exiting through a tangential nozzle at the tip of the column. The critical load for the combined fluid-structure system approaches that for Beck's column when the ratio of the fluid mass to the column mass approaches zero."

Interestingly, Païdoussis (1987) exclaims: "Indeed, the fact that a cantilever (that conveys fluid – i.e.) is not only a non-conservative problem similar to Beck's but also gyroscopic, explains the fascination it exerted and does so still, on applied mechanicians and mathematicians for the last twenty-five years. An additional advantage of this system is that it can readily be studied experimentally, unlike the original Beck problem which requires a rocket engine mounted to the free end of a beam column, or something similar!"

Païdoussis ((1998), p.170) also mentions: "In the case of a cantilevered beam with a tangential end-load at the free end, representing Beck's problem …, there is no simple way of minimizing the effect of fluid supply lines. Nevertheless, a successful experiment was conducted by Sugiyama et al. by attaching a solid-fuel rocket to the free end!… Also, not only the mass but the moment of inertia of the motor had to be taken into account. Agreement of experiment with theory is excellent, provided dissipation is ignored; once taken into account, viscoelastic damping of the column ($\alpha = 5.10^{-4}$) is found to diminish the theoretical critical thrust by a factor of 2 as compared to the undamped system. Thus rendering agreement rather poor. However, once the criterion for stability in finite time is used, the two sets of theoretical results come very close to each other, thus leading to very good agreement with experiment."

Due to the necessity to pick and choose the stability criterion, it appears worthwhile to quote from Lamb (1908) who stressed, nearly a century ago, "the difficulty of framing a definition of kinetic stability which shall be comprehensive and at the same time confirm to natural prepossession has long been recognized."

In accordance to Sugiyama and Païdoussis (1981), "cantilevers conveying fluid are rare examples of non-conservative systems with which one can verify theoretical predictions by corresponding experiments." "Certainly, some aspects of the problem have been known for a long time and are in almost everyone's common experience. Then the *buckling* (divergence) of a pipe with both ends supported, manifested by the large restraining force that must be exerted by those holding a fire-hose at high discharge rates, is also experienced albeit highly diminished, by one watering the lawn. The *flutter* of a cantilevered pipe, manifested by the thrashing, snaking motions of the fire-hose when released or by a garden-hose when dropped on the grass is well known to firemen and gardeners alike. In fact, these two phenomena are often irreverently but graphically referred to as the *fire-hose* and *garden-hose* instability, respectively."

As is clearly seen the justification or falsification of a theory may turn out to be a very difficult task. It is harder also to reach consensus since the interested parties often but not always try to "save" the candidate theory from falsification. The above general thoughts appear to be quite useful to bear in mind when confronted with *falsification* of numerical results associated with theoretical and/or experimental analyses in the literature on the "follower forces."

The experimental results on pipes conveying fluid are extensively described in a definitive monograph by Païdoussis (1998) and will not be described here. Namely, section 3.4.4 (pp. 103-111) describes experimental results on pipes simply supported at both ends, whereas numerous experiments associated with cantilevered pipes are discussed in section 3.5.6 (pp. 133-148).

It appears that the concept of "follower forces" has somehow escaped the process of model validation. It appears instructive to quote Zimmerman (2000) explaining the notion of model validation: "Model Validation is the qualitative part of the validation and verification process. It is a substantiation of the correctness of the model in the sense that it adequately represents the physics of a component or system. The physics are correctly represented when the mathematical form of the model *i.e.*, its constitutive equations of motion are capable of producing results which agree essentially with experimental data, given suitable parameter values."

It is important to keep in mind that,

1. "Model validation is a difficult process, and the majority of the effort is often adjusting the parameters of a model which do not have the correct physics;

2. appropriate experimental data must be available to identify the structure properly;

3.   exact, full analysis costs are typically too high, such that approximate analysis techniques must be available, and

4.   the selection of appropriate parameters to update is often difficult."

One may think, at the first glance, that model validation and the verification of any system must be a very difficult task, if all possible. Such a conclusion would however be opposite from being mature. Zimmerman (2000) demonstrates in his insightful article, that even large and complex structures are amenable to model validation and verification. Namely, he considers two large complex systems: The Russian Mir Station coupled to the U.S. Space Shuttle, and the P6 segment of the International Space Station in the launch configuration. If such systems lend themselves to model validation, why not to conduct detailed studies for the structures under follower forces? (Interesting material on matching analytical models with the experimental data is presented by Inman and Minas, 1990.)

The Effect of an Elastic Foundation on the Stability of a Pipe Conveying Fluid

As it was mentioned above an interest in the effect of an elastic foundation arose due to Smith & Herrmann's (1972) unexpected finding that for the uniform Beck's column on constant-modulus Winkler elastic foundation the critical flutter load is independent of the foundation modulus. This led to a strong criticism by (Voloshin and Gromov, 1977; and Panovko and Gubanova, 1987) who advocated that the paradox appeared due to setting the system to be elastic *ab initio*. Since the Beck's column is obtainable by neglecting the Coriolis force in a pipe conveying fluid, one could anticipate that if such a neglect is not performed, i.e., if one considers a real system, rather than purely statically applied "imaginary" system perhaps the paradoxical result would not appear.

Indeed, this is the case. As Lottati and Kornecki (1986) found the effect of elastic Winkler foundation is stabilizing, i.e., the flutter load of the pipe with elastic foundation is greater than that of the pipe without elastic foundation (see also papers by Roth, 1964, and Becker, 1980). According to Païdoussis (1998, p.149) "like gravity, the foundation provides an additional restoring force, which stabilized the system."

Becker, Hauger and Winzen (1978) studied also the effect of the rotary-foundation on the stability of the pipe conveying fluid. This amounts to introducing the term $-c\partial^2 w/\partial x^2$ in the equation of motion, leading to the reduced "effective" centrifugal force $\left(\rho A v^2 - c\right)\partial^2 w/\partial x^2$. This leads also to a stabilizing effect.

The effect of partial elastic foundation was considered in recent studies by Impollonia and Elishakoff (2000) and Elishakoff and Impollonia (2001). The former study dealt with an articulated pipe conveying fluid on either partial or full Winkler and/or Pasternak-type rotary foundation. It turned out that if the instability arises with divergence, the critical velocity of the pipe with elastic foundations may be smaller than that in the unsupported pipe. The introduction of a partial elastic foundation may reduce the critical velocity of the system, not unlike the findings by (Benjamin, 1961; and Gregory and Païdoussis, 1966), where the critical velocity reduction occurs by touching the pipe (without elastic foundations) with the finger. The divergence instability occurs when the elastic foundation goes beyond the first pipe. However, if the elastic foundation is confined within the first pipe, the system experiences flutter exclusively. *Full* elastic foundations of either kind increase the critical velocity in the case where stability is lost by flutter.

In the study by Elishakoff and Impollonia (2001) the effect of elastic Winkler and Pasternak-type rotary foundations on the stability of pipe conveying fluid. Both elastic foundations were partially attached to the pipe. For our purposes it ought to be stressed that the *single* foundation, either translatory nor rotatory, which is attached to the pipe along its *entire* length, *increases* the critical velocity. Such an intuitively anticipated strengthening effect is surprisingly *missing* for the elastic column on Winkler foundation subjected to so-called statically applied follower forces (Smith and Herrmann, 1972). Thus the sharp criticism expressed by Voloshin and Gromov (1977) and Panovko and Gubanova (1987) to the latter problem are inapplicable to the cantilevered pipe conveying fluid.

Djondjorov (2001) also studied the effect of partial Winkler elastic foundation on the stability of fluid conveying pipes. The objective was to "examine the influence of foundation length, position and rigidity on the critical velocities of such pipes." The Boobnov-Galerkin method was applied with a 10-term approximation. It was established that for small ratios of the masses of the fluid and the pipe, the maximal stabilizing effect of a Winkler foundation of certain length is achieved if it supports the free end of the pipe. For higher values of the mass ratio a foundation applied at a free end can either stabilize or distabilize the pipe depending on the foundation length. The author noted that "an elastic foundation applied at the midpoint is found to have a stabilizing effect on the pipe."

Djondjorov, Vasilev and Dzhupanov (2001) consider, *inter alia,* the elastic foundation with the stiffness

$$k(x) = \frac{EI}{L^4}k_0\left[4(1-\gamma)\left(\frac{x^2}{L^2}-\frac{x}{L}\right)+1\right] \tag{9}$$

where k_0 and γ are given constants. The elastic foundation supports a pipe conveying fluid. It is shown that a critical flow velocity of a straight cantilevered pipe on Winkler foundation increases together with the foundation stiffness $C_0 = k_0 EI/L^4$. Note that a parallel study on the effect of the foundation with the above modulus, *k(x),* was studied by Hauger and Vetter (1976). Djondjorov, Vassilev, and Dzhupanov (2001) also studied inhomogeneous elastic formulations of the polynomial type

$$k(x) = \frac{EI}{L^4}\left[k_0 + \sum_{n=1}^{6} k_n\left(\frac{x}{L}\right)^4\right] \tag{10}$$

Authors conclude: "obviously, strengthening of the foundation improves significantly (up to 82%) the stability of the pipe." They also summarize their work as follows: "Each foundation in this paper has a stabilizing effect on the straight cantilever pipes." The results derived by Elishakoff and Impollonia (2001), Djondjorov, Vassilev and Dzhupanov (2001), and Djornjorov (2001), appear, therefore, more realistic, than those derived by various investigators for the columns under "follower forces." This may again suggest that the follower force model may well constitute an oversimplification.

It appears to us that it is advisable to abandon the purely elastic Beck's column under purely statically applied follower forces, whose origin ought to look mysterious to a keen student of the subject.

A best simple model for the non-conservative stability problems appears to be the problem of the *pipe conveying fluid*. Indeed, as Païdoussis and Li (1993) write,

"... the pipe conveying fluid ... has established itself as a generic paradigm of kaleidoscope of interesting dynamical behavior.

Some of the reasons why this is so are the following:

(i)   it is a physically simple system, easily modeled by simple equations;
(ii)   it is a fairly easily realizable system, which thereby affords the possibility of theoretical and experimental investigation in parallel;
(iii)   this being a more general problem than that of the column and in some ways of the rotating shaft, yet including their essential characteristics, may be thought as complementing them;
(iv)   it is a problem in the large category of dynamical systems involving momentum transport, such as travelling chains and bands, chain-saw blades, etc....
(v)   being the simplest possible fluid-structure interaction system, it offers additional possibilities; e.g., examining the effects of fluid compressibility or viscidity on system dynamics."

This does not imply that there are no jet or rocket motor excited vibrations (see *e.g.* Kacprzynski and Kaliski, 1960; papers by Glaser, 1965; Feodosiev, 1965; Beal, 1965; Mladenov and Sugiyama, 1997; Kirillov and Seyranian, 1999). Yet these are not approximated, simplistically, by the purely elastic Beck's column. There appears nothing objectionable to producing follower forces by "smart" means (Chaudhry and Rogers, 1991; Tanaka and Idehara, 1999), or via automatic control (see *e.g.*, works by Horikawa, 1977; Wu, 1976a, 1976b; Park, 1985; Park and Mote, 1985).

The paradoxical result by Smith and Herrmann (1972) and the subsequent literature on this topic are discussed in detail by Dzhupanov (2001a, 2001b). The latter author devoted several articles to the Smith-Hermann paradox and gave spirited account of it, calling it a "great discussion."

Conclusion

It appears that the column subjected to so-called follower force does not represent a good approximation of the fluid-conveying pipe. This conclusion complements Koiter's (1996) recommendation "beware of unrealistic recommendation follower forces." There appear to be plenty of reasons to distance ourselves from characterizing so called follower forces as "realistic" (see Sugiyama, Langthjen, and Ryu, 1999). Likewise, the characterization of the "Beck's Column as the Ugly Duckling" (Sugiyama, Ryu, and Langthjem, 2002) from the fairy tales of Hans Christian Andersen appears to be beautiful but not (at least yet) substantiated. Indeed, Bolotin's (1999) statement appears to be not accidental: "I am not a partisan of *follower* forces. It seems that since 1961 when I had published my book, I never returned to the topic." Moreover, ".... most of the publications on the *follower* forces are of purely academic origin and seem artificial."

Still, "the fictitiously applied follower forces acting on a given system .... arbitrarily prescribed to depend on a certain manner on the deformation" (in the terminology of Herrmann, 1967) were not altogether negative.

It seems that they can be included in a course on elastic stability, as a didactic material, associated with curious paradoxes, if not in the engineering journals until there will be some novel development which may impart a new life to this nation. At the same time, one has to have in mind Arnold's (1998) statement: "It is important, that the simplest model be structurally stable, ie. that the conclusions derived from its investigation allow small variation of parameters and functions, that describe the model." Smith-Herrmann (1972) model vividly showed that the simplest model that exhibits a flutter instability – the column under so called follower forces – is structurally unstable. Fluid-conveying pipe, or a plate in supersonic flow therefore, appear to be a simplest useful models to exhibit the dynamic instability in question.

References

Anderson, G.L., 1975. The influence of rotary inertia, tip mass and damping on the stability of a cantilever beam on an elastic foundation. *Journal of Sound and Vibration*, **43**, 543-552.

Anderson, G.L., 1976. The influence of a Wieghardt type elastic foundation on the stability of some beams subjected to distributed tangential forces. *Journal of Sound and Vibration*, **44**(1), 103-118.

Arnold, V.I., 1998. "Hard" and "Soft" Mathematical Models. *Priroda*, **4**, 3-14 (in Russian).

Beal, T.R., 1965. Dynamic stability of s flexible missile under constant and pulsating thrusts. *AIAA Journal*, **3**, 486-494.

Beck, M., 1952. Die Knicklast des einseitig eingespannten tangential gedrückten stabes. *ZAMP*, **3**, 225-229 (in German).

Becker, M., Hauger, W., and Winzen, W., 1977. Influence of internal and external damping on the stability of Beck's column on an elastic foundation. *Journal of Sound and Vibration*, **54** (3), 468-472.

Becker, M., Hauger, W., and Winzen, W., 1978. Exact stability analysis of uniform cantilevered pipes conveying fluid or gas. *Archives of Mechanics* (Warsaw), **30**, 757-768.

Becker, O., 1980. Schwingungs- und Stabilitaets – Verhalten des durchstroemten Rohres. *Kernemergie*, **23**, 337-342 (in German).

Benjamin, T.B., 1961. Dynamics of a system of articulated pipes conveying fluid. I Theory, *Proceedings of the Royal Society of London*, Series A, 261, 487-499.

Blevins, R.D., 1990. *Flow-Induced Vibrations*. Van Nostrand-Reinhold, New York.

Bolotin,V.V., 1961. *Non-conservative problems on the theory of elastic stability*. Fizmatgit Publishers, Moscow (in Russian). (English translation, Pergamon Press, New York, 1964).

Bolotin, V.V., 1999. Dynamic instabilities on mechanics of structures. *Applied Mechanics Reviews*, 52(1), R1-R9.

Celep, Z., 1977. On the stability of a column on an elastic foundation subjected to a non-conservative load. Proceedings. Sixth National Science Congress, The Scientific and Technical Research Council of Turkey, 301-311.

Celep, Z., 1980. Stability of a beam on an elastic foundation subjected to a non-conservative Load. *Journal of Applied Mechanics*, 47, 111-120.

Chaudhry Z., and Rogers, C.A., 1991. Response of composite beams to an internal actuator force. Paper AIAA-91-1166-CP, 186-193.

Chen S.S., 1987. *Flow-Induced Vibration of Circular Structures*. Hemisphere, Washington D.C.

Crandall, S.H., 1995. The effect of damping on the stability of gyroscopic pendulums. *ZAMP* 46, special issue, S761-S780.

Djondjorov P.A., 2001. Dynamic stability of pipes partly resting on Winkler foundation. *Journal of Theoretical and Applied Mechanics*, (Sofia), 31 (3), 101-112.

Djondjorov, P., Vassilev, V., and Dzhupanov, V., 2001. Dynamic stability of fluid conveying cantilever pipes on elastic foundations. *Journal of Sound and Vibration*, 247 (3), 533-546.

Dzhupanov, V.A., 2001a. Twelve methodical notes on the paradoxical results in a class of dynamical problems, Part 1. *Journal of Theoretical and Applied Mechanics*, (Sofia), 31 (3), 74-100.

Dzhupanov, V.A., 2001b. Twelve methodical notes on the paradoxical results in a class of dynamical problems, Part 2. *Journal of Theoretical and Applied Mechanics*, (Sofia), 31 (4), 55-72.

Elishakoff, I., and Impollonia, N., 2001. Does a partial elastic foundation increase the flutter velocity of a pipe conveying fluid?. *Journal of Applied Mechanics*, 68, 206-212.

Elishakoff, I., and Wang, X.F., 1987. Generalization of Smith-Herrmann problem with the aid of computerized symbolic algebra. *Journal of Sound and Vibration*, 117(3), 537-542.

Feodosiev, V.I., 1965. On a problem of stability. *PMM*, 2, 391-392. (in Russian).

Glaser, R.F., 1965. Vibration and stability of analysis of compressed rocket vehicles. NASA TN D-2533.

Gregory, R.W., and Païdoussis, M.P., 1966a. Unstable oscillation of tubular cantilevers conveying fluid: II Experiments. *Proceedings of the Royal Society of London*, Series A, 293, 512-527.

Hauger, W., and Vetter, K., 1976. Influence of an elastic foundation and the stability of a tangentially loaded column. *Journal of Sound and Vibration*, 42, 296-299.

Herrmann, G., 1967. Stability of equilibrium of elastic systems subjected to nonconservative forces. *Applied Mechanics Reviews*, 20, 103.

Herrmann, G., and Nemat-Nasser, S., 1967. Instability modes of cantilevered bars induced by fluid flow through attached pipes. *International Journal of Solids and Structures*, 3, 39-52.

Horikawa, H., 1977. Active feedback control of an elastic body subjected to a non-conservative force. Ph. D. Dissertation, Princeton University.

Impollonia, N., and Elishakoff, I., 2000. Effect of elastic foundation on divergence and flutter of an articulated pipe conveying fluid. *Journal of Fluids and Structures*, **14**, 559-573.

Inman, D.J., and Minas, C., 1990. Matching analytical models with experimental data in mechanical systems. *Control and Dynamics Systems*, **37**, 327-363.

Jacoby, A., and Elishakoff, I., 1986. Discrete-continuous elastic foundation may leave the flutter load of the Pflüger column unaffected. *Journal of Sound and Vibration*, **108**, 523.

Kacprzynski, J., and Kaliski, S., 1960. Flutter of a deformable rocket in supersonic flow. Proc. Of the 2nd International Congress in the Aerospace Sciences, Zuerich, Sept. 12-16, 911-925.

Kar, R.C., 1980. Stability of a non-uniform viscoelastic cantilever beam on a viscoelastic foundation under the influence of a follower force. *Solid Mechanics Archives*, **4**, 457-473.

Kirillov, O.N., and Seyranian, A.P., 1999. Optimization and stability of a flying column. Proceedings, 3rd World Congress of Structural and Multidisciplinary Optimization (May 17-21, Buffalo, N.Y.).

Koiter, W.T., 1980. Buckling of a flexible shaft under torque loads transmitted by cardan joints. *Ingenieur-Archiv*, **49**, 369-373.

Koiter, W.T., 1985. Elastic stability. *Zeitschrift für Flugwissenschaften und Weltraumforschung*, **9**(4), 205-210.

Koiter, W.T., 1996. Unrealistic follower forces. *Journal of Sound and Vibration*, **194**, 636-638.

Lamb, H., 1908. *Proceedings of the Royal Society of London*. Series A, **80**, 168.

Lee, H.P., 1996. Dynamic stability of a tapered cantilever beam on an elastic foundation subjected to a follower force. *International Journal of Solids and Structures*, **33** (10), 1409-1424.

Lee, S.Y., and Yang, C.C., 1994. Non-conservative instability of non-uniform beams resting on an elastic foundation. *Journal of Sound and Vibration*, **169**, 433-444.

Lottati, I., and Kornecki, A., 1986. The effect of elastic foundation and dissipative forces on the stability of fluid conveying pipes. *Journal of Sound and Vibration*, **109**, 327-338.

Mladenov, K.A., and Sugiyama, Y., 1997. Stability of a jointed free-free beam under end rocket thrust. *Journal of Sound and Vibration* **199**, 1-15.

Nemat-Nasser, S., and Herrmann, G., 1966. Torsional instability of cantilevered bars subjected to non-conservative loading. *Journal of Applied Mechanics*, **33**, 102-121.

Païdoussis, M.P., 1987. *Flow-induced instabilities of cylindrical structures. Applied Mechanics Reviews*, **40**, 163-175.

Païdoussis, M.P., 1998. *Fluid-Structure Interaction, Slender Structures and Axial Flow*. Academic Press, London.

Païdoussis, M.P., and Li, G.X., 1993. Pipes conveying fluid: a model dynamical problem. *Journal of Fluid and Structures*, **7**, 137-204.

Panovko, Ya., G. and Gubanova, I.I., 1987. *Stability and vibrations of elastic systems*. Nauka Publishers, Moscow, 131 (in Russian, 4th edition).

Panovko, Ya., G., and Sorokin, S.V., 1993. The paradoxical influence of an increase of rigidity on buckling loads and natural frequencies for elastic systems. *Mechanics Research Communications*, **20** (1), 9-14.

Park, Y.P., and Mote, C.D., 1985. The maximum controlled follower force on a free-free beam carrying a concentrated mass. *Journal of Sound and Vibration*, **98** (2), 247-256.

Park, Y.P., 1985. Dynamic stability of a free–free Timoshenko beam under controlled follower force. Proceedings of the JSME Vibration Conference, **85**, 43-53.

Parnes, R., 1977. A paradoxical case in stability analysis. *AIAA Journal*, **15**(10) 1533-1534.

Roth., W., 1964. Instabilitaet durchstroemter Rohre. *Ingenieur-Archiv*, **33**, 236-263 (in German).

Sedov, L.I., 1980. *Thoughts About Science and About Scientists*. 50, "Nauka" Publishers, Moscow (in Russian).

Sinha, S.C., and Pawlowski, D.R., 1984. Stability analysis of a tangentially loaded column with a Maxwell type viscoelastic foundation. *Acta Mechanica*, **52**, 41-50.

Smith, T.E., and Herrmann, G., 1972. Stability of a beam on an elastic foundation subjected to a follower force. *Journal of Applied Mechanics*, **39**, 628-629.

Sugiyama, Y., Langthjem, M.A., and Ryu, B.-J., 1999. Realistic Follower Forces. *Journal of Sound and Vibration*, **225**(4), 779-782.

Sugiyama, Y., Ryu, S.-U., and Langthjem, M.A., 2002. Beck's Column as the Ugly Duckling. *Journal of Sound and Vibration*, **254**(2), 467-410.

Sugiyama, Y., and Païdoussis, M.P., 1981. Studies on stability of two-degree-of-freedom articulated pipes conveying fluid: the effect of characteristic parameter ratios. *Theoretical and Applied Mechanics*, **31**, 333-342.

Sundararajan, C., 1976. Stability of columns on elastic foundations subjected to conservative or non-conservative forces. *Journal of Sound and Vibration*, **37**, 79-85.

Tanaka, M., and Idehara, R., 1999. Robust shape design of non-conservative structural system with piezoelectric elements. Proceedings of the Third World Congress of Structural and Multidisciplinary Optimization (May 17-21, Buffalo, N.Y.).

Voloshin, I.I., and Gromov, V.G., 1976. Stability of cantilevered viscoelastic bar loaded by a follower force. *Mechanics of Solids*, **11** (4), 157-161.

Voloshin, I.I., and Gromov, V.G., 1977. On a stability criterion for a bar on an elastic base acted on by a follower force. *Mechanics of Solids*, **12**(4), 148-150.

Wahed, I.F.A., 1975. The instability of a cantilever on an elastic foundation under influence of a follower force. *Journal of Mechanical Engineering Science*, **33**, 219-222.

Wu, J.J., 1976a. Missile stability using finite elements – unconstrained variational approach. *AIAA Journal*, **14**, 313-319.

Wu, J.J., 1976b. On missile stability. *Journal of Sound and Vibration*, **49**(1), 141-147.

Zaslavsky, A., 1979. Increased length may enhance buckling load. *Israel Journal of Technology* **17**(3).

Zaslavsky, A., 1981. Discussion on destabilizing effect of additional restraint on elastic bar structures. *International Journal of Mechanical Sciences*, **23**(6), 383-384.

Zimmerman, D.C., 2000. Model validation and verification of large and complex space structures. *Inverse Problems in Engineering*, **8**, 93-118.

Engineering of Railway Tracks

Allan M. Zarembski, ZETATECH Associates, Inc.

Abstract

The railroad industry has been the subject of evolution, technological innovation, and engineering for nearly 200 years. This evolution of the railroad track, and its key components, has been paralleled by an evolution in railroad engineering. This paper presents an overview of railroad track engineering to include both design engineering and maintenance engineering. The paper also includes a discussion of the load environment to which track is subjected and which dominates railroad engineering today. Among the engineering perspectives addressed are the ability of the track and its major components to resist catastrophic failures such as a broken rail or track buckle, and the ability of the track to resist long term degradation or deterioration, such as through rail wear, fatigue, tie or ballast abrasion, and deterioration of geometry. A brief discussion of factors influencing future developments of railroad track engineering is also included.

Introduction

The railroad industry has been the subject of evolution, technological innovation, and engineering for nearly 200 years. The track structure, whose evolution is described in detail by Kerr (2003), likewise evolved through a combination of incremental improvements and technological innovation, such as the introduction of rolled steel sections into the rail manufacturing process which allowed the design and manufacture of the "T" rail section. The modern railway track structure, which consists of the "T" rail section, supported by discretely spaced crossties, was originally introduced in the mid-nineteenth century and continued to evolve through the introduction of more robust components, new materials, and improved component designs. Yet, many of these components continue to see service, not only on lightly used branch lines, where fifty-year-old rails and ties can still be found, but also on high density main-lines, where bridges built at the turn of the century continue to see active service. This evolutionary process is illustrated in Fig. 1 in the design of the rail section.

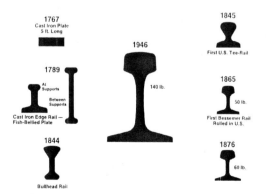

Fig. 1. Evolution of rail sections (Kerr, 2003)

This evolution of the railroad track, and its key components, has been paralleled by an evolution in railroad engineering. While nineteenth-century railroad engineering focused on "building" the railroads, with a strong emphasis on construction techniques, bridge and tunnel engineering and route alignment engineering, modern railroad engineering is focusing on improved analytical tools to improve the track structure's strength and ability to carry the ever increasing loads on the modern railway. Thus, while early railway engineers could get away with simple empirical formulas and rules of thumb to construct track that would support 8,000 lb. wheel loads, currently engineers must cope with 39,000 lb. wheel loads which generate stresses that approach or exceed the strength of the track materials. To do this efficiently and effectively requires sophisticated analytical tools and techniques to help develop design and maintenance standards for modern rail operations.

The focus of this paper is the modern railroad track structure, the environment in which it performs, and the tools used to optimize its performance.

Overview of the Railroad Track

The purpose of the railroad track structure is to support the loads of cars and locomotives and guide their movement (Hay, 1982). As part of this function, the track structure must withstand the loadings applied to it by both the vehicles and the environment and distribute these loads through a series of component elements, to a level that can be carried by the parent soil or subgrade beneath the track.

Fig. 2 presents an overview of the track. As can be seen, the rail is the structural member that contacts with and supports the wheels of the railway vehicles, and starts the process of load distribution by spreading each individual wheel load over a number of crossties. With static wheel loads of 36,000 lbs. and higher, the wheel/rail contact stresses generated are of the order of 100,000 psi, which can exceed the yield strength of the rail steel (though are generally below the steel's ultimate tensile strength). In addition, bending stresses of the order of 25,000 psi are also generated in the rail by the vertical loads, which when combined with lateral and longitudinal loading in the rail likewise can reach the strength of the rail steel.

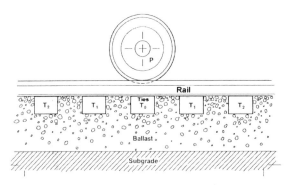

Fig. 2. Railroad track structure

As noted previously, the rail "beam" distributes the vertical wheel load to between 6 and 10 crossties[1] with contact stress levels of the order of 100 to 200+ psi at the surface of the tie. The tie, as illustrated in Fig. 2, in turn distributes this load to the top of the "rock" ballast layer, further reducing the level of stress to the order of 50 to 100 psi. Finally the ballast layer, with its associated sub-ballast layer (located between the ballast and subgrade) finishes this stress reduction process, bringing the level of stress to the order of 5 to 20 psi, a level that can be withstood by the parent soil material which most often makes up the final subgrade layer of the track. Fig. 3 further illustrates this load distribution behavior as a pyramid of stress reduction, with the corresponding AREMA[2] design limits defined for each component "layer" (AREMA, 2003).

*AREMA design limits

Fig. 3. Pyramid of bearing stresses

Engineering analyses of the railroad track structure must focus not only on the strength of the track and its components to resist catastrophic failure such as a broken rail or a track buckle, but also on its ability to resist long term degradation or deterioration, such as through rail wear, fatigue, tie or ballast abrasion, deterioration of geometry, etc. In fact, it is this maintenance of the track structure over extended periods of time (track component lives can extend over decades of severe service), which dominates railroad engineering today.

The Load Environment

In order to properly understand the issues of railroad engineering it is necessary to understand the load environment to which the track is subjected. This includes not only the vehicle induced loadings, vertical, lateral and longitudinal, but also any external loading such as the thermal loading of the rails generated by changes in ambient temperature.

---

[1] Tie plates are used between the rail and the crossties to increase the bearing area or the surface of the ties and reduce tie surface stresses.
[2] American Railway Engineering and Maintenance of Way Association, formerly known as the American Railway Engineering Association (AREA).

Vertical Load

The vertical wheel loading on the track structure represents a key load input and is often the primary load used in the engineering of the track. It has been the focus of most of the major engineering changes to the modern track structure. For example, the growth in vehicle weight and associated wheel loading has dominated the engineering of the track structure in the last century; with a quadrupling of vehicle wheel loads from the turn of the century (wheel loads of 8,000 lb) to today (wheel loads of 36,000 lb+) (Kerr, 2003). This growth is illustrated in Fig. 4, which shows the evolutionary growth in axle loads[3] from the 1920s through the year 2000.[4] The pace of this growth in axle load (and car weight), which is driven by the economics of heavier cars with improved carrying efficiency[5] (Zarembski and Blaze, 2003), has been set by the ability of the track structure to support these loads with minimum component failure and maximum component life. Thus it appears, at least to the track engineer, that no sooner has the design and maintenance of the track structure been brought to a level appropriate to the operation of the railroad, then the railroad makes use of this well engineered track structure to improve its operating efficiencies by increasing the weight of the cars.

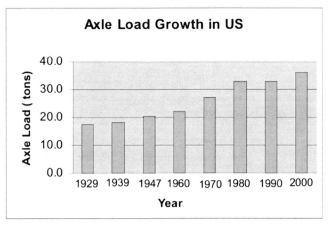

Fig. 4. Growth in axle loads from 1929-2000

According to Zarembski and Blaze (2003), Table 1, 36,000 lb. wheel loads represent the design loads for railroad operations in North America today. However, limited experience in the mining railroads of Western Australia together with extensive research and testing in the U.S. is already opening the way for a future increase in this load limit to 39,000 lbs. While European railways have tried to keep their load levels to 24,000 – 27,500 lbs., the demands of modern railroad economics are forcing them likewise to increased loadings, with recent introduction of 33,000 wheel loads for iron ore freight operations in Sweden leading the way.

---

[3] Axle load is double the static wheel load based on the standard two-wheel axle.
[4] Based on growth of carloads as defined in AAR Railroad Facts, 2002 edition.
[5] Car carrying efficiency is defined in terms of net to tare ratio, the ratio of goods carried in the car to the "dead" weight of the car itself.

Table 1 Static Wheel Loads – Worldwide
(Zarembski and Blaze, 2003)

Axle Load		Gross Weight of Cars		Status
Tonnes	Tons	kg	lb	
22	24	88,000	193,600	Common European Limit
25	27.5	100,000	220,000	UK and Select European limit
30	33	120,000	263,000	North American free interchange limit
				BV (Sweden) limit on Malmbanan Ore Line
32.5	36	130,000	286,000	Current HAL weight for North American Class 1 RR
35.5	39	142,000	315,000	In use in Australia
				Currently under test in US
				Limited use in double stack operations

The static load, however, is only a portion of the picture. Dynamic augments to the static loads, due to dynamic effects of track geometry imperfections, rail or wheel surface defects, increased operating speeds, stiffness transitions, etc., can dramatically increase these load levels. According to Scott[6] (1986), until recently, rare impact loads of the type illustrated in Fig. 5, with a dynamic impact factor of almost 4, had the ability of causing severe track damage and associated component failure. Only recently has the use of wayside force measurement systems allowed for the monitoring and control of these very high forces. With the use of these monitoring systems, railroads in North America recently introduced a 90,000 lb. dynamic wheel load limit, which still represents a factor of almost 3 times the static wheel load. While these load levels are quite rare, field measurement of dynamic wheel loads have found that between 0.1% and 0.5% of all freight car wheels on high density corridors that see 286,000 tons cars (with 36,000 lb. static wheel load) experience dynamic load levels exceeding 75,000 lbs, more than double the static load level.

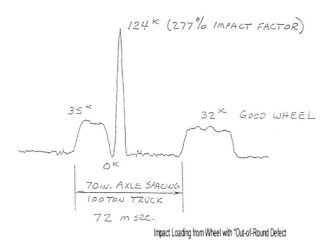

Fig. 5. Impact loads from wheel with "out-of-round" defect (Scott, 1986)

---

[6] The same behavior is noted by Talbot (1980) and Kerr (2003).

To represent this dynamic behavior, several analytical approaches have been introduced over the years, including empirical "quasi-static" dynamic augment equations and sophisticated multi-degree of freedom vehicle track interaction models. Fig. 6 (Doyle, 1980) presents several of the more common quasi-static impact formulas that have been developed worldwide. The most common one used in the US today, the AREA (now AREMA) formula (AREMA, 2003) is

$$V_{dynamic} = V_{static}\ (1+33\ v/100D) \tag{1}$$

Where

v	=	speed in mph
D	=	wheel diameter in inches
$V_{dynamic}$	=	quasi-static dynamic load
V_{static}	=	static wheel load

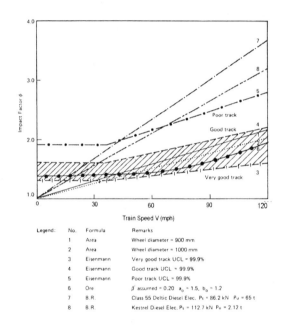

Fig. 6. Comparison of impact factor formulae (Doyle, 1980)

Alternatively, one of the more common representations of this dynamic load is the P1 and P2 impact load shown in Fig. 7, where a high impact short duration P1 force is followed by a longer duration lower amplitude impact force P2 (Jenkin, 1974; Ahlbeck, 1980). While the higher frequency P1 is generally most damaging to the rail and tie components, the lower frequency P2 force causes ballast damage and corresponding track degradation. The P1 force is given by

$$P_1 = P_0 + 2\alpha v \sqrt{\frac{K_h M'}{1+\dfrac{M'}{M_u}}} \tag{2}$$

P_1 =   Dynamic Wheel Load (lbs)
P_0 =   Static Wheel Load (lbs)
α =   Rail Joint Dip Angle (rad)
v =   Vehicle Speed (in/sec)
K_h =   Hertzian Contact Stiffness (lb/in)
M' =   Effective Mass of Rail and Tie (lb-sec^2/in)
M_u =   Unsprung Vehicle Mass (lb-sec^2/in)

The P_2 forces are

$$P_2 = P_0 + \left[1 - \pi\xi^2 \frac{M_t}{M_u + M_t}\right][2\alpha V]\sqrt{K_t M_u}\left[\frac{M_u}{M_u + M_t}\right]^{\frac{1}{2}} \qquad (3)$$

The variables in the above equation are defined as follows:

P_2 =   Dynamic Wheel Load (lbs)
P_0 =   Static Wheel Load (lbs)
ξ =   Track Effective Damping Ratio
M_t =   Effective Track Mass (lb-sec^2/in)
M_u =   Unsprung Vehicle Mass (lb-sec^2/in)
α =   Rail Joint Dip Angle (rad)
v =   Vehicle Speed (in/sec)
K_t =   Effective Track Stiffness (lb/in)

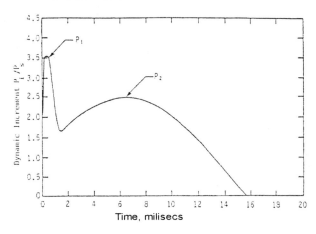

Fig. 7. Dynamic wheel/rail force vs. time (Jenkin, 1974)

Finally, multi-degree of freedom vehicle–track interaction models have been used to define a truer dynamic loading environment. The most common of these models; NUCARS (US), VAMPIRE (UK), A'GEM (Canada), etc. are often used in the definition of the dynamic load environment for a specific wheel/rail dynamic interaction situation. According to Zarembski and Abbott (1978), for broader analyses, statistical distributions of loading from large-scale wayside tests, such as illustrated in Fig. 8 for general interchange traffic are often used.

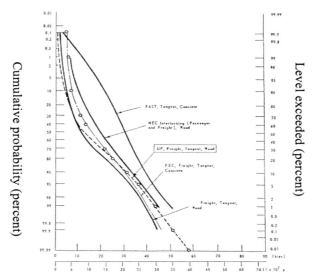

Fig. 8. Peak vertical wheel-rail load (Zarembski and Abbott, 1978)

Lateral Load

The lateral load environment is one that is often underestimated in railroad engineering, however it represents a major load condition, particularly in curves. Because railway vehicles have rigid axles, i.e., no differential that allows for independent turning of each wheel, during curving there is often a condition of lateral and longitudinal slip that takes place. Although railway wheel treads have a conicity, which is designed to provide a lateral steering mechanism around curves, this is usually effective only for shallow curves. For medium to severe curves there is generally flanging of the wheels and associated high wheel/rail lateral forces. This is illustrated in Fig. 9, which shows a flanged wheel with classic two point contact, one contact point between the wheel flange and the side or gauge face of the rail and the second contact point between the wheel tread and the top of the rail head.

Fig. 9. Two-point wheel-rail contact

Furthermore, this lateral flanging force is not just the steady state curving force, but also includes a large dynamic or transient augment to the steady state centrifugal force, due to the dynamics of the wheel negotiating the curve. Since most railway axles are configured as two or three axle trucks (or bogies), which are usually constrained to remain parallel to each other as they go around a curve,[7] as shown in Fig. 10, there is a resulting angle of attack between the wheel and the rail, which contributes to this high dynamic force. Fig. 11 illustrates the combined steady state and transient dynamic components for a rail vehicle negotiating a moderate curve according to Koci and Swenson (1978). Figure 12 presents a statistical distribution of lateral wheel forces as a function of occurrences per 1,000 miles of travel. As can be seen from this figure, lateral loads in the 30,000+ range have been measured on a low probability of occurrence basis, while loads in the 15,000+ range occur on a more common basis[9].

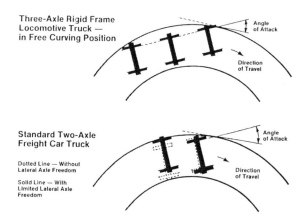

Fig. 10. Angle of attack during turning

---

[7] Trucks which allow the axles to become radial to the curve are generally more expensive to manufacture and maintain and as such are not common in freight operations. They are common in passenger operations where ride comfort is an important consideration.

[9] Based on test data from AAR/TTCI and FAST sources.

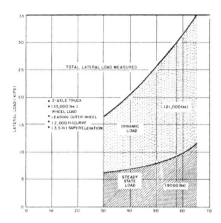

Fig. 11. Steady state and dynamic lateral loads (Koci and Swenson, 1978)

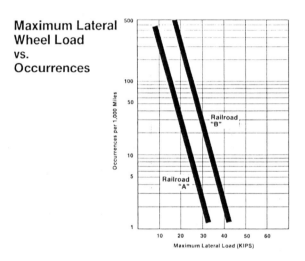

Fig. 12. Maximum lateral wheel load vs. occurrences

As can be seen in Fig. 9, these lateral loads are transmitted to the railhead both at the top of the head and at the side or gauge face of the rail. Furthermore, they act concurrently with the previously defined vertical loads thus generating a severe load environment on moderate to sharp curves where both the lateral and vertical loads are present.

As in the case with the vertical loads, the lateral loads are also transmitted through the rail, acting as a beam, to multiple crossties. This lateral load is then distributed to the ballast, where it is usually resisted by the shoulder of the ballast, the area outside the end of the crosstie, as well as by friction between the base and side of the tie and the ballast in the cribs (between the ties) and under the ties.

Longitudinal Loads

Longitudinal forces are input into the track structure through two distinct mechanisms; mechanical forces through train action and thermal forces through changes in ambient temperature (Flamanche, 1904).

Mechanically induced longitudinal forces are directly related to longitudinal train handling and operations. This includes train acceleration, specifically at the traction wheels of the locomotives and train deceleration or braking, either in the traction wheels only (dynamic braking) or throughout all of the wheels in the train consist (train air braking). Maximum mechanical forces of up to 60,000 lbs. per rail have been recorded, though more typically these forces are in the range of 20,000 lbs. per rail (Hiltz et al., 1937).

Thermally induced longitudinal rail forces are caused by the change in ambient (and corresponding rail) temperature from the "neutral" or "force free" temperature of the rail. These forces can be either tensile or compressive in nature (depending on whether the rail temperature is above or below the force free temperature) and can reach significant levels as illustrated in Fig. 13. In curves, this longitudinal effect can also result in significant lateral forces, with corresponding lateral movement of the curves. As can be seen in Fig. 13, a 100 F temperature change can generate 250,000 lbs. of longitudinal force in 132 RE rail. While this is a very high change in temperature, changes in rail temperature of the order of 50 to 60 degrees are quite realistic.

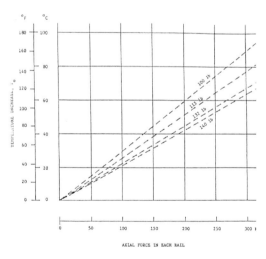

Fig. 13. Rail temperature increase vs. axial force in rail (Kerr, 1978)

Railroad Engineering

Railroad engineering, as it has evolved and emerged into current practice can be divided into two broad categories: design based engineering and maintenance based engineering. While there are, of course, overlaps between the two, there is a major difference in focus and approach, with railroad maintenance personnel being primarily concerned with the latter and component and system designers being primarily concerned with the former.

Design Based Engineering

Design based engineering is concerned with the areas of new designs of entire railroad track systems, subsystems, or individual components. As such, the "standardized" tools presented by the AREMA Manual for Railroad Engineering (AREMA, 2003) and summarized in the pyramid of stress shown in Fig. 3 provide an overview of the design approach stated earlier, to define the stress environment at each level of the track and each of its components.

All "modern" railroad engineering design starts with Beam On Elastic Foundation (BOEF) theory. This is the approach that is used to define the behavior of the track under vertical loads, and treats the track structure as a rail beam attached to a continuous elastic foundation, which represents the crossties, ballast and subgrade (Kerr, 2003; Hay, 1982). This model was first applied to longitudinal tie track by Winkler (1867) and later extended to crosstie track by Flamanche (1904), Talbot (1918), and Timoshenko (1915). The corresponding analytical model is presented in Fig. 14 and is defined by the equation:

$$EI\frac{d^4w(x)}{dx^4} + kw(x) = q(x) \tag{4}$$

If the general load distribution is replaced by a single wheel (or axle if the beam represents both rails), then the formulation becomes solvable using classical BOEF theory (Hetenyi, 1946). The resulting rail deflection is expressed as

$$w(x) = \frac{P\beta}{2k}e^{-\beta x}\left[\cos(\beta x) + \sin(\beta x)\right] \tag{5}$$

and the corresponding rail bending moments are

$$M(x) = \frac{P}{4\beta}e^{-\beta x}\left[\cos(\beta x) - \sin((\beta x)\right]$$

where

$$\beta = \sqrt[4]{\frac{k}{4EI}}$$

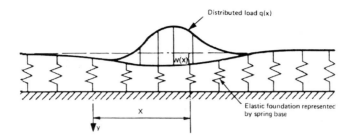

Fig. 14. Representation of a continuously supported infinite beam on an elastic foundation
subjected to load q (x)

This analytical approach is very much dependent on the definition of the track modulus or stiffness (also referred to as the *k* or u in different formulations). As a result, significant testing has been performed, both in the field and in the laboratory to try to define this value as well as to develop a standardized approach to its measurement. Extensive testing by AREA's Special Committee on Stresses in Railroad Track (Talbot et al., 1980), the Association of American Railroads (Zarembski and Choros, 1979) and others (Kerr, 1983, 2003) has determined track modulus values for wood tie track which range from 500 psi under poor support conditions to over 5000 psi for excellent support conditions. In the case of frozen roadbed, concrete tie track or slab track, the modulus values can increase to levels of 10,000 psi or even higher (Redden et al., 2002)

One of the major shortcomings of the BOEF theory is that it treats the tie-ballast-subgrade system as a linear spring and then tries to relate actual track load-deformation behavior to this linear spring value. While numerous studies have addressed the development of testing techniques to calculate this stiffness value *k*, (Kerr, 2003; Talbot et al., 1980; Zarembski et al., 1979), questions remain as to the validity of this linear stiffness assumption with various field tests showing linear, bi-linear, or non-linear load-deflection characteristics. (Kerr, 2003; Talbot et al., 1980; Zarembski et al., 1979; Selig and Waters, 1994).

In order to overcome this deficiency, variations on the base response model have been introduced with different types of foundation models such as the one presented in Fig. 15a and 15b which introduces a rotational resistance effect between the tie and the ballast (Kerr, 1979, 2003) with the resulting equation:

$$EI\frac{d^4w(x)}{dx^4} - s\frac{d^2w}{dx^2} + kw(x) = q(x) \qquad (6)$$

with the solution for a single (one) axle load P being

$$w(x) = \frac{P\beta^2}{2k\alpha\kappa}e^{-\alpha|x|}\left[\kappa\cos(\kappa|x|) + \alpha\sin(\kappa|x|)\right]$$

where

$$\beta = \sqrt[4]{\frac{k}{4EI}} \quad \text{and} \quad \alpha, \kappa = \pm\sqrt{\beta^2 \pm \frac{s}{4EI}}$$

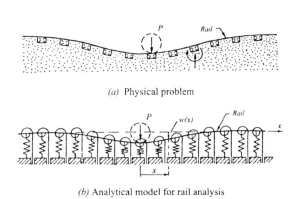

(a) Physical problem

(b) Analytical model for rail analysis

Fig. 15(a) and (b). Improved analytical model for rail analyses with rotational resistance (Kerr, 2003)

A second approach is illustrated in Fig. 15c using a two-dimensional Pasternak foundation (Kerr, 2003) which includes a shear layer G_P as well as a linear spring layer k_P. In this case the governing equation is:

$$EI\frac{d^4w(x)}{dx^4} - G_P\frac{d^2w}{dx^2} + k_Pw = q(x) \tag{7}$$

which is similar in form to the equation with rotational resistance shown above.

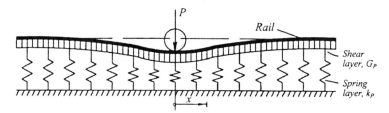

Fig. 15(c). Rail on a two-dimensional Pasternak base (Kerr, 2003)

All of the formulations discussed above use a quasi-static (time independent) load to represent the vehicle applied vertical loading. In reality, however, the actual applied load is dynamic in nature, as discussed previously. Therefore, a more exact formulation of the BOEF equation under dynamic loading would be that for a moving load on an infinite beam resting on an elastic base of the form[10]:

---

[10] This formulation does not include inertia of the base.

$$EI \frac{\partial^4 w(x,t)}{\partial x^4} + m(x) \frac{\partial^2 w(x,t)}{\partial t^2} + k(x)w(x,t) = q(x,t) \qquad (8)$$

where

$w(x,t)$	=	vertical deflection of the beam
EI	=	flexural stiffness of beam
$m(x)$	=	track mass
$K(x)$	=	varying vertical track modulus
$q(x,t)$	=	weight and vertical inertia of moving object

For a moving wheel load of a railroad vehicle, the resulting applied load is as follows:

$$q(x,t) = \left[P - M \frac{\partial^2 w(x,t)}{\partial t^2} \right] \delta(x,t) \qquad (9)$$

where

P	=	applied wheel load
M	=	concentrated mass of applied load
δ	=	Dirac delta function
(x,t)		

A related problem is one where the supporting track stiffness is not uniform, but rather varies with distance. While the solution of this problem dynamically is very complex, the quasi-static version of this problem, as illustrated in Fig. 16 can be formulated with three differential equations as follows (Zarembski et al., 2001):

Differential Equations:

$$EI \frac{d^4 w_1(x_1)}{dx_1^4} + k_1 w_1(x_1) = 0 \qquad -\infty < x_1 \leq 0 \qquad (10)$$

$$EI \frac{d^4 w_2(x_2)}{dx_2^4} + k_1 w_2(x_2) = 0 \qquad 0 < x_2 \leq l$$

$$EI \frac{d^4 w_3(x_3)}{dx_3^4} + k_2 w_3(x_3) = 0 \qquad 0 < x_3 \leq \infty$$

where

EI	=	properties of the rail
P	=	wheel load (factored for dynamics)
k_1	=	track modulus of the parent track structure

k_2 =   track modulus of the crossing track structure
x_i =   longitudinal coordinate for the three zones
l =   location of the load

For other related formulations, the reader is referred to (Kerr, 2003; Doyle, 1980; Timoshenko and Langer, 1932; or Hetenyi, 1946).

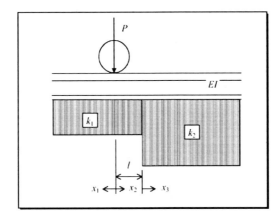

Fig. 16. Stiffness transition model (Zarembski et al., 2001)

As already noted, the above BOEF and related formulations all treat the ties, ballast, and subgrade as a consolidated support condition term(s). While the distribution of load to the individual ties can be calculated from BOEF theory[11] the distribution of stress through the tie, onto the ballast, through the ballast and onto the subgrade cannot be addressed using this generalized approach. Rather, additional analyses are required.

In the case of crosstie analysis, a major issue is that of the support of the tie by the ballast. This tie support can range from completely uniform, to support under the rail seats to center bound (where the tie is supported at the center only - an undesirable support condition that can result in failure of the tie at its center due to bending). While many formulations have evolved for this analysis, including BOEF type analysis, the more simplified analyses that are commonly used assume partial support of the tie on the ballast. Thus, the AREMA design manual (AREMA, 2003) assumes the tie is supported on the outside thirds of the tie length with no support in the center. Using beam-bending theory it is thus possible to develop tie-bending guidelines such as illustrated in Fig. 17 (Hay, 1982).

---

[11] Using the track deflection w, track modulus or stiffness k and the tie spacing.

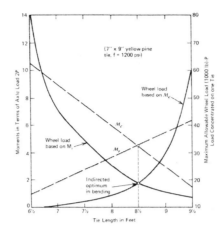

Fig. 17. Bending moments and wheel loads versus tie length (Talbot et al., 1980)

Using this simplified approach, the resulting tie-ballast bearing area is thus $A_b = 2b*L/3$
Where b is the tie width and L is the tie length.

The corresponding ballast pressure is then the individual tie load F_i divided by the tie bearing area A_b. Based on AREMA guidelines, this value must be less than 85 psi (65 psi for poorer quality ballast/wood ties) (AREMA, 2003), (see Fig. 3).

The analysis of the load transmission through the ballast layer includes both simplified and more complex analyses as well as empirically derived analytical formulations.

Certainly the most simplistic approach is the "load spread" approach illustrated in Fig. 18, where the tie-ballast pressure is distributed over a constant area that increased with depth. Both a 1:1 and 1:2 load spread has been used in various applications.

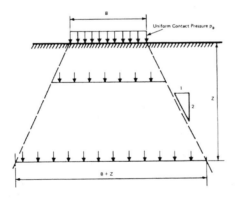

Fig. 18. Vertical stress transmission by means of the 2:1 distribution (Doyle, 1980)

Another commonly used approach uses Boussinesq Elastic Theory to calculate the distribution of load through a semi-infinite homogeneous elastic half space (Doyle, 1980; Boussinesq, 1885). The resulting stresses calculated for the Boussinesq application of a uniformly loaded circular area at the surface of a semi-infinite elastic medium (Doyle, 1980), (Bathurst & Kerr, 1999) is

$$\sigma_z = p_a\left[1 - \frac{z^3}{\left(a^2 + z^2\right)^{.5}}\right] \quad \text{for the vertical stress} \tag{11}$$

$$\sigma_x = \sigma_y = \frac{p}{2}\left[\left(1 + 2v\right) - \frac{2\left(1 + v\right)z}{\left(a^2 + z^2\right)^{0.5}} + \frac{z^3}{\left(a^2 + z^2\right)^{.5}}\right] \quad \text{for the horizontal stresses}$$

and

$$\tau_{max} = \frac{\sigma_z - \sigma_x}{2} \quad \text{for the maximum shear stress}$$

where

p_a = average uniform pressure over the loaded area

a = radius of the circular loaded area and v is Poisson's ratio.

Fig. 19 shows the results of the Boussinesq vertical stress analysis, as a function of ballast depth, compared to the load spread results (for both 1:1 and 1:2 spread ratios).

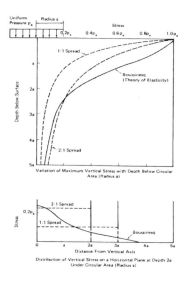

Fig. 19. Comparison of the vertical stress distribution under a uniformly loaded circular area based on Boussinesq equations and 1:1 and 2:1 distributions (Doyle, 1980)

Another ballast pressure distribution approach was proposed by Schramm (Schramm). This analytical approach calculated maximum vertical pressure at a depth beneath the crosstie as a function of the angle of internal friction of the ballast, which varies from 30 to 40 degrees. The resulting equation for maximum vertical subgrade pressure is

$$\sigma_z = p_a \left[\frac{1.5(l-g)B}{\{3(l-g)+B\}z \tan \theta} \right] \qquad (12)$$

where: p_a = average uniform contact pressure under rail seat

l = tie length, g = distance between rail centers, B = width of crosstie, z = depth of ballast layer

and θ = angle of internal friction of the ballast.

Still another relationship is an empirical equation developed by the AREA Talbot Committee (Talbot et al.) in the period 1918-1940 which was incorporated in the AREA design manual for many years. This relations is

$$p_c = \left[\frac{16.8 p_a}{(h)^{1.5}} \right] \qquad (13)$$

where p_c is the ballast pressure underneath the center of the tie at depth h and p_a is the applied tie/ballast pressure

Fig. 20 presents a comparison of these different analyses showing vertical stress as a function of ballast depth. It should be noted that the actual material properties of the ballast and subgrade are quite complex and have been simplified in these approaches in order to allow for an engineering design formulation for the load transmission through the ballast to the subgrade. Thus, as noted in Fig. 3, the ultimate stress at the ballast subgrade interface must be less than 20 psi (or lower depending on the actual bearing strength of the material). For a more comprehensive discussion of ballast and subgrade material behavior and properties refer to (Selig and Waters, 1994).

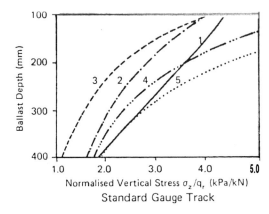

Legend:	Curve No.	Researcher	Method
	1	Boussinesq	Simplified for case of a circular area
	2	Eisenmann	Strip Load
	3	Clarke	Load Spread
	4	Schramm	Load Spread
	5	Talbot	Empirical

Fig. 20. Comparison of theoretical, semi-empirical and empirical stress distributions with ballast depth (Doyle, 1980)

Alternately, more complex system analysis models such as Finite Element models have been used to analyze the track structure and its stress levels. Such a Finite Element model is illustrated in Fig. 21 where each layer of the track is in turn modeled as a separate series of elements. These FE models include such models as ILLITRACK (Robnett et al., 1975), developed by the University of Illinois in the 1970s, GEOTRACK (Chang et al., 1980) and KENTRACK (Huang et al., 1984). The comparative merits of these models are discussed by Selig (Selig and Waters, 1994). However, in general, the FE models have not come into wide use because of the difficulty in calibrating them to actual conditions and the limited advantage they give as compared to the more simplified analytical approaches already discussed.

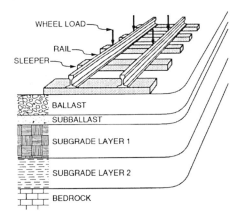

Fig. 21. Configuration of Geotrack (Chang et al., 1980)

Maintenance Based Engineering

Maintenance based engineering is concerned with the area of existing track and how to optimize its performance, particularly in the long term railroad environment of increasing loads that was presented previously. While there are applications on a system basis, most of the focus here is usually on specific components or subsystems. As in the case of design engineering, many of the "standards" are presented in the AREMA Manual for Railroad Engineering (AREMA, 2003), however these tend to be more in the way of generalized guidelines rather than specific engineering approaches.

While many maintenance practices have evolved empirically over the years, there has also emerged engineering analyses and studies that have worked with and have led this empirical development of maintenance practices.

Perhaps the most important area of the track where this maintenance engineering approach has evolved is in the rail, which is the portion of the track that carries the direct wheel load and which, for many heavy axle load operations represents the highest maintenance and replacement cost area for the track structure (as well as a major area of safety concern). As shown by Kerr (2003) in Fig. 22, the rail stresses is a complex one involving multiple loading stress environments. It is further complicated by the presence of lateral loads, which similarly generate bending and contact stresses.

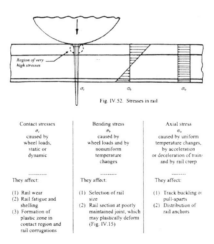

Fig. 22. Summary of rail stresses and their effects (Kerr, 2003)

The rail stresses include the bending stresses discussed previously as well as wheel/rail contact stresses and the longitudinal stresses such as from the thermal environment. These stresses are additive as illustrated in Fig. 23, particularly in the longitudinal direction. While the bending stresses play an important factor in the rail design process, the contact and longitudinal stresses are most important in the maintenance engineering activity, where in-track maintenance policies and practices are strongly affected by these stresses and the resulting failure modes that they generate. As noted in Fig. 22, these include fatigue related problem, both surface and subsurface in the rail, wear related problems, rail pull-apart problems, etc.

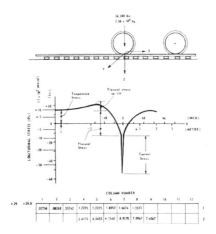

Fig. 23. Combined stress components (Zarembski and Abbott, 1978)

In order to define the proper maintenance engineering response, the correct formulation of these stresses, and the related understanding of those parameters that affect them, becomes very important.

In the case of the longitudinal stresses, these are calculated from the thermally induced rail forces, as distributed over the cross-sectional area of the rail. Thus

$$\sigma_o = \frac{F_o}{A} = \frac{EA\alpha\Delta T}{A} \tag{14}$$

where: F_0 = Thermally induced longitudinal rail force
E = Modulus of Elasticity of the rail steel
A = rail cross-sectional area α = coefficient of thermal expansion
and ΔT = change in temperature from neutral (force free) temperature

Fig. 24 presents the relationship between longitudinal stress and temperature change.

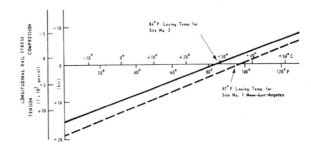

Fig. 24 – Temperature – stress curves (Zarembski and Abbott, 1978)

Knowing the level of longitudinal stress (based on the neutral temperature which can be field adjusted by the railroad) can assist in determining the potential for failure, through either pull-aparts or track buckling.

The wheel/rail contact stress, which is generally defined using Hertzian Contact stress theory or some variation from this theory (Kerr, 2003; Zarembski and Abbott, 1978), is directly related to the local interface geometry of the wheel and the rail, together with the loading and material properties. This contact can be centrally located on the rail head, as illustrated in Fig. 25, two point contact to include wheel flange contact on the side of the rail head (Fig. 9) or can contact at the gauge corner of the rail as shown in Fig. 26. Depending on these factors, the magnitude of the stresses can vary significantly. This contact stress however is very local to the surface of the railhead, and as such decreases rapidly away from the surface as shown in Fig. 27. As a result many of the problems that are caused by this high contact stress condition are local to the surface of the railhead or occur just subsurface at the point of maximum shear stress (see Fig. 27). By changing the shape or profile of the rail head through such means as rail grinding, it is possible to control the location and shape of the wheel/rail contact zone and the associated contact stresses. This in turn allows for the "engineering" and maintaining of an optimum profile(s) for curves.

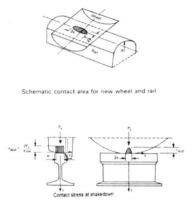

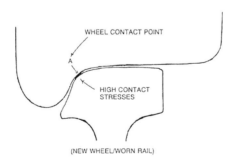

Fig. 25. Contact surface between rail and wheel (Doyle, 1980)

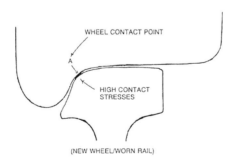

Fig. 26. Single point contact between gauge corner of rail and wheel flange throat

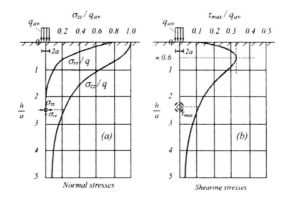

Fig. 27. Distribution of stresses and the loaded surface (Kerr, 2003)

The combination of these stresses, in turn, can produce internal defects, which accumulate over time (or total traffic tonnage) and result in the required replacement of the rail. This accumulation of defects with traffic is illustrated in Fig. 28, which shows the rate of accumulation of these defects with traffic. This rate of defect accumulation has been shown to follow a Weibull logarithmic distribution, which is defined by the equation:

$$PD(MGT) = 1 - e^{-\left(\frac{MGT}{\beta}\right)^{\alpha}} \qquad (15)$$

where α and β are the Weibull slope and intercept parameters describing the characteristics of defect formation. The rate of defect occurrence (defect/rail/MGT), can be determined from the Weibull rate equation:

$$\lambda(MGT) = \frac{\alpha MGT^{\alpha-1}}{\beta^{\alpha}} \qquad (16)$$

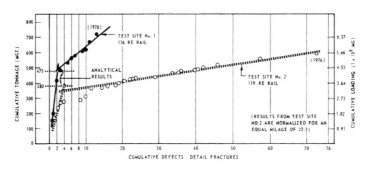

Fig. 28. Accumulation of rail defects (Zarembski and Abbott, 1978)

Rail also wear under the loads and stresses defined above. As can be seen in Fig. 29, this includes wear both at the top of the railhead (h) and at the side or gauge face of the rail (g). Rail wear has been the subject of numerous empirical studies, with the rate of wear being defined as a function of key track, traffic and environmental parameters. One such generalized wear equation is of the form:

$$Railwear = \frac{f(C,H,L,S)R'J'}{g(D)h(G)F(P,S)}$$

(17)

where:

C = curvature	J' is a function of welded vs. jointed track
H = rail hardness	D = traffic density
L = level of lubrication	G = grade
S = train speed	and P = axle load
R' is a function of rail size	

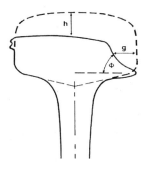

Fig. 29. Rail wear

Wear and fatigue represent separate failure (and replacement) mechanisms for rail. Which one is dominant varies with track and traffic (vehicle) conditions. As can be seen in Fig. 30, under heavier wheel loads (and higher stresses) fatigue failure occurs first, with fatigue being the dominant failure mechanism. Under lighter axle loads and/or severe curvature, rail will wear out before fatigue failure occurs.

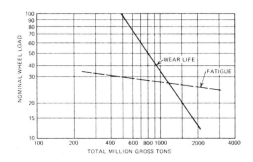

Fig. 30. Fatigue life versus wear life for various wheel loadings (Zarembski, 1979)

Track Stability

Finally, no discussion of railroad engineering can be complete without a discussion of Track Buckling. This analysis approach can be considered to fall into either design based engineering or the maintenance based engineering category, since it can be applied in either approach. The evolution of this analysis approach is explained in detail by Kerr (2003, 1978, 1980, 1974) and will only be summarized here.

As noted previous, there is a class of rail force directly associated with the change in temperature of the rail (which is directly related to the ambient air temperature) the distribution of which is illustrated in Fig. 31. As can be seen in this Figure, there is a zone of constant thermal force, and there are two end zones where the force reduces to zero, as a function of the longitudinal resistance of the track r_o. When the force in the rails exceed the lateral resistance of the track system (to include the rail/fastener/tie/ballast interfaces), the track can buckle laterally as illustrated in Fig. 32. The buckling zone, likewise experiences a reduction in axial force as illustrated in Fig. 32b.

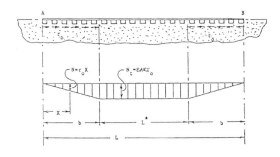

Fig. 31. Axial force distribution in a track of length L caused by uniform temperature change
(Kerr, 2003)

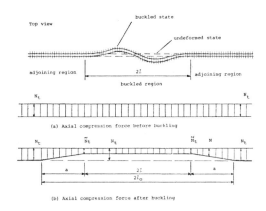

Fig. 32. Distribution of axial compression forces before and after buckling (Kerr, 2003)

The analyses presented by Kerr (2003,1980) show three equilibrium branches for the track subject to longitudinal thermal loading. These branches are illustrated in Fig. 33 for tangent (straight) track, where branches 1 and 3 are stable and branch 2 is unstable. Thus for rail temperatures above T_L there exists a stable deformed state where track buckling can occur.

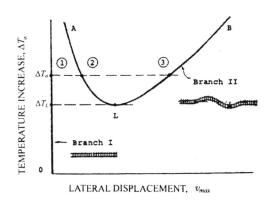

Fig. 33. Typical equilibrium branches for a heated straight track (Kerr, 2003)

The results of the buckling analysis yields maximum allowable temperature increase above the neutral temperature of the rail to avoid track buckling. This is then used both in setting initial track installation temperatures (track design) and also in the ongoing monitoring of the track condition (track maintenance). For details refer to Kerr (2003).

Future Directions

As noted previously, the railway track structure has evolved over the last 100 years, keeping pace with increased car weights, increased traffic congestion, and more complex operations. The factors that are most likely to influence the development of railroad track engineering are:

1. Continuing increased axle loads. As shown in Fig. 4, railroad car weight and wheel load has increased significantly over the last decade and the trend is clearly continuing. As railroads learn to cope with the effect of the 286,000 lb. cars (36 ton axle loads), they will move to the next generation 315,000 lb cars (39 ton axle loads) which are already in use in Western Australia and which have been extensively tested in the U.S. Even heavier cars are on the horizon with the Western Australian mining roads looking at 41+ ton axle loads as well.

2. High-speed passenger operations. Europe and parts of Asia (Japan, Korea, Taiwan) already focus on high-speed passenger rail operations with the design of the track structure geared to the precise control of geometry needed for train speeds of 150 to 200 mph. The U.S. is also starting to move in that direction, with Amtrak operating at 150 mph and much discussion on high-speed corridors, in many case overlaid onto heavy axle load freight corridors. This latter problem poses a major engineering challenge since the tight geometric tolerances needed for high-speed passenger operations degrade rapidly under heavy axle load freight operations for conventional track structures.

3. Economics. Railways worldwide are finding that their operating and maintenance budgets are being cut and railway engineering personnel are more and more looking to optimize the cost of track maintenance. This has already led to such innovations as rail profile grinding, lubrication-grinding management, and improved fastener configurations. In the future introduction of new component or system designs will be strongly tied into the cost of maintenance.

It is expected that the track structure will continue to evolve with a focus on the "weak spots" that fail under traffic. This would involve development of improved components and or materials such as

- Improved rail steels with fewer inclusions (cleaner steel) and better wear and fatigue resistance. The industry is even now starting to look at rail steels with a bainitic microstructure instead of the currently used pearlitic steel.
- Plastic crossties, lighter than concrete with long lives and little environmental degradation.
- Improved welding processes, particularly in-track welding to replace current thermit welding processes, which have high failure rates after relatively short lives.
- Improved turnout designs to allow for improved vehicle/track dynamics, higher speeds, and lower wheel/rail forces.
- Improved rail lubricants with greater temperature performance and little or no environmental impacts (such as the new generation soy based lubricants being tested).
- Improved maintenance practices such as stone blowing to provide better geometry correction and profile grinding to reduce wheel/rail contact stresses and associated rail degradation.

Finally the potential for development of new improved track systems such as slab tracks or track laid on asphalt underbeds exists. This has the potential for improving overall performance, however in the short term, this may be confined to new construction such as for transit systems or short track rebuilds. Since most of the main line track in the U.S. and Europe is currently in a maintenance mode, the potential for extensive rebuild of existing track is limited.

However, as trains continue to get heavier and there exists a need to design economical high strength long lived components and systems, the role of railroad track engineering will continue to take an important place in the rail transportation environment.

References

Ahlbeck, D.R., 1980. An Investigation of Impact Loads Due to Wheel Flats and Rail Joints. *American Society of Mechanical Engineers*, **80**-WA/RT-1.

American Railway Engineering and Maintenance of Way Association, 2003. Manual for Railway Engineering.

Bathurst, L.A., and Kerr, A.D., 1999. An Improved Analysis for the Determination of Required Ballast Depth. Proc. of the Annual AREMA Conference.

Boussinesq, J., 1885. Application des Potentiels a l'Etude de l'Equilibre et du Movement des Solides Elastiques. Gautheir-Villars, Paris, France.

Chang C.S., Adegoke, C.W., and Selig E.T., 1980. The GEOTRACK Model for Railroad Track Performance. *Journal of the Geotechnical Engineering Division*, ASCE, **106**,(GT11).

Doyle, N.F., 1980. Railway Track Design: A Review of Current Practice, Australian Government Publishing Service, Canberra.

Flamanche, A., 1904. Researches on the Bending of Rails. *Bulletin, International Railway Congress Association*, **18**.

Hay, W.W., 1982. *Railroad Engineering*, Second Edition, John Wiley and Sons.

Hetenyi, 1946. *Beams on Elastic Foundations*. University of Michigan Press, Ann Arbor, 1946.

Hiltz, J.P. Jr. et al., 1937. Measurement Under Traffic of Dynamic Rail Creepage Forces. *Bulletin of the American Railway Engineering Association*, **38**.

Huang, Y.H., Lin C., and Deng, X., 1984. Hot Mix Asphalt for Railroad Track Beds—Structural Analysis and Design. *Association of Asphalt paving Technologists*.

Jenkin, H.H. et al., 1974. The Effect of Track and Vehicle Parameters on Wheel/rail Vertical Dynamic Forces. *Railway Engineering Journal*.

Kerr, A.D., 1974. The Stress and Stability Analyses of Railroad Tracks. *Journal of Applied Mechanics*, **41**, (4).

Kerr, A.D., 1978. Thermal Buckling of Straight Tracks; Fundamentals, Analyses and Preventive Measures. *Bulletin of the American Railway Engineering Association*, **669**.

Kerr, A.D., 1979. Improved Stress Analysis for Cross-tie Tracks. *Journal of the Engineering Mechanics Division*, Proc. ASCE, **4**.

Kerr, A.D., 1980. An Improved Analysis for Thermal Track Buckling. *International Journal of Non-Linear Mechanics*, **15**.

Kerr, A.D., 1983. A Method for Determining the Track Modulus using a Locomotive or Car on Multi-Axle Trucks. In Proc. AREA, **84**.

Kerr, A.D., 2003. *Fundamentals of Railway Track Engineering*. Simmons Boardman Press, Omaha.

Koci, L., and Swenson, C., 1978. Locomotive Wheel and Rail Loadings – A Systems Approach. *Heavy Haul Railways Conference*. Perth Western Australia.

Redden, J.W.P., Selig, E.T., and Zarembski, A.M., 2002. Design Considerations in Stiff Track Modulus Environments. *Railway Track & Structures Magazine*.

Robnett, Q.L., Thompson, M.R., Knutson, R.M., and Tayabji S.D., 1975. Development of a Structural Model and Materials Evaluation Procedure. *US DOT Report DOT-FR-30038*.

Schramm, G., 1961, Permanent Way Technique and Permanent Way Economy. Otto Elsner, *Verlagsgesellschaft Darmstadt*, (English translation by Hans Lange).

Scott, J.F., 1986. Evaluation of Concrete Turnout Ties and Bridge Ties. *American Railway Engineering Association*, **706**.

Selig, E.T., and Waters, J.M., 1994. *Track Geotechnology and Substructure management*. Thomas Telford Press, London.

Talbot, A.N. et al., 1980. Report on the Special Committee on Stresses in Railroad Track. *Bulletin of the American Railway Engineering Association*, **205**, **224**, **253**, **275**, **319**, **358** and **418**, 1918-1940, Reprinted as Stresses in Railroad Track – The Talbot Reports, AREA.

Timoshenko, S., and Langer, B.F., 1932. Stresses in Railroad Track. *Transactions of the American Society of Mechanical Engineers*, **54**, APM 54-26.

Winkler, E., 1867. *Die Lehre von der Elasticitat und Festigkeit* (Elasticity and Strength in German), Verlag von H. Dominicus, Prague.

Zarembski, A.M., 1979. Effect of Rail Section and Traffic on Rail Fatigue Life. American Railway Engineering Association, 78th Annual Technical Conference, Chicago, IL.

Zarembski, A. M., and Abbott, R. A., 1978. Fatigue Analysis of Rail Subject to Traffic and Temperature Loading. *Association of American Railroads report,* **315**.

Zarembski, A.M., and Blaze, J., 2003. The Economics of Heavy Axle Loads: Costs and Benefits. *Heavy Haul: The Solution for Europe's Future*, Paris.

Zarembski, A. M., and Choros, J., 1979. On the Measurement and Calculation of Vertical Track Modulus. *Bulletin of the American Railway Engineering Association*, **81** (675).

Zarembski, A.M., Palese, Joseph W., and Katz, L., 2001. Reduction of Dynamic Wheel/Rail Impact Forces at Grade Crossings Using Stiffness Transitions. American Society of Mechanical Engineers, 2001 ImechE Congress, New York, NY.

Zarembski, A.M., Resor, R.R., and Patel, P., 2003. Economics of Wayside Inspection Systems. ASME International Mechanical Engineering Congress and Exposition, Washington, DC.

Design Ice Forces and Fracture Scaling

John P. Dempsey, Clarkson University

Abstract

The scale dependence on the fracture energy G_c^{scale} of freshwater and sea ice is discussed. Next, via highly idealized radially cracked hollow cylinder and sphere models, the nominal ice pressure induced during ice structure indentation is shown to be dependent on the ice thickness h in the form $h^{-1/2}\sqrt{EG_c^{scale}}$. Through considerations concerning the scale dependence of G_c^{scale} on the degree of stable cracking (prior to spalling or flaking), the offshore platform design ice force is predicted to depend on ice thickness as $h^{-\eta}$, in which $0<\eta<1/2$, in good agreement with design pressures currently used by ice engineers. Finally, the bearing capacity of radially cracked ice sheets in the presence of crack closure under extension and bending is ascertained.

In-Plane Ice Forces: Offshore Ice-Structure Interactions

The design of offshore structures and marine transportation systems, principally for petroleum exploration and production, has driven much of the ice mechanics research over the last three decades to be focused on a physical range in scale from a fraction of a meter to several hundred meters. Much of this ice mechanics research has focused on the interaction of intact ice sheets, ridges, rubble fields, and fragmented ice covers with fixed and moving structures (Croasdale, 1988). In the literature concerning ice forces on structures, the effective pressure is defined as the total interaction force divided by the contact area, which is usually taken to be the product of structure width and thickness. The measured effective pressures versus contact area from small-scale tests show a large degree of scatter in the data, whilst the data from full-scale measurements reveal a dramatic decrease with increasing nominal contact area (Masterson and Spencer, 2001). There is considerable interest, on the part of the design engineer, in determining just how far ice loads can be lowered, safely. Unfortunately, there is not enough confidence on the part of design engineers in being able to make direct use of current indentation theories and small-scale ice properties. Essentially, the design ice force has always been much less than small-scale strengths would predict. The need exists for an increased recognition of, and emphasis on, both scale and velocity effects as regards the topic of ice forces on structures.

When ice sheets interact with structures the failure (compressive) strength displays a pronounced scale-effect (Sanderson, 1988; Blanchet, 1998). Plots presenting peak strengths regardless of such factors as failure mode, ice type, and loading rate have been in existence since the early seventies. Sanderson's plot is the most well known. There is now no doubt that a family of pressure-area or pressure-width curves will prove to be necessary, depending on the application, ice type, ice velocities, etc. One real difficulty, however, is that this pressure-area/width dependence, while so important, depends on peak pressures under compression. Compressive failure strengths, however, depend on the boundary conditions and scale. While a rather large number of failure mechanisms and loading situations must be analyzed, it does not seem rational to try to deterministically model every brittle mode of failure.

At the present time, the relative importance of contact area, ice sheet velocity, and the structural width versus ice thickness ratio on the global design ice force is not sufficiently well understood. Once gained, this knowledge needs to be incorporated in ice-structure interaction models and global ice load models. A very important question concerns the role of brittle failure processes in reducing effective pressures at higher ice sheet velocities versus the dependence on contact area implicit in the Sanderson plot, or versus width. Important sub-problems include the nucleation of cleavage cracks, spalling and flaking, the size and failure of individual crushing zones, and the ductile-to-brittle transition speed and its dependence on temperature. The hypothesis that the effective pressure generated during ice-structure interaction depends primarily on the contact area is not universally accepted: the influence of ice velocity (or indentation speed) and surrounding ice extent is perhaps just as important.

During the edge indentation of an ice sheet, if the ice moves very slowly, the force builds up relatively slowly, and contact develops over the substantial portion of the ice thickness. Near-horizontal cleavage cracks parallel to the ice surface at first spread stably outward from the contact zone, but later become unstable and grow much longer, though they remain parallel to the surface (Hirayama et al., 1974; Kry, 1981). Ultimately a region on one side of the crack becomes unstable, the crack turns towards a free surface, and a spall or flake breaks away. Any relative vertical movement (related, for example, to tide or to large-scale flexural buckling) makes spalling more likely. Repeated spalling leads to a sequence of wedge-shaped profiles.

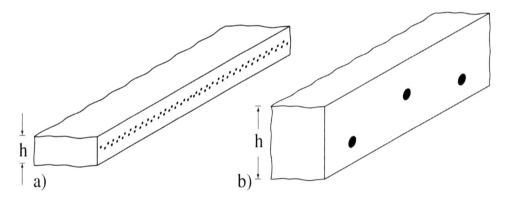

Fig. 1. a) Line-like and b) localized distribution of high pressure zones.

If the ice moves somewhat faster, the distribution of contact force alters. The largest contact pressures occur close to mid-thickness, along a *line-like high-pressure zone*, shown schematically in Fig. 1a. The zone is composed of myriad small contact regions, *high-pressure zones (HPZs)*, which continually form and fade away, each contact region centered on a local asperity left by an earlier fracture. An individual HPZ may only carry a large force for a very short time, perhaps hundredths of a second. Together, however, the HPZs transmit a force that is almost uniformly distributed across the contact breadth (though not through the contact thickness). At any instant, the distribution of contact force looks like a rather smooth range of rounded hills. The ice literature calls this *simultaneous contact*, and calls the failure mode *ductile crushing.*

If the edge indentation speed surpasses some critical speed, the *transition velocity*, the local contact pressure is much less uniform, and the ice pressure (averaged over the nominal contact area, the ice thickness multiplied by the contact breadth) is distinctly lower. Although the HPZs are still forming and fading rapidly across the breadth, they are only weakly correlated in space and time. At any instant, the distribution of contact force looks like a jagged mountain range, roughly centered on the mid-thickness of the ice, with many sharp peaks. Fig. 1b is an idealization. The ice literature describes this scenario as *non-simultaneous* contact, and the mode as *brittle crushing*.

In contacts with compliant structures, there is an intermediate indentation speed range in which the mode alternates cyclically between a load-up phase (associated with simultaneous contact and a reduced relative indentation speed) and a briefer extrusion phase that follows failure of the ice. In the extrusion phase the structure rebounds and the relative indentation speed is much higher. A plot of force against time looks like the teeth of a ripsaw. This is *intermittent crushing*. If the structure or indenter may be characterized as rigid, there is one transition velocity and the ice contact will switch from simultaneous to non-simultaneous

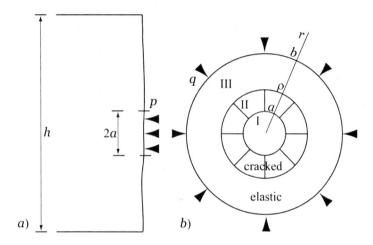

Fig. 2. a) High-pressure zone on the contact face of an ice sheet; b) radially cracked cylinder or sphere.

(ductile to brittle crushing). If the structure or indenter is compliant, there are two transition velocities: one for the switch from ductile to intermittent crushing, and the other for the switch from intermittent to brittle crushing. During the penetration of an indenter into an ice sheet, cracking on different scales is activated. These include the grain-scale grain boundary facet cracks, across column cracks, and transgranular cracks. With increasing pressure, this cracking will progressively localize into partially and then fully developed process zones, followed by the stable growth of a system of macrocracks. The indentation pressure reaches a maximum when the crack growth becomes unstable, this happening because the cracks "feel" the presence of a remote boundary, in the sense that the boundary affects the strain energy release rate. A few of the cracks then propagate unstably towards the boundary. In the context of ice-structure indentation, this remote boundary is most often the upper and lower surfaces of the ice sheet (for the indentation of floating ice sheets). There are no doubt a number of complicated deformation mechanisms active in the contact zone during ice-structure interactions. The formation and spatial charactcristics of the high-pressure zones seem to be of central importance to the whole subject. The objective of this treatment is to elucidate the factors that control both the size ("vertical width" if line-like, and "diameter" if localized) and peak pressures of the high-pressure zones that form. This is done via models that are deliberately highly idealized.

For instance, Fig. 2a depicts a line-like HPZ on one plane face of a uniform ice sheet. It carries a local contact pressure p across a zone breadth $2a$, small by comparison with the sheet thickness h. Fig. 2b shows a model that represents conditions close to the HPZ, a cylinder centered at the mid-height of the HPZ. The ice is idealized as a uniform linear-elastic/brittle material characterized by Young's modulus E, Poisson's ratio v and critical strain energy release rate G_c. Position is referred to cylindrical coordinates r, θ, z, radial displacement is denoted u, and the stress components are σ_{rr}, $\sigma_{\theta\theta}$ and σ_{zz}. The cylinder in Fig. 2b is divided into three cylindrical regions, as follows: in region I, $0<r<a$, $\sigma_{rr} = \sigma_{\theta\theta} = -p$; in region II, $a<r<\rho$, n equally-spaced radial cracks extend to a radius ρ and release the circumferential stress, so that $\sigma_{\theta\theta}$ is zero; in region III, $\rho<r<b$, the ice remains intact and elastic. For now let the outer boundary of the cylinder at b be loaded by a pressure q.

One of the questions to be explored is when does the internal pressure p acting over the vertical width $2a$ become aware of the scale at hand? In this model, the scale at hand is the outer radius b. In the case of edge indentation, the scale at hand is the shorter of the distance from the high-pressure zone to the top and bottom surfaces. As shown by Dempsey et al. (2001), the energy release rate associated with the radial cracking induced by a line-like high pressure is given by

$$G = \frac{p^2 a^2}{bE'} f(C, \rho/b) \qquad (1)$$

in which $C = qb/pa$. The function $f(C, \rho/b)$ is dimensionless. The maximum pressure that may be observed under line contact conditions is therefore given by

$$p_{max} \sim \Psi \frac{b^{1/2}}{a} \sqrt{E' G_c^{scale}} \qquad (2)$$

in which E' is the short-time modulus and G_c^{scale} is the critical-energy-release-rate for the scale at hand. The free-surface correction function Ψ has been introduced because one must modify the hollow cylinder idealization and cut the cylinder in half (lengthwise). Note that one could re-write Eq. (2) in the form

$$p_{max} \sim \Psi \frac{h^{1/2}}{a} K_Q \qquad (3)$$

in which b has been loosely equated with the ice thickness h, and the equality $(EG_c^{scale})^{1/2} \equiv K_Q$ has been assumed, K_Q being the scale dependent apparent fracture toughness. The force per unit length (along the width) associated with the line-like high pressure zone follows from the maximum pressure in the first term of Eq. (2) multiplied by the internal diameter ($2a$) of the hollow cylinder in Fig. 4b. The associated nominal pressure p_{nom} follows by dividing the latter force by the ice thickness h. Thus

$$p_{nom} \sim 2\Psi h^{-1/2}\sqrt{E'G_c^{scale}} \quad \text{(line-like HPZ)} \tag{4}$$

The force exerted via a localized high-pressure zone, on the other hand, follows from the maximum pressure in a radially cracked hollow sphere, with the maximum pressure being multiplied by the effective loaded area πa^2. Suppose that there are N active localized high-pressure zones over a width equal to the ice thickness: the nominal pressure follows directly as

$$p_{nom} \sim \pi \Omega N h^{-1/2}\sqrt{EG_c^{scale}} \quad \text{(localized HPZ)} \tag{5}$$

At this point, there is the need for the compilation of test data and measurements. Evidence suggests that G_c^{scale} should increase with ice thickness and the degree of stable crack growth (Mulmule and Dempsey, 2000). In the latter paper, which used the sea ice fracture experimental results described earlier in the paper (Dempsey et al., 1999), G_c^{scale} was found to vary as $h^\beta (0 < \beta < 1/2)$. It follows then from Eqs. (4) and (5) that the nominal pressure would vary as $h^{-1/2+\beta/2}$. In this regard, it is remarkable to note that the design ice force for large aspect ratios (aspect ratio being equal to the ratio of the structure width to the ice thickness) specifies that $p_{nom} \sim h^{-0.174}$ (Masterson and Spencer, 2001).

To truly predict design ice forces for thicker ice and wider structures, it is clear that the scale dependence on the fracture of ice needs to be known. The next section addresses this topic.

Fracture Scaling, G_c^{scale} and K_Q

In a far-sighted statement, Weeks and Assur (1972) stated that "We feel that an understanding of the scale effect in ice testing is essential before a thorough scientific basis can be developed for the utilization of small-scale testing in engineering design problems." Scale effects on the tensile strength over the size range of 0.1-100 m have now been measured (Dempsey et al., 1999), and geophysical tensile strengths can be predicted at or less than the scale of 1000 m, depending on the effective crack lengths one expects at these larger scales (Dempsey, 1996). If one adopts the viewpoint that scale effects in the compressive strength of sea ice are but a reflection of the scale effect on the tensile strength, one would expect the geophysical stress resultants to also become scale invariant at length scales of a similar order. Certain constitutive quantities examined at both lab- and structural-scale reveal no scale effect (Cole and Dempsey, 2001); it is clearly time to obtain a greater understanding of the scale-invariant versus scale dependent parameters.

At lab-scale (less than 0.5 m), issues such as inhomogeneity and polycrystallinity are especially important to the fracture behavior of ice. Because of the large grain sizes that one can encounter in ice sheets, it is essential that the effects of sample size on the fracture behavior be determined. In other words, are small-scale (lab-scale) results applicable at larger scales (at the scale of ice-structure interactions, for instance)? In an attempt to answer some of these questions (Dempsey, 1996), full thickness lake and sea ice edge-notched square plates of dimension L (Dempsey et al.,

1999) were tested over the size range of (0.34<L<28.64 m) and (0.5<L<80 m), respectively (see Fig. 3). The results were rather interesting.

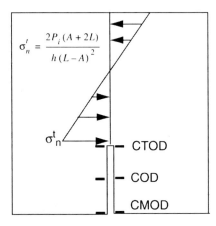

$$\sigma_n' = \frac{2P_i\,(A + 2L)}{h\,(L - A)^2}$$

Fig. 3. Edge-notched square plate test geometry of dimension L, crack length A and thickness h (normal to the above plan view); note the definition of the nominal tensile peak stress at crack initiation, σ_n^t, which figures prominently in the scaling discussion below.

The lab- to structural-scale (0.34<L<28.64 m) fracture tests conducted on S1 freshwater lake ice at Spray Lakes, Alberta, used the base-edge-notched reverse-tapered plate geometry and covering a size range of 1:81. A Bažant-type size effect analysis (Bažant, 1984) of the measured fracture strengths (which do reveal a significant dependence on scale) was unexpectedly clouded by the fact that the data collected violates the associated scatter requirements, even though the size range tested is large. Moreover, via Hillerborg's fictitious crack model, large fracture energies were back-calculated (of order 20 N/m), but for miniscule process zone sizes; in addition, not all of the measured deformations for each test could be matched simultaneously. Apparently, these very warm S1 macrocrystalline lake ice experiments were dominated by nonlocal deformation and energy release rate mechanisms, in all likelihood brought about by grain boundary sliding. The reduced effectiveness of both the Bažant-type size effect analysis and Hillerborg's fictitious crack model (Mulmule and Dempsey, 1997, 1998) is due mainly to the lack of crack growth stability achieved in the experiments. These unstable fractures truncated the fracture process. Given the irregular and large grain structure, the very warm ice temperatures, and the non-negligible grain boundary surface energy, there was a marked dependence on specimen size and distinctly non-unique pre-failure processes occurred. These observations have spurred an interest in fracture size effects versus polycrystalline inhomogeneities. If grain boundary sliding can occur with some inelasticity, apparently one encounters a situation in which the test specimen must incorporate not hundreds but thousands of grains.

The lab- to structural-scale (0.5<L<80 m) in-situ full thickness (1.8 m) fracture tests were conducted on first-year sea ice at Resolute, N.W.T. using self-similar (plan view) edge-cracked square plates. With a size range of 1:160, the data has been used, via size effect analyses, to evaluate the influence of scale effects on the fracture behavior of sea ice over the range 0.1 m (laboratory) to 100 m and to predict the scale effect on tensile strength up to 1000 m. The influence of scale on the ice strength and fracture toughness is dramatic, as shown in Fig. 4. For the thick first-year sea ice tested, the size-independent fracture toughness is of order 250 kPa√m, not the 115 kPa$\sqrt{m}$ that has been commonly used. The number of grains spanned by the associated test piece was 200, much larger than the number 15 typically quoted for regular tension-compression testing. The size-independent fracture energy was 15 N/m, while the requisite LEFM test size for the edge-cracked square plate geometry (for loading durations of less than 600 s and an average grain size of 1.5 cm), was 3 m square. Size effect analyses of

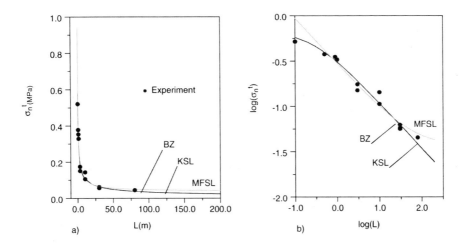

Fig. 4. Size effect laws for fits carried out over a range of 0.5-80 m. Nominal tensile strength vs specimen size, (b) Log (nominal strength) vs log (specimen size).

sub-ranges of the data show that unless the specimen sizes tested are themselves sufficiently large, the true nature of the scale effect is not revealed. This was a concern raised by 30 years ago: *"A particularly dangerous aspect noted of the size effect is that it may not occur unless member sizes are sufficiently large and consequently may not appear in scaled-down laboratory testing"* (Leicester, 1973). In other words, the predictive capability of the various size effect laws showed a hitherto unemphasized dependence on the actual subset used. The goodness of fit was highly dependent on the subset of the size range covering the major area of change. All size effect laws are ad hoc, and all can in general be made to fit the data—this matter is pursued in greater detail in Dempsey et al. (1999). In the case of the fracture tests reported in the latter paper, based on the lab-scale and field-scale strength data measured between 0.1 and 3 m and using Bažant's size effect law, it is possible to accurately predict the tensile strengths for all of the remaining tests, up to and including 80 m.

By direct measurements, the fracture scaling of first-year sea ice 1.8m thick at Resolute, N.W.T. has been established from lab-scale to structural-scale (0.1 to 100 m). During later experiments, the microstructure of sea ice has been studied on the scale of the ice sheet thickness, and the dramatic skeleton sub-structure of the brine drainage networks (Cole and Shapiro, 1998) has been made much clearer. At this stage, one is afforded the following length scales in sea ice: the platelet spacing, grain size, brine channel diameters, brine channel spacing, ice sheet thickness, thermal crack spacing, floe size, aggregate scale, shear band widths, and soon. The spacing of the brine drainage channels is possibly a significant factor in the 3m transitional size found above. The evolution of this structure with ice thickness needs to be understood, as does the influence of thermal cracks at larger scales (between 100m and several km). Apparently, thermal cracks are spaced on the order of 200 m apart. Thermal cracking may weaken large floes (larger than 1 km, especially), or perhaps this is caused more by thickness variations in the thick ice sheet combined with the thin ice of refrozen leads. The expressions deduced by Dempsey (1996) for full thickness tensile strengths via several size effect laws coincide with geophysical strengths at the scale of 1 km. As noted therein, the strengths predicted to occur at the scale of 1 km were predicted to lie between 11 and 38 kPa. For 1.8m thick ice, this is the same as strengths per unit thickness to lie between 20 and 70 kN/m. Note that Coon et al. (1998) set the unconfined tensile limit (based on a great deal of data) for sea ice at 50 kN/m.

The tensile fracture of ice is complicated by the presence of creep, even for short times. The time dependent fracture mechanics necessary to describe the tensile fracture of sea ice has now been developed (Mulmule and Dempsey, 1998). The fracture of sea ice has been modeled using a viscoelastic fictitious crack (cohesive zone) model. At the outset, both the stress-separation law active in the cohesive zone and the creep compliance function for the bulk material were unknown and had to be back-calculated through load versus crack opening displacements at the crack mouth, an intermediate location, and at the physical crack tip. The scale-invariant fracture energy (at structural scales, at least) has been determined to be 15 N/m. The scale effect on strength can now be predicted as a function of any crack length and for any geometry. The scale dependence on the apparent fracture toughness K_Q or G_c^{scale} can now be evaluated (Mulmule and Dempsey, 1999, 2000).

Vertical Ice Forces: Bearing Capacity

In cold regions, floating ice sheets are often used as roads, airfields, parking lots (especially by ice fishing enthusiasts), and construction platforms (Gold, 1971; Ashton, 1986; Masterson and Smith, 1991). For loads of short duration, under cold conditions, the ice cover may be assumed to behave elastically. As revealed by Kerr's comprehensive reviews of the literature (Kerr, 1976, 1996), most analyses have presumed 1) the validity of classical thin-plate theory, 2) a strength-based failure criterion, 3) breakthrough coincides with the formation of the first crack system, and 4) given that the radial cracks form a system of wedges, one may proceed by analyzing one floating wedge. This led to design formulas expressed in terms of the maximum stress and the thickness of the ice squared (h^2), with no recognition of scale effects or the consideration of fracture mechanics. Moreover, as stated by Kerr (1984), this method has shortcomings, "since the number of observed radial cracks varies widely, and also there are interaction forces between the wedges that are not included in these analyses."

As shown by a number of field tests, the formation of the "first crack system" is not associated with the peak bearing capacity and failure. Subject to an increasing vertical load, radially oriented surface cracks initiate at the bottom of the plate. These cracks propagate up through-the-thickness as well as radially. Ultimately, circumferential cracking ensues, initiated by tension on the top surface of the plate. The crack face interaction that occurs after a number of radial cracks have "popped in" produces a wedging action that allows the plate to carry an additional load. This wedging action, or crack face interaction, should be more evident for thicker sheets. This crack face interaction is a complicated three-dimensional contact problem. The contact pressure distribution is unknown and acts over an unknown area. The constraint that the contact pressure be positive (compressive) or zero, thus excluding tensile tractions on the crack faces, in itself makes the problem nonlinear (even for small deformations and linearly elastic material behavior).

Slepyan (1990) studied the radial cracking of a floating plate without including the influence of crack closure, and determined that the bearing capacity would vary as $h^{13/8}$. Conversely, in a rigorous treatment of radial cracking with closure, Dempsey et al. (1995) determined that the bearing capacity varies instead as $h^{3/2}(EG_c^{scale})^{1/2}$.

In the latter paper, the progressive radial cracking of a clamped non-floating plate subjected to crack-face closure was studied. The material behavior was assumed to be elastic-brittle. The cracks were assumed to be relatively long in the sense that the three-dimensional contact problem could then be described via a statically equivalent two-dimensional idealization. The number of cracks was supposed large enough to permit a quasi-continuum approach rather than one involving the discussion of discrete sectors. The formulation incorporated the action of both bending and stretching as well as closure effects of the radial crack face contact. In this type of problem, since the contact scenario envisaged is of the receding type, the contact area changes discontinuously from its initial to its loaded extent and shape on application of the first increment of load. Further, if the nature of the loading does not change but increases in magnitude only, and if the cracked zone radius does not change, the extent and shape of the closure contact does not change. The final special property of receding contacts is especially important in the context of the problem under consideration: the intensity of the closure stress distribution will increase, without change in form, in direct proportion to the load. Fracture mechanics was used to explore the load-carrying capacity and the importance of the role of the crack-surface-interaction. For a given crack radius, the closure contact width was assumed to be constant. Under this condition, a closed-form solution was obtained for the case of a finite non-floating clamped plate subjected to a concentrated central force. Crack growth stability considerations predicted that the system of radial cracks would initiate and grow unstably over a significant portion of the plate radius. The closure stress distribution was determined exactly in the case of narrow contact widths and approximately otherwise.

The strength-based bearing capacity predictions (Kerr, 1976) predict a nominal failure stress P_{max}/h^2 that is independent of ice thickness. Given that G_c^{scale} can vary as h^β with ($0<\beta<1/2$), the fracture mechanics treatment suggests that P_{max}/h^2 should vary as $h^{-1/2+\beta/2}$. That is, the bearing capacity should exhibit a size effect on the ice thickness. Bažant and Kim (1998) reached the same conclusion. As revealed by multiple discussions on the issue of bearing capacity given radial cracking with closure (Kerr, 1996; Dempsey et al., 2000), there is the need for careful experiments over a larger range in ice thickness, with detailed measurements and observations.

Conclusions

Knowledge concerning the influence of scale on the tensile fracture behavior of ice is summarized. This knowledge is then used to infer the influence of ice thickness on in-plane design ice forces and bearing capacity.

Acknowledgments

This material is based upon work supported in part by the National Science Foundation under Grant No. 0338226 and in part by the U.S. Army under Grant No. DAAD19-00-1-0479.

References

Bažant, Z.P., 1984. Size effect in blunt fracture: concrete, rock, metal. *ASCE Journal of Engineering Mechanics*, **110**, 518-535.

Bažant, Z.P., Kim, J.J.H., 1998. Size effect in penetration of sea ice plate with part-through cracks. I:Theory. II: Results. *Journal of Engineering Mechanics*, **124**, 1310-1315 and 1316-1324.

Blanchet, D., 1998. Ice loads from first-year ice ridges and rubble fields. *Canadian Journal of Civil Engineering*, **25**, 206-219.

Cole, D.M., and Dempsey, J.P., 2001. Influence of scale on the constitutive behavior of sea ice. In: IUTAM Symposium on Scaling Laws in Ice Mechanics and Ice Dynamics (Eds. J.P. Dempsey and H.H. Shen), Kluwer Academic Publishers, Dordrecht, The Netherlands, 251-264

Cole, D.M., and Shapiro, L.H., 1998. Observations of brine drainage networks and microstructure of first-year sea ice. *Journal of Geophysical Research*, **103**, 21739-21750.

Coon, M. D., Knoke, G.S., and Echert, D.C., Pritchard, R.S., 1998. The architecture of an anisotropic elastic-plastic sea ice mechanics constitutive law. *Journal of Geophysical Research*, **103**(C10), 21915-21925.

Croasdale, K.R., 1988. Ice forces: current practices. Proceedings of the 7th International OMAE Conference IV, 33-151.

Dempsey, J.P., 1996. Scale effects on the fracture of ice. The *Johannes Weertman Symposium*, (Eds. R.J. Arsenault, D.M. Cole, T. Gross, G. Kostorz, P. Liaw, S. Parameswaran, and H. Sizek), The Minerals, Metals and Materials Society (TMS), Warrendale, Pennsylvania, 351-361.

Dempsey, J.P., Palmer, A.C., and Sodhi, D.S., 2001. High-pressure zone formation during compressive ice failure. *Engineering Fracture Mechanics*, **68**, 1961-1974.

Dempsey, J.P., DeFranco, S.J., Adamson, R.M., and Mulmule, S.V., 1999. Scale effects on the in-situ tensile strength and fracture of ice. I: Large grained freshwater ice at Spray Lakes Reservoir, Alberta. & II: First-year sea ice at Resolute. *N.W.T. International Journal of Fracture*, **95**, 325-345 and 347-366.

Dempsey, J.P., Sodhi, D.S., Bažant, Z.P., and Kim, J.J. H., 2000. Discussion of "Size effect in penetration of sea ice plate with part-through cracks. I:Theory. II: Results." *Journal of Engineering Mechanics*, **126**, 438-442.

Hirayama, K. -I., Schwarz, J., and Wu, H.C., 1973. Model technique for the investigation of ice forces on structures. Proceedings of the 2nd International POAC Conference, 332-344.

Kerr, A.D., 1976. The bearing capacity of floating ice plates subjected to static or quasi-static loads – A critical survey. *Journal of Glaciology*, **17**, 229-268.

Kerr A.D., 1984. Mechanics of ice cover breakthrough. Proceedings of a Workshop on Penetration Technology, Hanover, NH, 245-262.

Kerr, A.D., 1996. Bearing capacity of floating ice covers subjected to static, moving, and oscillatory loads. *Applied Mechanics Reviews*, **49**, 463-476.

Kry, P.R., 1981. Scale effects in continuous crushing of ice. Proceedings of the 6th IAHR Symposium on Ice II, 565-580.

Leicester, R.H., 1973. Effect of size on the strength of structures. CSIRO Australian Forest Products Laboratory, Division of Building Research Technological Paper No. **71**, 1-13.

Masterson, D.M., and Spencer, P.A., 2001. Ice force calculation for large and small aspect Ratios. In: IUTAM Symposium on Scaling Laws in Ice Mechanics and Ice Dynamics (Eds. J.P. Dempsey and H.H. Shen), Kluwer Academic Publishers, Dordrecht, The Netherlands, 31- 42.

Mulmule, S.V., and Dempsey, J.P., 1997. Stress-separation curves for saline ice using the fictitious crack Model. *ASCE Journal of Engineering Mechanics*, 123,870-877.

Mulmule, S.V., and Dempsey, J.P., 1998. A viscoelastic fictitious crack model for the fracture of sea ice. *Mechanics of Time-Dependent Materials*, **1**, 331-356.

Mulmule, S.V., and Dempsey, J.P., 1999. Scale effects on sea ice fracture. *Mechanics of Cohesive-Frictional Materials*, **4**, 505-525.

Mulmule, S.V., and Dempsey, J.P., 2000. LEFM size requirements for the fracture testing of sea Ice. *International Journal of Fracture*, **102**, 85-98.

Sanderson, T.J.O., 1988. *Ice Mechanics, Risks to Offshore Structures*. Graham & Trotman Limited, UK.

Weeks, W.F., and Assur, A., 1972. Fracture of lake and sea ice. In: *Fracture* (Ed. H. Liebowitz), VII, 879-978.

Fracture Analysis and Size Effects in Failure of Sea Ice

Zdeněk P. Bažant, Northwestern University

Abstract

This study,[1] dedicated to professor Arnold Kerr of the University of Delaware, is based on the premise (recently validated by Dempsey's in-situ tests) that large-scale failure of sea ice is governed by cohesive fracture mechanics. The paper presents simplified analytical solutions for (1) the load capacity of floating ice plate subjected to vertical load and (2) the horizontal force exerted by an ice plate moving against a fixed structure. The solutions clarify the fracture mechanism and agree with the previous numerical simulations based on cohesive fracture mechanics. They confirm the presence of a strong deterministic size effect. For the case of vertical load, the size effect approximately follows the size effect law proposed in 1984 by Bažant. In the case of an ice plate moving against a fixed obstacle, radial cleavage of the ice plate in the direction opposite to ice movement causes a size effect of structure diameter which follows linear elastic fracture mechanics for small enough diameters but becomes progressively weaker as the diameter increases. The present solutions contradict the earlier solutions based on material strength or plasticity theories, which exhibit no size effect.

Introduction

Based on the classical studies of Kerr (1975) and many others (see Kerr, 1996, for a review), the failure of sea ice has until recently been analyzed exclusively according to the material strength criteria, in the form of either plasticity or elasticity with a strength limit. Recently, however, it transpired that this classical approach, which is known to agree with small-size field tests and normal laboratory experiments, is adequate only for small-scale failure, but not for large-scale failure (e.g., failure of a floating ice sheet more than about 0.5 m thick). The reason is that the classical approach exhibits no size effect on the nominal strength of structure.

Until recently, whenever a size effect was observed in tests, it was explained by Weibull theory of strength randomness. However, such an explanation of size effect is dubious because the maximum load in ice failure is usually not reached at the initiation of fracture but only after large stable crack growth (e.g., Bažant and Planas, 1998; Bažant, 1997a; Bažant and Chen, 1997; Bažant, 2002a; Bažant, 2004; RILEM, 2004). Rather, the explanation must be sought in quasibrittle fracture mechanics. Recently, fractality of fracture surfaces or microcrack distributions was suggested as a source of structural size effects, but this idea also does not pass rigorous scrutiny (Bažant, 1997b; Bažant and Yavari, 2004).

Various recent experiments (Dempsey, 1991; DeFranco and Dempsey, 1994; DeFranco, Wei, and Dempsey, 1991), especially Dempsey's in-situ tests of record-size specimens (Dempsey et al., 1999a,b; Mulmule et al., 1995; Dempsey et al., 1995), indicate that sea ice does follow

---

[1] A major part of this paper is reproduced from the paper Bažant (2001), and a smaller from Bažant (2000). Thanks are due to the editors of these works for their permission.

fracture mechanics and on scales larger than about 10 m is very well described by linear elastic fracture mechanics (LEFM). Consequently, the size effects of fracture mechanics (Bažant and Planas, 1998; Bažant and Chen, 1997; Bažant, 2001) must get manifested in all the failures in which large cracks grow stably prior to reaching the maximum load. This includes two fundamental problems: (1) vertical load capacity of floating ice plate (penetration fracture), and (2) the maximum horizontal force exerted on a fixed structure by a moving ice plate. The former has been analyzed by fracture mechanics at various levels of sophistication in several recent studies; see Bažant and Li (1994), Li and Bažant (1994), Dempsey et al. (1995), Bažant and Kim (1998), of which the last presents rather realistic numerical simulation confirming a strong deterministic size effect. Acoustic observations also suggest a size effect (Li and Bažant,1998).

Quasibrittle fracture analysis of another problem, namely the large-scale thermal bending fracture of floating ice (Bažant,1992), also indicated strong size effect, following, however, a different law. So did the analysis of ice plate failure subjected to a vertical line load (Bažant, 2001b; Bažant and Guo, 2002). The line load problem is very instructive because an accurate solution can be obtained from one-dimensional differential equation for a beam on elastic foundation and can be cast in terms of explicit formulas. These two problems, however, are beyond the scope of this paper.

The purpose of this article is to summarize the recent numerical studies of size effect at Northwestern University and briefly outline a simplified fracture analysis of size effect which is based on the technique of asymptotic matching, an approach that leads to explicit formulae. The present brief exposition (based on a workshop in Fairbanks, Bažant, 2000), is extended in much more detail in a separate journal by Bažant (2002), which also includes detailed discussions of previous ice fracture and scaling studies (Ashton, 1986; Atkins, 1975; Dempsey et al., 1995, 1999a; Goldstein and Osipenko, 1993; Palmer, 1983; Ponter, 1983; Slepyan, 1990; etc.).

Review of Numerical Analysis of Vertical Penetration

Bažant and Kim (1998) conducted a detailed numerical analysis of vertical load penetration, for which the typical fracture pattern is shown in Fig. 1a. The radial cracks at maximum load penetrate through only a part of ice thickness (Dempsey et al., 1995; Bažant and Li, 1995); Fig. 1b,c. The radius of each crack is divided by nodes into vertical strips in each of which the crack growth obeys Rice and Levy's (1972) nonlinear "line-spring" model relating the normal force N and bending moment M in the cracked cross section to the relative displacement Δ and rotation θ (Fig. 1b). The following ice characteristics have been assumed: tensile strength $f_t' = 0.2$ MPa, fracture toughness $K_c = 0.1$ MPa $\sqrt{m}$, Poisson ratio $\upsilon = 0.29$, and Young's modulus $E = 1.0$

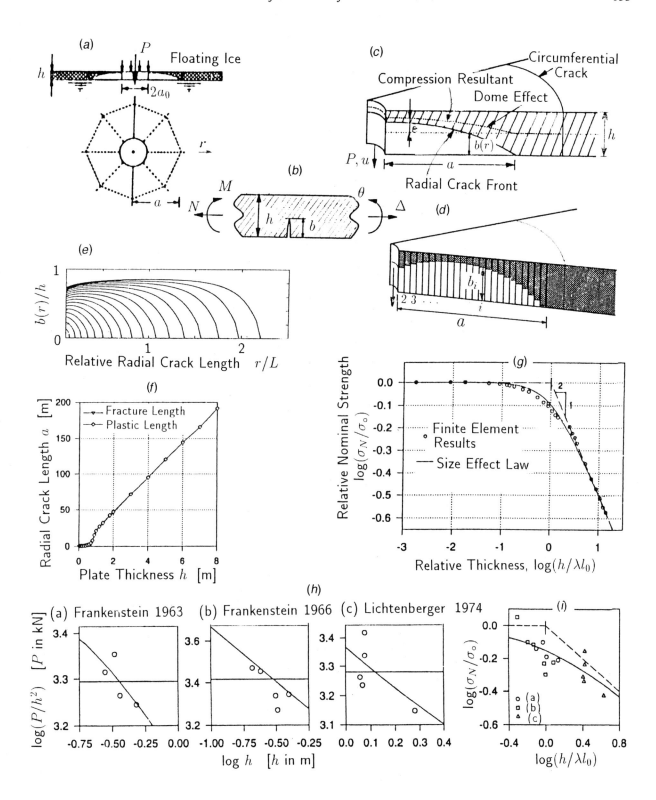

Fig.1. Vertical penetration fracture problem analyzed by Bažant and Kim (1998), main numerical results, and comparison with field tests of Frankenstein (1963, 1966) and Lichtenberger (1974).

GPa, with the corresponding values: fracture energy $G_f = K_c^2/E = 10\,\text{J/m}^2$, and Irwin's fracture characteristic length $l_0 = \left(K_c/f_t'\right)^2 = 0.25$ m.

The analysis is based on a simplified version of the cohesive crack model in which the vertical crack growth in each vertical strip is initiated according to a strength criterion. The cross section behavior is considered to be elastic-plastic until the yield envelope in the (N,M) plane is crossed by the point (N,M) corresponding to fracture mechanics. For ease of calculations, a non-associated plastic flow rule corresponding to the vector (dΔ,dθ) based on fracture mechanics is assumed. To suppress moment singularity under concentrated load P, the load is replaced by a distributed load along a small circle centered at the load point.

Fig. 1e displays, with a strongly exaggerated vertical scale, the calculated crack profiles at subsequent loading stages. Fig. 1f shows the numerically calculated plot of the radial crack length a versus the ice thickness h ("fracture length" means the radial length of open crack, and "plastic length" the crack length up to the tip of plastic zone). This plot reveals that, except for very thin ice, the radial crack length

$$a \approx c_h h \tag{1}$$

where $c_h \approx 24$.

The size effect is understood as the dependence of the nominal strength σ_N on the structure size, which is here represented by the ice thickness, h. For the vertical penetration problem, we define

$$\sigma_N = P/h^2 \tag{2}$$

where P = load. The data points in Fig. 1g show, in logarithmic scales, the numerically obtained size effect plot of the normalized σ_N versus the relative ice thickness. The initial horizontal portion, for which there is no size effect, corresponds to ice thinner than about 20 cm.

Since the model of Bažant and Kim includes plasticity, it can reproduce the classical solutions with no size effect. The ice thickness at the onset of size effect depends on the ratio of ice thickness to the fracture characteristic length, h/l_0. For realistic ice thicknesses h ranging from 0.1 m to 6 m, the computer program would yield perfectly plastic response with no size effect if the fracture characteristic length l_0 were at least 100× larger, i.e., at least 25 m. This would, for instance, happen if either were at least 10× smaller ($f_t' \leq 0.01$ MPa) or K_c at least 10× larger ($K_c \geq 10$ MPa $\sqrt{m}$). The entire diagram in Fig. 1g would then be horizontal.[2]

---

[2] Larger values of l_0 are of course possible in view of statistical scatter, but nothing like 100× larger. For example, by fitting Dempsey et al.'s (2000b) size effect data from in-situ tests at Resolute, one gets $K_c \approx 2.1$ MPa $\sqrt{m}$, and with $f_t' \approx 2$ MPa one has the fracture characteristic length $l_0 = (K_c/f_t')^2 = 1$ m. But this larger value would not make much difference in the size effect plot in Fig. 1g. The reason that these values were not used in the plot in Fig. 1g was that they correspond to long-distance horizontal propagation of fracture, rather than vertical growth of fracture.

The curve in Fig. 1g is the optimum fit of the numerically calculated data points by the generalized size effect law proposed in Bažant (1985). The final asymptote has slope $-1/2$, which means that the asymptotic size effect is $\sigma_N \propto h^{-1/2}$, the same as for LEFM with similar cracks, and not $h^{-3/8}$ as proposed by Slepyan (1990) and by Bažant and Li (1994). The $-3/8$ power scaling would have to apply if the radial cracks at maximum load were full-through bending cracks. The $-1/2$ power scaling may be explained by the fact that during failure the bending cracks are not full-through and propagate mainly vertically, which is supported by the calculated crack profiles in Fig. 1e (for thermal bending fracture, though, exponent $-3/8$ is valid; Bažant, 1992).

By fitting of the size effect data in Fig. 1g, the following generalized size effect law (Bažant, 1985; Bažant and Planas, 1998) has been calibrated (see the curve in Fig. 1g):

$$P_{\max} = \sigma_N h^2 \quad , \quad \sigma_N = B f_t' \left[1 + \left(h/\lambda_0 l_0\right)^r\right]^{-1/2r} \tag{3}$$

where $B = 1.214$, $\lambda_0 = 2.55$, $m = 1/2$, $r = 1.55$ and $l_0 = 0.25$ m ($f_t' = 0.2$ MPa in Fig. 1g). The test data available for checking this formula are very limited. The data points in Fig. 1h represent the results of the field tests by Frankenstein (1963, 1966) and Lichtenberger (1974), and the curves show the optimum fits with the size effect formula verified by numerical calculations. After optimizing the size effect law parameters by fitting the data in the three plots in Fig. 1h, the data and the optimum fit are combined in the dimensionless plot in Fig. 1i.

Interesting discussions of Bažant and Kim's (1998) study were published by Dempsey (2000) and Sodhi (2000) and rebutted by the authors. Sodhi criticized the neglect of creep in Bažant and Kim's analysis. Intuition suggests that the influence of creep might be like that of plasticity, which tends to increase the process zone size, thereby making the response less brittle and the size effect weaker. But the influences of creep and plasticity are very different. This is documented by studies of concrete (e.g., Bažant and Gettu, 1992; Bažant et al., 1993; Bažant and Planas, 1998; and especially Bažant and Li, 1997; and Li and Bažant, 1997) which show that creep always makes the size effect due to crack growth stronger. In the plot of $\log\sigma_N$ versus $\log$(size), a decrease of loading rate causes a shift to the right, toward the LEFM asymptote, which means that the size effect is intensified by creep, contrary to the opinion of Sodhi. The physical reason, clarified by numerical solutions with a rate-dependent cohesive crack model (Li and Bažant, 1997), is that the highest stresses in the fracture process zone get relaxed by creep, which tends to reduce the effective length of the fracture process zone. The shorter the process zone, the higher is the brittleness of response and the stronger is the size effect. It thus transpires that, to take creep into account, it suffices to reduce the value of fracture energy and decrease the effective length of the fracture process zone.

Approximate Analysis of Vertical Penetration

An ice plate floating on water behaves exactly as a plate on Winkler elastic foundation (Fig. 1a), with a foundation modulus equal to the specific weight of water, ρ. Failure under a vertical load is known to involve formation of radial bending cracks in a star pattern (shown in a plan view in

Fig. 1a). These radial cracks do not reach through the full ice thickness before the maximum load is reached (Dempsey et al., 1995; Bažant and Li, 1995; Bažant and Kim, 1998). Rather, they penetrate at maximum load to an average depth of about *0.8h* and maximum depth *0.85h* where *h* is the ice thickness (Fig. 2a). The maximum load is reached when polygonal (circumferential) cracks, needed to complete the failure mechanism, begin to form (dashed lines in Fig. 1a).

Sea ice is not sufficiently confined to behave plastically (this is, for example, confirmed by the absence of yield plateau apparent in the load-deflection diagrams measured for instance by Sodhi, 1998). Sea ice is a brittle material (e.g., Dempsey, 1991; DeFranco and Dempsey, 1992, 1994; DeFranco et al., 1991; Bažant, 1992a,b; Bažant and Li, 1994; Li and Bažant, 1994; Bažant and Kim, 1998), and so the analysis must be based on the rate of energy dissipation at the crack front and the rate of energy release from the ice-water system.

Dimensional analysis, or alternatively a transformation of the partial differential equation for the bending of a plate on Winkler foundation to dimensionless coordinates, shows that the behavior of the plate is fully characterized by the characteristic length

$$L = \left(D / \rho\right)^{1/4} \tag{4}$$

where $D = E\,h^3\,/12\,(1 - v^2) =$ cylindrical bending stiffness of the ice plate; $v =$ its Poisson ratio. According to Irwin's relation, the energy release rate is

$$G = K_I^2 / E' = N^2 g(\alpha) / E'h. \tag{5}$$

Here $E' = E / (1 - v^2)$ and $g(\alpha)$ is a dimensionless function:

$$g(\alpha) = \pi\alpha \left[6F_m(\alpha)e/h + F_N(\alpha)\right]^2 \qquad (\alpha = a/h) \tag{6}$$

which is obtained by superposing the expressions for the stress intensity factor K_I (given in handbooks), characterized by functions $F_M(\alpha)$ and $F_N(\alpha)$ for loading by bending moment M and normal force N; $a =$ crack depth, $e = -M/N =$ eccentricity (positive when the compression resultant is above the mid-plane).

To relate M and N to vertical load P, let us consider element 12341 of the plate (Fig. 1a and 2e,f,g), limited by a pair of opposite radial cracks and the initiating polygonal cracks (with zero depth at initiation). Since the cracks must form at the location of the maximum radial bending moment, the vertical shear force on the planes of these cracks is zero. The distance R of the polygonal cracks from the vertical load P may be expected to be proportional to the characteristic length L, and so we may set $R = \mu_R L$ where μ_R is assumed to be a dimensionless constant (Irwin's characteristic length of fracture is expected to have no effect because the initiation of polygonal cracks is governed by a strength criterion).

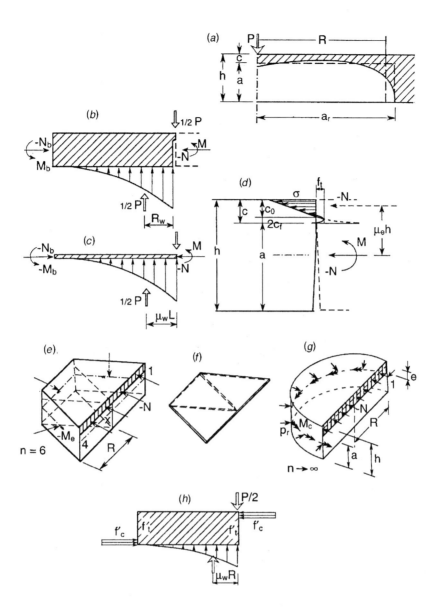

Fig. 2. Analysis of vertical penetration fracture: (a) Crack profile and (b-h) forces acting on element 123401.

In each narrow radial sector, the resultant of the water pressure due to deflection w (Fig. 2b,c) is located at a certain distance r_w from load P; r_w must be proportional to L because its solution must depend on only one parameter, L. Integration over the area of element 12341, taken as a semi-circle of radius r_w, yields the vertical resultant of water pressure acting on the whole element 12341. Again, the distance of this resultant (whose magnitude it $P/2$) from load P must be proportional to L, i.e., may be written as

$$R_w = \mu_w L \tag{7}$$

where the factor μ_w is dimensionless and is assumed to be a constant (which is justified when crack tip plasticity is negligible).

For simplicity, we assume N and M along the radial as well as polygonal cracks to be uniform. The condition of equilibrium of horizontal forces acting on element 12341 in the direction normal to a radial crack then requires N on the planes of the polygonal cracks to be equal to the N acting in the radial crack planes (this becomes clear upon noting that the loading by N along the entire circumference of element 12341 is in equilibrium with a two-dimensional hydrostatic stress in the horizontal plane).

The axial vectors of the moments M_c acting on the polygonal sides are shown in Fig. 2e,g by double arrows. Summing the projections of these axial vectors from all the polygonal sides of the element, one finds that their moment resultant with axis in the direction 14 is $2R\,M_c$, regardless of the number n of radial cracks. So, upon setting $R = \mu_R L$, the condition of equilibrium of the radial cracks with the moments about axis 14 (Fig. 2b,c,e,g) located at mid-thickness of the cross section may be written as:

$$2(\mu_R L)M + 2(\mu_R L)M_c - \frac{1}{2}P(\mu_w L) = 0 \tag{8}$$

Consider now the initiation of the polygonal cracks. It occurs when the normal stress σ reaches the tensile strength f_t' of the ice. A layer of distributed microcracking, of some effective constant thickness D_b that is a material property, will have to form at the top ice before the polygonal crack could (cf. Bažant and Li, 1996, for concrete). The polygonal cracks may be assumed to initiate when the average stress in layer D_b reaches the tensile strength f_t', and since the average stress is roughly the elastically calculated stress for the middle of layer D_b, the criterion of bending crack initiation may simply be written as

$$\sigma_b + N/h = f_t' \tag{9}$$

where

$$\sigma_b = \left(12M_c/h^3\right)(h - D_b)/2 = \left(6M_c/h^2\right)(1 - D_b/h) \tag{10}$$

This criterion, however, can be correct only when h is sufficiently larger than D_b , i.e., asymptotically for $h/D_b \to \infty$. The case $h < D_b$ is physically meaningless. For $h \approx D_b$, M can be reasonably approximated as the plastic bending moment, which may be approximately taken as 1.5× larger than the elastic bending moment for the same material strength. This condition is satisfied by replacing the aforementioned initiation criterion with:

$$\frac{6M_c}{h^2 q(\xi)} + \frac{N}{h} = f_t', \qquad q(\xi) = \frac{2+\xi}{1+\xi}, \qquad \xi = \frac{h}{D_b} \tag{11}$$

This replacement is justified by noting the large-size asymptotic expansion

$$\begin{aligned} 1/q(\xi) &= (1 + D_b/h)/(1 + 2D_b/h) \\ &= (1 + D_b/h)(1 - 2D_b/h + (\cdot)/h^2 + ...) \\ &= 1 - D_b/h + (\cdot)/h^2 + (\cdot)/h^3 + ... \end{aligned} \tag{12}$$

and realizing that factor *1/q(ξ)* and the original factor *(1 –D_b/h)* are both valid only up to the first two terms of the asymptotic expansion in powers of *1/h*, which are common to both (cf. Bažant and Chen, 1997). The approximation by function *q(ξ)* achieves the asymptotic matching of the size effects at very large and very small sizes.

Now substitute

$$M = -Ne = N\mu_e h \tag{13}$$

into Eq. (8) (*N* is negative when compressive); $\mu_e = e/h \approx$ = constant < 0.5 (according to the numerical results of Bažant and Kim (1998), $\mu_e \approx 0.45$, since the average crack depth *a* at maximum load is about 0.8 *h*). Then M_c may be expressed from Eq. (8) and substituted into Eq. (11). Furthermore, one must take into account the condition of vertical propagation of the radial bending cracks, which may be written as $G=G_f$ where G_f is the fracture energy of ice. Thus, the critical value of normal force (compressive, with eccentricity *e*) may be written as

$$N = -\sqrt{E'G_f h / g(\alpha)} \tag{14}$$

Algebraic rearrangements eventually lead to the following equation:

$$\sigma_N = \frac{2\mu_R}{3\mu_w}\left\{[6\mu_e + q(\xi)]\sqrt{\frac{E'G_f}{hg(\alpha)}} + f_t' q(\xi)\right\} \tag{15}$$

To decide the value of α, note that a finite fracture process zone (FPZ) of a certain characteristic depth $2c_f$ which is a material property must exist at the tip of vertically propagating radial crack in ice. This zone was modeled in the numerical simulations of Bažant and Kim (1998) as a yielding zone. The tip of the equivalent LEFM crack lies approximately in the middle of the FPZ, i.e., at a distance c_f from the actual crack tip, whose location is denoted as a_0. In structures of different sizes, the locations of the actual crack tip are usually geometrically similar, i.e., the value of $\alpha_0 = a_0/h$ may be assumed to be constant when ice plates of different thicknesses h are compared (Bažant and Planas, 1998). Thus, denoting $g'(\alpha_0) = dg(\alpha_0)/d\alpha_0$, one may introduce the approximation

$$g(\alpha) \approx g(\alpha_0) + g'(\alpha_0)(c_f/D) \tag{16}$$

Substituting this into Eq. (15) and rearranging, one gets for the size effect the formula:

$$\sigma_N = \frac{4\mu_R}{\mu_w}\left[\mu_e \frac{q(\xi)}{6}\right]\sqrt{\frac{E'G_f}{hg(\alpha_0) + c_f g'(\alpha_0)}} + \frac{\mu_R}{3\mu_w}q(\xi)f_t' \tag{17}$$

Consider now the special case in which the size dependence of $q(\xi)$ is neglected, i.e. $q(\xi) = 1$ (this is justified when $\xi = h/D_b$ is large because $q(\xi)$ approaches 1 as $1/h$, which is much faster than the asymptotic trend of Eq. (18), which is $1/\sqrt{h}$). In this special case, Eq. (17) reduces to the classical size effect law with non-zero residual strength σ_r (proposed by Bažant, 1987):

$$\sigma_N = \sigma_0\left(1 + \frac{h}{h_0}\right)^{-1/2} + \sigma_r \tag{18}$$

Note, however, that σ_r appeared to be negligible in Bažant and Kim's (1998) numerical simulations, and in that case the formula reduces to the size effect law proposed in Bažant (1984). Eq. (17) reduces to this law with $\sigma_r = 0$.

in which $\quad \sigma_0 = \frac{4\mu_R\mu_e}{\mu_w}\sqrt{\frac{E'G_f}{c_f g'(\alpha_0)}}$, $\quad h_0 = c_f \frac{g'(\alpha_0)}{g(\alpha_0)}$, $\quad \sigma_r = \frac{\mu_e}{3\mu_w}f_t'$ $\quad$ (19)

This formula was shown to agree with the numerical simulations by Bažant and Kim (1998), which in turn were shown not to disagree with the experimental data that exist (Fig. 1h,j).

The size effect could be absent only if the bending moments in the radial cracks as well as polygonal cracks corresponded to the birectangular plastic stress distribution. Denoting the tensile and compressive yield strengths as f_t and f_c, and taking the moment equilibrium condition of element 12341 (Fig. 2) about line 14, one can show that the nominal strength would in that case be

$$\sigma_N = (4\mu_R/\mu_w)(f_c^{-1} + f_t^{-1})^{-1} \tag{20}$$

which exhibits no size effect.

Horizontal Force Exerted on Obstacle by Moving Ice

Another important problem of scaling is the force P that a moving ice plate of thickness h exerts on a fixed structure, idealized as a rigid circular cylinder of diameter d. The nominal strength of the structure may in this case be defined as

$$\sigma_N = P / hd. \tag{21}$$

Several mechanisms of break-up are possible.

One is the elastic buckling of the floating plate (e.g., Slepyan, 1990), shown in Fig. 3a. Although the present interest is in fracture, it may be mentioned that the elastic buckling failure exhibits a reverse size effect of the type

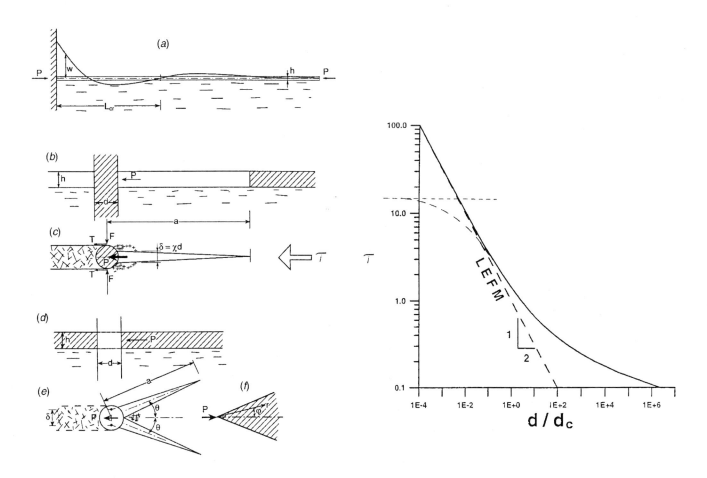

Fig. 3. (a) Elastic buckling of an ice plate moving against an obstacle; (b – c) radial cleavage crack of a plate pressing on an obstacle and, on the right, the corresponding size effect when the cohesive zone length is ignored; (d – f) diverging V-cracks.

$$\sigma_{Ncr} = \kappa \left(d/h \right) \sqrt{\rho E'} \; \sqrt{h} \qquad (22)$$

(Bažant, 2000) where κ is a dimensionless parameter depending on d/h. Thus, only very thin floating plates can fail by buckling.

Global Failure Due to Cleavage Fracture

Consider now a long radial cleavage crack in the ice plate, propagating against the direction of ice movement (Fig. 3b,c). The ice exerts on the structure a pair of transverse force resultants F and a pair of tangential forces T in the direction of movement; $T = F \tan \varphi$ where φ = effective friction angle. Considering the ice plate as infinite, we have

$$K_1 = \left(F/h \right)\sqrt{2/\pi a} \qquad (23)$$

(e.g., Tada et al.'s handbook, 1985). The energy release rate is

$$\mathrm{G} = \frac{1}{h}\left[\frac{\partial \prod{}^{*}}{\partial a} \right]_{F} = \frac{1}{h}\frac{d}{da}\left[\frac{1}{2}C(a)F^{2} \right] = \frac{F^{2}}{2h}\frac{dC(a)}{da} \qquad (24)$$

where a = crack length ((Fig. 3b,c) and $C(a)$ = load-point compliance of forces F. Using Irwin's relation, we have

$$\mathrm{G} = K_{I}^{2}/E = 2F^{2}/\pi Eh^{2}a. \qquad (25)$$

Equating this to Eq. (24), we thus get

$$dC(a)/da = 4/\pi Eha. \qquad (26)$$

This expression may now be integrated from $a = d/2$ (surface of structure, Fig. 3b,c) to a. Thus one gets $C(a)$, and from it the opening deflection

$$\delta = C(a)F = \left(4F/\pi Eh \right)\ln\left(2a/d \right). \qquad (27)$$

Now note that there is likely to be at least some amount of local ice crushing at the structure, and so the relative displacement between the two flanks of the crack must be less than d. Setting $\delta = \chi d(\xi<1)$, we obtain from the foregoing expression for δ the relation

$$a = \frac{1}{2}d \exp\left(\pi Eh\chi d/4F \right) \qquad (28)$$

(note that a/d is not constant but increases with d; hence, the fracture modes are not geometrically similar, and so the LEFM power scaling cannot be expected to apply). Now we substitute this into

$$K_I = (F/h)\sqrt{2\pi/a} \tag{29}$$

set

$$K_I = K_c = \sqrt{EG_f} \tag{30}$$

where K_c = fracture toughness of ice, and write

$$\sigma_N hd = P = 2T = 2(F\tan\varphi) \tag{31}$$

where φ = friction angle. After some rearrangements, this yields the size effect relation

$$\frac{2\sqrt{\pi}\,F}{h\sqrt{EG_f d}} = \exp\left(\frac{Eh\chi d}{8\pi F}\right) \tag{32}$$

The pair of forces F is related to load P on the structure ($P = 2T$, Fig. 3c) by a friction law, which may be written as

$$P = 2F\tan\varphi \tag{33}$$

where φ is the friction angle. Now we may substitute $F = P/2\tan\varphi$ and $P = \sigma_N hd$ into Eq. (32), and solve the resulting equation

$$d = \frac{(\tan\varphi)^2}{\pi}\frac{EG_f}{\sigma_N^2}\exp\left(\frac{E\chi\tan\varphi}{2\pi\sigma_N}\right) \tag{34}$$

or

$$\frac{d}{d_c} = \left(\frac{\sigma_c}{\sigma_N}\right)^2 e^{\sigma_c/\sigma_N} d/d_c = \tau^{-2}e^{1/\tau} \tag{35}$$

in which $\tau = \sigma_N/\sigma_c$ and d_c and σ_c are constants defined as

$$d_c = 4G_f/\pi E\chi^2, \qquad \sigma_c = \frac{1}{2}\pi E\chi\tan\varphi. \tag{36}$$

Eq. (34), plotted in Fig. 3 (right), represents the law of radial cleavage size effect in an inverted form. Note that, for $d \ll d_c$, $\sigma_N \approx \sqrt{d_c / d}$, which is the LEFM scaling for similar cracks.

So far, the length $2c_f$ of the cohesive zone of the radial cleavage crack was considered negligible compared to a. If this is not so, then the log-log size effect plot (dashed curve) is found to start from the left with a horizontal asymptote and then gradually approach the curve in Fig. 3 (right) (see the formula in Bažant, 2001).

$$\text{for } d \ll d_c : \quad \sigma_N \approx \sqrt{d_c / d}; \tag{37}$$

To figure out the asymptotic size effects, we may rewrite Eq. (34) as $\ln(d/d_c) = 2\ln(1/\tau) +)1/\tau)$. In this sum, the first term dominates when τ is very large (and d very small), while the second term dominates when τ is very small (and d very large). This rigorously follows (with the notation $\xi = 1/\tau$) from the limits:

$$\lim_{\xi \to 0} \frac{2\ln\xi + \xi}{2\ln\xi} = 1, \qquad \lim_{\xi \to \infty} \frac{2\ln\xi + \xi}{\xi} = 1 \tag{38}$$

Thus the following asymptotic behaviors transpire:

$$\text{for } d \ll d_c : \quad \sigma_N \approx \sqrt{d_c / d}; \qquad \text{for } d \gg d_c : \quad \sigma_N \approx 1/\ln(d/d_c) \tag{39}$$

At intermediate sizes, there is a smooth transition between these two simple scaling laws. As seen in Fig. 3 (right), the size effect is getting progressively weaker with increasing d (although no horizontal asymptote is approached). The reason is that the cracks are dissimilar, i.e., the ratio a/d increases with d.

Comments on Diverging V-Shaped Cracks

Another observed mechanism (e.g., Sanderson, 1988, ch. 7) consists of diverging V-shaped cracks; Fig. 3d,e. To estimate in a crude manner the complementary energy Π^*, we may consider only the stresses within the wedge between the cracks (Fig. 3f). From a well-known solution (Timoshenko and Goodier, 1970),

$$\sigma_r = -Pk_\theta \cos\varphi / rh, \sigma_\varphi = \sigma_{r\varphi} = 0 \tag{40}$$

where σ_r, σ_φ and $\sigma_{r\varphi}$ are the stress components in polar coordinates r, φ and

$$k_0 = 1 / \left(\theta + \frac{1}{2}\sin 2\theta \right), \tag{41}$$

θ being the inclination angle of the cracks (Fig. 3e). The displacement at $r = d/2$ (structure surface) is

$$u = \int_{d/2}^{\infty} \left(\sigma_r / E\right) dr = \left(P k_\theta / Eh\right) \ln\left(2a/d\right). \tag{42}$$

Then

$$\prod{}^{*} = \frac{1}{2} Pu = \left(P^2 k_\theta / 2Eh\right) \ln\left(2a/d\right) \tag{43}$$

The complementary energy before fracture may be estimated as the value of $\prod{}^{*}$ for $\theta = \pi$, i.e.

$$\prod{}_0^{*} \approx \left(P^2 / 2\pi Eh\right) \ln\left(2a/d\right) \tag{44}$$

The total energy release due to V-cracks in the ice plate is

$$\Delta \prod{}^{*} = \prod{}^{*} - \prod{}_0^{*} \tag{45}$$

The condition

$$\left[\partial \Delta \prod{}^{*} / \partial a\right]_P = 2h G_f \tag{46}$$

yields

$$\sigma_N = \frac{P}{hd} \approx \frac{2}{d} \sqrt{\frac{E G_f}{\pi^{-1} - k_\theta}} \sqrt{a} \tag{47}$$

To determine crack length a and angle θ, one could use two conditions: (a) the opening displacement at the crack mouth, δ, must be equal to $\chi d/(2 \cos \theta)$, which means that the load-point displacement of force P must be

$$u = \left(\chi d / 2\right) \tan \theta \tag{48}$$

and (b) P should be minimized with respect to θ. However, the solution is quite complicated and will not be pursued here. Besides, there is also the question of a possible simultaneous axial (radial) cleavage crack, and the question of simultaneous ice crushing. Unlike the radial cleavage fracture, the V-shaped crack mechanism cannot accommodate continuous movement of the ice and can occur only from time to time.

A further possible break-up mechanism is the compression fracture of the ice plate in contact with the obstacle. This mechanism also leads to a pronounced size effect. A simplified formula for it was derived in Bažant and Xiang (1997) and Bažant (2000) (also Bažant, 2001, 2002).

Finally, the ice floe can fracture globally, upon impact. It can be approximately treated as a deep beam loaded by distributed forces representing the inertia forces, and by a concentrated reaction from the obstacle. This type of fracture of quasibrittle materials has been treated in various works, and a strong size effect due to energy release has been shown to exist (for a review, see Bažant and Chen, 1997; Bažant, 1999, 2001, 2002).

Closing Remarks

The present fracture analysis, of course, does not mean that the classical strength theory (based either on plastic limit analysis or elastic analysis with allowable stress) is rendered useless. The classical theory, which exhibits no size effect, remains to be useful in two ways: (1) for analyzing small enough ice structures, for example, vertical penetration of sea ice plates less than about 50 cm thick; and (2) for providing a small-size asymptotic anchor to the size effect solution according to the cohesive crack model (or crack band model).

Acknowledgment

Partial financial support under grant N00014-91-J-1109 from the Office of Naval Research and grant CMS-0301145 from the National Science Foundation, both to Northwestern University, is gratefully acknowledged.

References

Ashton, G. (ed.), 1986. *River and Lake Ice Engineering.* Water Resources Publications.

Atkins, A.G., 1975. Icebreaking modeling. *J. of Ship Research,* **19**(1): 40-43.

Barenblatt, G.I., 1979. *Similarity, self-similarity and intermediate asymptotics.* Consultants Bureau (Plenum Press), New York, N.Y. (trans. from Russian original, 1978).

Barenblatt, G.I., 1987. *Dimensional analysis.* Gordon and Breach, New York.

Bažant, Z.P., 1984. Size effect in blunt fracture: concrete, rock, and metal. *J. of Engrg. Mech., ASCE,* **110**: 518-535.

Bažant, Z.P., 1985. Fracture mechanics and strain-softening in concrete. Preprints, U.S.- Japan Seminar on Finite Element Analysis of Reinforced Concrete Structures, Tokyo, **1**, 47-69.

Bažant, Z.P., 1987. Fracture energy of heterogeneous material and similitude. Preprints, SEM-RILEM Int. Conf. on Fracture of Concrete and Rock (held in Houston, Texas), ed. by S.P. Shah and S.E. Swartz, publ. by SEM (Soc. for Exper. Mech.), 390-402.

Bažant, Z.P., 1992. Large-scale thermal bending fracture of sea ice plates. *J. of Geophysical Research,* **97** (C11): 17,739-17,751.

Bažant, Z.P., 1992. Large-scale fracture of sea ice plates. (Proc. 11th IAHR Ice Symposium, Banff, Alberta), June (ed. by T.M. Hrudey, Dept. of Civil Engineering, University of Alberta, Edmonton), **2**, 991-1005.

Bažant, Z.P., 1993. Scaling laws in mechanics of failure. *J. Engrg. Mech. ASME,* **119** (9), 1828-1844.

Bažant, Z.P., 1997a. Scaling of quasibrittle fracture: asymptotic analysis. *Int. J. of Fracture,* **83** (1): 19-40.

Bažant, Z.P., 1997b. Scaling of quasibrittle fracture: Hypotheses of invasive and lacunar fractality, their critique and Weibull connection. *Int. J. of Fracture,* **83** (1): 41-65.

Bažant, Z.P., 1999. Size effect on structural strength: a review. *Archives of Applied Mechanics* (Ingenieur-Archiv, Springer Verlag), **69**: 703-725.

Bažant, Z.P., 2000. Scaling laws for brittle failure of sea ice. *Preprints* distributed at IUTAM Workshop on Scaling Laws in Ice Mechanics and Ice Dynamics, held at University of Alaska, Fairbanks, June 2000, J.P. Dempsey, H.H. Shen, and L.H. Shapiro, eds., 1-23.

Bažant, Z.P., 2001a. *Scaling of Structural Strength.* Hermes Scientific Publications, Oxford and Paris.

Bažant, Z.P., 2001b. Scaling of failure of beams, frames and plates with softening hinges. *Meccanica* (Kluwer Acad. Publ.), 36, 67-77, 2001 (special issue honoring Giulio Maier).

Bažant, Z.P., 2001. Scaling laws for sea ice fracture. *Proc., IUTAM Symp. on Scaling Laws in Ice Mechanics and Ice Dynamics*, (held in Fairbanks, June 2000), ed. by J.P. Dempsey and H.H. Shen, Kluwer Academic Publ., Dordrecht, 195-206.

Bažant, Z.P., 2002. Scaling of Sea Ice Fracture. I. Vertical Penetration. II. Horizontal Load from Moving Ice. *ASME Journal of Applied Mechanics,* 69 (Jan.), 11-18 and 19-24.

Bažant, Z.P., 2002a. *Scaling of Structural Strength.* Hermes Penton Science (Kogan Page Science), London, (280 + xiii pages, monograph); and French translation 2004.

Bažant, Z.P., 2004. Scaling theory for quasibrittle structural failure. *Proc., National Academy of Sciences,* 101, in press.

Bažant, Z.P., and Cedolin, L., 1991. *Stability of structures: Elastic, inelastic, fracture and damage theories.* Oxford University Press, New York.

Bažant, Z.P., & Chen, E.-P., 1997. Scaling of structural failure. *Applied Mechanics Reviews,* ASME, **50** (10): 593-627.

Bažant, Z.P., & Gettu, R., 1991. Size effects in the fracture of quasi-brittle materials. in Cold Regions Engineering (Proc., 6th ASCE International Specialty Conference, held in Hanover, NH, Feb. 1991), D.S. Sodhi (ed.), ASCE, New York, 595-604.

Bažant, Z.P., and Guo, Z., 2002. Size effect on strength of floating sea ice under vertical line load. *J. of Engrg. Mechanics*, **128** (3), 254-263.

Bažant , Z.P., and Kim, J.-K., 1985. Fracture theory for nonhomogeneous brittle materials with application to ice. Proc. *ASCE Nat.* Conf. on Civil Engineering in the Arctic Offshore – ARCTIC 85, San Francisco, L. F. Bennett (ed.), ASCE, New York, 917-930.

Bažant, Z.P., and Kim, Jang-Jay H., 1998. Size effect in penetration of sea ice plate with part-through cracks. I. Theory. *J. of Engrg. Mechanics*, ASCE, **124** (12): 1310-1315; with discussions and closure in **126** (4): 438-442 (2000).

Bažant, Z.P., and Kim, Jang-Jay H., 1998. Size effect in penetration of sea ice plate with part-through cracks. II. Results. *J. of Engrg. Mechanics* ASCE, **124** (12): 1316-1324; with discussions and closure in **126** (4): 438-442 (2000).

Bažant, Z.P., Kim, J.J., and Li, Y.-N., 1995. Part-through bending cracks in sea ice plates: Mathematical modeling. ICE MECHANICS—1995, J.P. Dempsey and Y. Rajapakse (eds.), ASME AMD, **207**, 97-105.

Bažant, Z.P., and Li, Y.-N., 1994. Penetration fracture of sea ice plate: Simplified analysis and size effect. *J. of Engrg. Mech.* ASCE, **120** (6): 1304-1321.

Bažant, Z.P., and Li, Y.-N., 1995. Penetration fracture of sea ice plate. *Int. J. Solids Structures,* **32** (No. 3/4): 303-313.

Bažant, Z.P., and Planas, J., 1998. *Fracture and size effect in concrete and other quasibrittle materials.* CRC Press, Boca Raton, Florida.

Bažant, Z.P., and Yavari, A., 2004). Is the cause of size effect on structural strength fractal or energetic-statistical?. *Engrg. Fracture Mechanics,* 43, in press.

Bažant, Z.P., and Xiang, Y., 1997. Size effect in compression fracture: splitting crack band propagation. *J. of Engrg. Mechanics* ASCE, **123** (2): 162-172.

Dempsey, J.P., 1991. The fracture toughness of ice. *Ice Structure Interaction.* S.J. Jones, R.F. McKenna, J. Tilotson and I.J. Jordaan, eds., Springer-Verlag, Berlin, 109-145.

Dempsey, P.P., 2000. Discussion of Size effect in penetration of ice plate with part-through cracks. I. Theory, II. Results. by Z.P. Bažant and J.J.H. Kim. *J. of Engrg. Mech.,* **126** (4): 438; with authors' rebuttal, 438-442.

Dempsey, J.P., Adamson, R.M., and Mulmule, S.V., 1995. Large-scale in-situ fracture of ice. Proc., FRAMCOS-2, Wittmann, F.H., ed., AEDIFICATIO Publishers, D-79104 Freiburg, Germany, 675-684.

Dempsey, J.P., Adamson, R.M., and Mulmule, S.V., 1999b. Scale effects on the *in situ* tensile strength and fracture of ice: Part II. First-year sea ice at Resolute, N.W.T. *Int. J. of Fracture,* **95**: 346-378.

Dempsey, J.P., DeFranco, S.J., Adamson, R.M., and Mulmule, S.V., 1999a. Scale effects on the *in situ* tensile strength and fracture of ice: Part I. Large grained freshwater ice at Spray Lakes Reservoir, Alberta. *Int. J. of Fracture,* **95**: 325-345.

Dempsey, J.P., Slepyan, L.I., and Shekhtman, I.I., 1995. Radial cracking with closure. *Int. J. of Fracture,* **73** (3): 233-261.

DeFranco, S.J., and Dempsey, J.P., 1992. Nonlinear fracture analysis of saline ice: Size, rate and temperature effects. Proc. of the 11th IAHR Symposium, Banff, Alberta, **3**, 1420-1435.

DeFranco, S.J., and Dempsey, J.P., 1994. Crack propagation and fracture resistance in saline ice. *J. Glaciology,* **40**: 451-462.

DeFranco, S.J., Wei, Y., and Dempsey, J.P., 1991. Notch acuity effects on fracture of saline ice. *Annals of Glaciology,* **15**: 230-235.

Frankenstein, E.G., 1963. Load test data for lake ice sheet. *Technical Report 89,* U.S. Army Cold Regions Research and Engineering Laboratory, Hanover, New Hampshire.

Frankenstein, E.G., 1966. Strength of ice sheets. Proc., Conf. on Ice Pressures against Struct.; Tech. Memor. No. 92, NRCC No. 9851, Laval University, Quebec, National Research Council of Canada, Canada, 79-87.

Goldstein, R.V., and Osipenko, N.M., 1993. Fracture mechanics in modeling of icebreaking capability of ships. *J. of Cold Regions Engrg. ASCE ,* **7** (2): 33-43.

Li, Y.-N., and Bažant, Z.P., 1994. Penetration fracture of ice plate: 2D analysis and size effect. *J. of Engrg. Mech.* ASCE, **120** (7): 1481-1498.

Li, Z., and Bažant, Z.P., 1998. Acoustic emissions in fracturing sea ice plate simulated by particle system. *J. of Engrg. Mechanics* ASCE, **124** (1): 69-79.

Lichtenberger, G.J., Jones, J.W., Stegall, R.D., and Zadow, D.W., 1974. Static ice loading tests Resolute Bay—Winter 1973/74. APOA Proj. No. 64, Rep. No. 745B-74-14, (CREEL Bib # 34-3095), Sunoco Sci. & Technol., Rechardson, Texas.

Kerr, A.D., 1975. The bearing capacity of floating ice plates subjected to static or quasi-static loads – A critical survey. Research Report 333, U.S. Army Cold Regions Research and Engineering Laboratory, Hanover, New Hampshire.

Kerr, A.D., 1996. Bearing capacity of floating ice covers subjected to static, moving, and oscillatory loads. *Appl. Mech. Rev., ASME Reprint,* **49** (11): 463-476.

Mulmule, S.V., Dempsey, J.P., and Adamson, R.M., 1995. Large-scale in-situ ice fracture experiments - part II: modeling efforts, in ice mechanics, 1995. ASME Joint Applied Mechanics and Materials Summer Conference, AMD – MD, 1995. University of California, Los Angeles, June, 28-30.

Nevel, D.E., 1958. The theory of narrow infinite wedge on an elastic foundation. *Transactions, Engineering Institute of Canada*, 2(3).

Palmer, A.C., Goodman, D.J., Ashby, M.F., Evans, A.G., Hutchinson, J.W., and Ponter, A.R.S., 1983. Fracture and its role in determining ice forces on offshore structures. *Annals of Glaciology*, **4**: 216-221.

Ponter, A.R.S., Palmer, A.C., Goodman, J., Ashby, M.F, Evans, M.F., and Hutchinson, J.W., 1983. The force exerted by a moving ice-sheet on an offshore structure. 1. The creep mode. *Cold Regions Sci. & Tech.*, **8**: 109-118.

Rice, J.R. and Levy, N., 1972. The part-through surface crack in an elastic plate. *J. Appl. Mech. ASME,* **39**: 185-194.

RILEM Committee QFS, 2004. "Quasibrittle fracture scaling and size effect." Materials and Structures (Paris), 37, in press.

Sanderson, T.J.O., 1988. *Ice Mechanics: Risks to Offshore Structures*. Graham and Trotman Limited, London.

Schulson, E.M., 1990. The brittle compressive fracture of ice. *Acta Metall. Mater,* **38**(10): 1963-1976.

Schulson, E.M., 2001. Brittle failure of ice. *Engineering Fracture Mechanics*, 68: 1839-1887.

Sedov, L.I., 1959. *Similarity and dimensional methods in mechanics*. Academic Press, New York.

Slepyan, L.I., 1990. Modeling of fracture of sheet ice. *Mechanics of Solids.* (transl. of Izv. AN SSSR Mekhanika Tverdoga Tela), 155-161.

Sodhi, D.S., 1995a. Breakthrough loads of floating ice sheets. *J. Cold Regions Engrg. ASCE,* **9** (1): 4-20.

Sodhi, D.S., 1995b. Wedging action during vertical penetration of floating ice sheets. *AMD-Vol.207, Ice Mechanics*, Book No. H00954, 1995, 65-80.

Sodhi, D.S., 1996. Deflection analysis of radially cracked floating ice sheets. 17th Int. Conf. OMAE Proceedings, Book No. G00954, 1996, 97-101.

Sodhi, D.S., 1998. Vertical penetration of floating ice sheets. *Int. J. of Solids and structures,* **35** (31-32), 4275-4294.

Sodhi, D.S., 2000. Discussion of size effect in penetration of ice plate with part-through cracks. I. Theory, II. Results. by Z.P. Bažant and J.J.H. Kim. *J. of Engrg. Mech.*, **126** (4): 438-440; with authors' rebuttal, 438-442.

Tada, H., Paris, P.C., and Irwin, J.K., 1985. *The Stress Analysis of Cracks Handbook*, 2nd ed., Paris Productions, Inc., St. Louis, MO.

Timoshenko, S.P., and Goodier, J.N., 1970. *Theory of elasticity*. 3rd ed., McGraw Hill, New York, 110.

Weeks, W.F., and Mellor, M., 1984. Mechanical properties of ice in the Arctic seas. *Arctic Technology & Policy*, I. Dyer & C. Chryssostomidis (eds.), Hemisphere, Washington, D.C., 235-259.

Weeks, W.F., and Assur, A., 1972. Fracture of lake and sea ice. *Fracture*, H. Liebowitz, ed., Vol. II, 879-978.

Weibull, W., 1939. The phenomenon of rupture in solids. *Proc., Royal Swedish Institute of Engineering Research* (Ingenioersvetenskaps Akad. Handl. (Stockholm)), **153**: 1-55.

Follow That White Concrete Road:
The Evolution of Concrete Pavement Analysis

Anastasios M. Ioannides, University of Cincinnati

Abstract

This paper traces the evolution of analytical methods for concrete pavements over the period from the publication of Westergaard's early work in the 1920s until today. It is demonstrated that this development was accomplished in parallel with the birth and coming of age of geotechnical engineering, and was likewise pioneered by a number of colorful and distinguished personalities. Westergaard's analysis had been based on a series of limiting assumptions, which have defined the task for generations of pavement engineers. The eventual elimination of these shortcomings has been possible only through a combination of road tests and laboratory studies, jerry-rigged equipment and sophisticated electronic tools. As a new era of mechanistic-empirical pavement design is ushered in, it behooves the profession to undertake a concerted and serious effort to eliminate the most crucial remaining stumbling block to their efforts: Miner's fatigue hypothesis. Fracture mechanics, similitude concepts and the principles of dimensional analysis hold the brightest promise of a future worthy of the pioneers celebrated herein.

Pavement Definition and Classification

A pavement is a complex engineering system, whose analysis and design involves the interaction of three equally important components, namely, the natural supporting layers; the constructed layers; and the geometry of the applied loads (Ioannides, 1991a). The importance of the first two components is self-evident; the third is dictated by the prevailing use of linear elasticity, which robs the level of the applied load of its natural significance. Pavement engineers are invariably aware of this definition, albeit unconsciously at times, since it defines precisely the three main categories of input data required in any analytical exercise. The conventional flowchart for the so-called mechanistic-empirical method of pavement design, whose popularity has been increasing exponentially in the last few years, also recognizes this definition, while adding climate as a distinct input category (Fig. 1); the effect of the latter, of course, is accommodated through its impact on the properties and the boundary conditions of the pavement layers. In contrast, a more functional, if also conventional, definition of a pavement as a prepared surface for rubber-tired vehicles providing a safe, comfortable and economic ride, is of little help to those interested in formulating the differential equations describing a given phenomenon and proceeding with their solution. Moreover, it gives rise to the fallacy that the pavement is no more than an æsthetic coat (to which, indeed, the German words for pavement, *der Belag*; *das Pflaster*, literally translate), which has been responsible for many a blunder over the years.

171

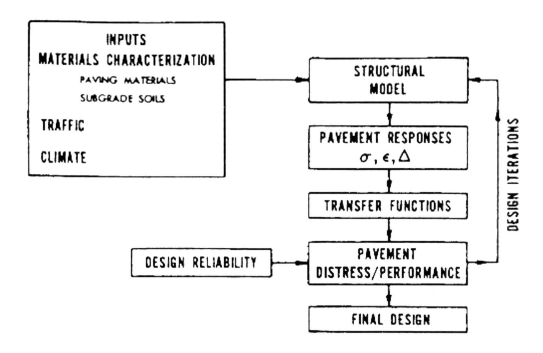

Fig. 1. The Mechanistic-Empirical Design Procedure (after Thompson, 1996)

Suffice it to mention here the scant attention to the subgrade as a design factor in the *Interim Guides for the Design of Flexible/Rigid Pavement Structures* of the American Association of State Highway Officials (AASHO, 1961, 1962). In contrast, an engineering definition of the pavement as a system allows pavement engineers to recapture their honored position among the practitioners of structural and soil mechanics, revealing the parallel development and indebtedness of their trade to those of their fellow professionals.

As an engineering system, a pavement cannot be characterized merely with reference to its constituent materials, especially since terms like "rigid" or "flexible" hardly ever mean what a dictionary or an engineer would define these words as. It would be more descriptive to classify pavements as "black" and "white," which is in essence what we do when we speak in terms of "asphalt" and "concrete" pavements, respectively. Conversely, the definition of a pavement as an engineering system leads to a classification of pavements that is not only meaningful, but also comprehensive, since it focuses instead on response and behavior. Restricting ourselves to the prevailing practices, we may immediately eliminate the geometry of the applied loads from consideration at this time, since circular loads are conventionally assumed to be applied to the pavement surface, without unduly compromising the validity of our solutions (Baron, 1942). Consequently, two categories may be recognized with respect to the constructed layers, namely, those that obey the restrictions of plate theory, and those that follow the generalized principles of linear elasticity, exhibiting compressibility through their thickness. With regard to the natural supporting layers, commonly idealized as a single material, the conventional characterizations are those of the Winkler dense liquid and the Boussinesq elastic solid (Winkler, 1867; Boussinesq, 1885). Table 1 shows the four major pavement classes that result from such a classification, along with the names of the pioneers who established the corresponding analytical procedures.

TABLE 1 - Pavement Classification

| | | Constructed Layers | |
		ELASTIC LAYER	ELASTIC PLATE
Natural Supporting Layers	ELASTIC SOLID	*Class I* (Burmister, 1943)	*Class II* (Losberg, 1960)
	DENSE LIQUID	*Class III* (Van Cauwelaert, 1990)	*Class IV* (Westergaard, 1926)

It is apparent that despite the fact that *Class I* has been traditionally applied to asphalt pavements, and that *Class IV* has been conventionally employed for concrete pavements, no Class is exclusively associated with a particular pavement type, nor is any pavement type invariably to be identified with a given analytical Class.

The Pioneers

Just as Karl Anton von Terzaghi (1883-1963) is the "father of soil mechanics," Harald Malcolm Westergaard (1888-1950) must be recognized as the person to whom modern pavement mechanics owes its birth. In his fertile mind germinated the seeds planted by a series of distinguished thinkers who preceded him. The methodology he brought to life in the early 1920s has been the progenitor of every significant and worthwhile development in pavement engineering since then. This is the primary reason why, in the interest of time and space, the presentation below will follow primarily the "white concrete road." That is not to undervalue the contributions of the other pillar of the profession, namely Donald Martin Burmister (1895-1981), who some twenty years later produced the theory and analytical solution that has formed the backbone of the efforts of those who have followed the alternate "black asphalt route."

Westergaard's first paper on the analysis of concrete pavements was published in Danish in 1923; his last publication (in English) appeared merely twenty-five years later, in 1948 (Westergaard, 1923, 1948). Both treated a slab-on-grade as a plate on a bed of springs, subject to a series of restrictive assumptions. Table 2 provides an estimate of the error in the calculated critical bending stress each of these assumptions precipitates, as well as the name of the investigator who first addressed each one effectively. The correlation between the magnitude of the error and the date of its resolution is by no means accidental. Nor is it surprising that it was Westergaard himself that provided the first two most significant links in the chain of progress represented in Table 2, viz., through his solutions of the temperature curling problem, published in 1927, and of the load transfer problem, contained in the DSc thesis of his last graduate advisee, Mikhail Skarlatou Skarlatos (b. 1922), deposited at Harvard University in 1949 (Westergaard, 1927; Skarlatos, 1949). Consideration of these two solutions can shed light on some interesting aspects of Westergaard's *modus operandi*.

TABLE 2 - Westergaard's Limiting Assumptions

	Typical Associated Error (%)	Resolved By	Year of Resolution	Governing Dimensionless Variables
Uniform Support: No Curling	50	Westergaard	1927	$\forall$)T
One Slab: No Load Transfer	25	Skarlatos	1949	f
Single Wheel Load: No Multiple Wheel Gears	15	Pickett and Ray	1951	s/P
Single Placed Layer: No Base	10	Childs, et al.	1957	k_{eff}/k
Infinite Slab: Interior, Edge and Corner Loading Conditions	5	Ioannides	1984	L/P
Semi-infinite Foundation: No Rigid Bottom	2	Pending	Pending	Pending

The Curling Problem

In attempting to address the uniform contact limitation, Westergaard (1927) retained all six restrictions noted in Table 2, while introducing the following four additional assumptions: (7) Infinite slab self-weight; (8) Applicability of the principle of superposition with regard to the summation of load-induced and thermal stresses; (9) Adequacy of considering a linear temperature variation through the slab thickness; (10) Slab response under night-time conditions is a mirror image of the corresponding behavior under day-time conditions. At first glance, these new restrictions appear to doom any solution obtained to being hopelessly unrealistic. The consequence of the infinite self-weight assumption is that the slab remains in contact with the subgrade, which seems hardly justifiable if the whole purpose of Westergaard's new analysis is the elimination of the uniform contact restriction. In addition, the principle of superposition can only be accepted if boundary and support conditions remain unchanged, a condition ensured by assumption (7), but

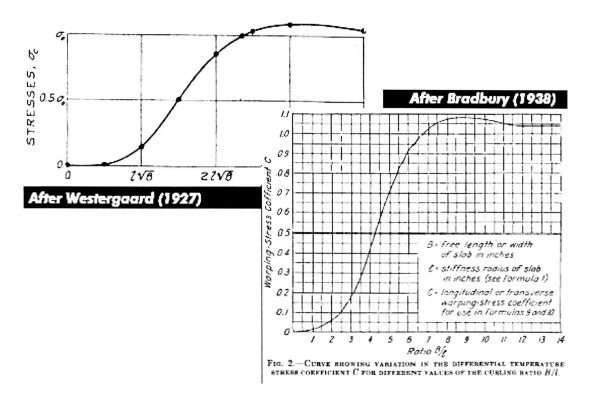

Fig. 2. C-Coefficient for Curling Analysis

hardly reflecting the slab field behavior responsible for the deterioration spawning engineering interest in the curling phenomenon in the first place. It is no consolation that recent experience gained on the basis of the finite element method, provides additional evidence undermining Westergaard's assumptions (9) and (10), as well.

Westergaard (1927) offered only scant discussion of these self-evident shortcomings of his new analysis, but proceeded quite confidently with the development of practical design tools, intended to complement the tool chest he had already started to compile for load-induced stresses. The new tools he derived were in the form of the now well-known equations for curling stresses at the interior and at the edge of a pavement slab, coupled with the widely used chart of the C-coefficient (Fig. 2). The latter is necessary for accommodating the pronounced effect of slab size in curling stress calculation. Furthermore, Westergaard (1927) stipulated that superposition be used in calculating critical combined (load-induced plus thermal) stresses under interior and edge loading conditions, during daytime and nighttime conditions.

On account of their manifestly "weak foundation," it is quite conceivable that these new tools could have been considered as unlikely to be useful in practical applications and their developer himself could have been dismissed as a dweller in the proverbial academic ivory tower. Indeed, relatively few practicing engineers today would recognize these widely used tools as Westergaard's, let alone appreciate their feeble technical basis. Instead, it is quite natural see the aforementioned curling equations and C-coefficient chart ascribed to Bradbury (1938), even in such respected textbooks as those by Yoder and Witczak (1975), and by Huang (1993). In this manner, Westergaard became the first of a long list of unsung heroes in the area of curling analysis, whose only consolation may be found in the thought that uncited imitation is the highest form of flattery. To Bradbury's credit, however, it must be freely admitted that his verbatim reproduction of Westergaard's work lent it the credibility that it sorely needed at the time. This promoted its adoption by the industry, a task to which a respected practitioner was much better suited than a university professor.

The Load Transfer Problem

In his 1948 paper, Westergaard provided some preliminary considerations for the solution of the edge load transfer problem. These, however, were limited by his implicit assumption that the load transfer in terms of deflection was identical to the load transfer in terms of stress, a postulate disproved by more recent finite element investigations (Ioannides and Korovesis, 1990, 1992). Thus, on the prominent issue of load transfer, in which pavement engineers were greatly interested, it had been the common understanding that Westergaard's last contribution was the very cursory treatment contained in his 1948 paper. A recent examination of the archives of the Airfields and Pavements Division of the Geotechnical Laboratory at Waterways Experiment Station, however, has brought to light a consulting report submitted to the Corps by Westergaard in June 1949 that contains a detailed examination of the load transfer problem. This report, cited herein as Skarlatos (1949), had been prepared under Contract W-33-017-eng-4240 from the Ohio River Laboratories of the Corps of Engineers, and describes analytical investigations by M.S. Skarlatos under the supervision of Westergaard. As far as could be ascertained (Skarlatos, 1994: personal communication), the findings of this report had never been published elsewhere, although at least one citation of this work has been encountered in the literature (Woodhead and Wortman, 1973). Because of the considerable significance of this work, both from the scientific and the historical perspectives, a paper co-authored by Skarlatos has since been published outlining the major findings of what appears to have been one of Westergaard's last and most important projects (Skarlatos and Ioannides, 1998).

To expedite dissemination of the information acquired, a second paper by Ioannides and Hammons (1996) presented the results of an investigation sponsored by the Federal Aviation Administration, which sought to extend and refine the load transfer solution by Skarlatos (1949) using modern computational tools. A comparison of this solution to earlier finite element solutions was also presented. Following the same approach as Skarlatos and Westergaard, closed-form equations have been derived for the maximum deflection and maximum bending stress occurring on the unloaded side of an edge capable of load transfer. Together with Westergaard's 1948 edge loading equations, the two formulae derived by Ioannides and Hammons (1996) effectively extend Westergaard's solutions to the case of the load transfer problem. In the process of this investigation, the genius of Westergaard became apparent once again: the entire solution to the load transfer problem was already contained in its essence in his original Danish paper (Westergaard, 1923). Its effective utilization is now possible owing to the development of modern electronic computational tools (software and hardware)!

Westergaard's accomplishments came about as a result of the work of a number of previous investigators, most prominent among them being Arthur Newell Talbot (1857-1942), his own academic mentor and eventual father-in-law. Talbot is well known for his analytical work in railway engineering, for which he idealized the rail as a beam on a Winkler medium. The evolution of railway track analysis and of methods for defining the rail-support parameter has been summarized by Kerr (1987). The extension of Talbot's 1-D solution to 2-D by Westergaard is, therefore, seen to be a natural development, whose time had come with the transition from the train to the automobile as the primary mode of transportation. Winkler, himself, had published his paper "On Elasticity and Stiffness" in 1867, and owing to its versatility, the foundation model he had proposed found quickly an application in the analysis of floating ice sheets by Hertz (1884), although it had been arguably employed earlier by rival dense liquid "parent," the Swiss-Russian academician Nikolai Fuss (1755-1826), as well as by the celebrated Leonhard Euler (1707-1783). A comprehensive review of the history and state-of-the-art of floating ice cover analysis is presented by Kerr (1996). As Selvadurai (1979) points out, however, the roots of the Winkler approach run much deeper, stretching to ancient Greece, since the dense liquid model "represents a simple consequence" of the principle of buoyancy, first articulated by Archimedes (287-212 BC), in his memorable sprint on the white marble road of ancient Syracuse, shouting "Eureka! Eureka!"

During his tenure at the University of Illinois, Malcolm Westergaard continued his friendship and professional association with Karl Terzaghi, who derided him on account of the low salary that academics, as opposed to consultants, were willing to accept (Goodman, 1999). The development of soil and of pavement mechanics proceeded in parallel throughout the first few decades of their evolution, exemplified by the ground breaking pavement engineering papers that appear in the first few International Conferences on Soil Mechanics and Foundation Engineering (e.g., Morton, 1936), Westergaard's analytical solution for a varved clay (Westergaard, 1938), as well as Terzaghi's classic paper on the determination of the subgrade modulus, k (Terzaghi, 1955). Westergaard, however, was primarily a structural engineer, and concrete was his material of choice, at least in terms of research. His contribution to the construction of the mammoth Boulder (later renamed to Hoover) dam (1933-1935) is documented by the plaque bearing his name at that site, which provides but one example of how the histories of concrete and of pavement engineering also paralleled each other. Nowhere is that more evident than in the friendly competition that gradually developed between the University of Illinois and Purdue University regarding concrete fatigue, whose main application was in concrete pavements (Ioannides, 1997b). The "Boilermakers" led the way in that arena for the first forty years or so, catapulted to the front runner status by the pioneering work of Professor William K. Hatt, who introduced such popular (if subsequently questioned) ideas as the existence of an endurance limit of $(\Phi/M_R) = 0.50$, in which Φ is the maximum bending stress arising in the concrete slab, and M_R is its modulus of rupture.

The Early Years Prior to the Computer and the AASHO Road Test

In the years following the publication of Westergaard's pioneering papers, i.e., that on the calculation of pavement responses in general (Westergaard, 1926), and that on the elimination of the flat-slab (no curling) assumption (Westergaard, 1927), additional investigators attempted to refine his solutions and to eliminate the rest of his restrictive assumptions. Prominent among these was Gerald Pickett, who focused his efforts on eliminating the single-wheel load assumption. The breakthrough came from an unexpected neighbor to Westergaard, namely from Nathan M. Newmark (1910-1981), who had occupied at the time an office in the same building as the concrete pavement pioneer. Newmark is familiar to the structural mechanics community for his manifold contributions, especially those in earthquake engineering (e.g., Zeevaert and

Newmark, 1956), and to every student of soil mechanics on account of his celebrated Charts, used for the determination of the vertical stress distribution in an elastic half-space under an arbitrarily shaped load (Newmark, 1942). On the basis of the latter, Pickett and Ray (1951) ingeniously produced the corresponding charts for the bending stress distribution in a slab on a dense liquid foundation, under a multiple-wheel load. Through his association with researchers from India, such as S. Badaruddin and S.C. Ganguli who were keen in applied mathematics, Pickett also developed a series of Charts pertaining to the slab on an elastic solid (Boussinesq) subgrade (Pickett, et al., 1955; Pickett and Badaruddin, 1957).

The elastic solid foundation always exerted an almost irresistible appeal to pavement engineers, in view of its perceived more realistic representation of the soil medium, particularly in light of Terzaghi's early work in footing and mat foundation design (Terzaghi, 1943). In contrast, the Winkler subgrade modulus, k, has always been elusive to determine, exhibiting undue sensitivity to the test conditions prevailing during its estimation, notably the plate size, so that it could hardly be considered an intrinsic soil property (Teller and Sutherland, 1943). Recognizing this, Westergaard (1926) had suggested that the k-value should be backcalculated from existing slabs-on-grade, and persisted in its use since he found the elastic solid foundation to be simply unsuitable for the examination of the critical edge and corner load conditions (Ioannides, 1991b). In the early 1920s, with the experiments at the Bureau of Public Roads' Arlington, Virginia Experiment Station (1918-1923) by A.T. Goldbeck and at the Bates Road Test (1921-1924) in Illinois by Clifford S. Older, edges and corners of concrete slabs had been recognized as the loci of initiation of several distress types, including transverse or corner cracking, pumping and faulting. Kerr and Shade (1984), for example, cite the repair of concrete slab blowups in Delaware in 1925, as reported by *Engineering News Record*.

For this reason, layered elastic theory, developed in 1943 by Donald M. Burmister, was never considered applicable to concrete pavements, for which it predicted unreasonably high subgrade stresses along the perimeter of the slab. Burmister's solution dispenses with the restrictions of plate theory and treats both the constructed and the natural layers as compressible elastic solids. In many ways, the new derivation was analogous to Westergaard's; if it took two additional decades to develop, this can be ascribed to the fact that "unlike the Winkler model, where the governing equations are of a differential form, problems associated with the elastic continuum model generally require the solution of integral or integrodifferential equations" (Selvadurai, 1979). The lack of effective computational tools for this purpose was largely responsible for the time lag between the plate-on-liquid solution by Westergaard (1926), and that of the layer-on-solid solution by Burmister (1943, 1945). In fact, the first mention of the use of modern computers in structural engineering could well be that by Burmister (1956): "Mr. Robert L. Schiffman did most of the computation work on the I.B.M. computing machines." Schiffman (1923-1997), who subsequently excelled as a professor of soil mechanics specializing in the phenomenon of consolidation (Schiffman, 2000), continued to elaborate on Burmister's method, extending its applicability first to three- and eventually to n-layered systems (Shiffman, 1957, 1962).

It is interesting to note at this point, that the approach followed by Burmister was essentially one of deriving a correction factor to a simpler solution obtained previously. According to Boussinesq, for example, the deflection, w, at the surface of a half space under a circular load can be written as:

$$w = \frac{1.5pa}{E_s} \tag{1}$$

assuming that the material Poisson ratio, $\mu_s = 0.5$, and denoting the elastic modulus by E_s, the pressure by p, and the radius of the applied load by a. To extend this solution to a two-layered system, Burmister (1943, 1945) wrote:

$$w = \frac{1.5pa}{E_s} * F_w \left[\frac{h}{a}, \frac{E}{E_s} \right] \tag{2}$$

where h is the thickness and E the modulus of the constructed layer, whose Poisson's ratio is assumed again to be $\mu = 0.5$. Note that the three "equally important interacting components" of the pavement system are clearly represented in this expression. In other words, Burmister's solution of the two-layered problem involves two dimensionless ratios, $(h/a, E/E_s)$; for its part, Westergaard's solution for the same problem involves only a single dimensionless ratio, (a/l), where l denoted the characteristic length of the system, defined as:

$$l = \sqrt[4]{\frac{Eh^3}{12(1-\mu^2)k}} \tag{3}$$

The concept of a characteristic length is familiar to researchers of the mechanics of diverse structural systems, including beams on elastic foundation and foundation piles. Westergaard (1926) called l the radius of relative stiffness of the slab-subgrade system, because it is of a linear dimension and because it reminded him of the radius of gyration, employed in the buckling of columns.

The elegance of the Westergaard and Burmister formulations prompted a series of analysts to pursue their amicable combination, even though such a development would probably be applicable only to the interior loading condition of concrete pavement slabs, i.e., when the loads are placed away from any edges or corners. Hogg (1938) and Holl (1938) presented the fundamental steps for such a solution to the plate on elastic solid problem, apparently independently yet practically simultaneously. This has led to the speculation that A.H.A. Hogg may be merely an academic pseudonym for Dio Lewis Holl (1895-1954), whose career as a professor at Iowa State University is well documented. Evidently on account of the lack of computational tools, the solution remained in the form of multiple infinite integrals until 1960, when Anders Losberg of Chalmers Tekniska Högskola, Sweden completed their numerical evaluation and presented a series of equations in a form analogous to those first proposed by Westergaard. The new governing independent variable is (a/l_e), where:

$$l_e = \sqrt[3]{\frac{Eh^3(1-\mu_s^2)}{6(1-\mu^2)E_s}} \tag{4}$$

Notice that Losberg (1960) presented only interior loading solutions. The corresponding edge and corner loading solutions were developed by Ioannides (1988) on the basis of 2-D finite element results.

The Single Placed Layer Problem

The question of the contribution of the base layer to the response and performance of a concrete pavement system (Ioannides and Khazanovich, 1994) continued to generate concern among cement and concrete suppliers, who hoped that a reduction in slab thickness would make their products more competitive during the period of the construction of the Eisenhower interstate highway system (after 1956). The analytical basis for a solution of this problem was provided by Odemark (1949), who suggested an ingenious scheme for extending Burmister's solution to multi-layered systems, without the need for prohibitively demanding computations. Nil Odemark's suggestion was based on the method of equivalent thicknesses, and involved the reduction of the multi-layered system to a two-layered one, namely of a single placed layer resting on a subgrade of increased elastic modulus, E_s. Despite the fact that extrapolating the method to an augmented value of the modulus of subgrade reaction, k, is unwarranted (recall that k cannot be considered an intrinsic soil property), a series of such solutions have been proposed, starting with one that resulted from experiments conducted by the Potland Cement Association (PCA) in the 1950s (Childs et al., 1957). A more robust and theoretically rigorous solution has been presented by Ioannides et al. (1992), on the basis of the method of transformed sections.

The Computer Age: The Finite Element Method and Dimensional Analysis

The AASHO Road Test (1958-1960) constitutes a watershed event in the history of pavement engineering. Its purpose was to generate the data necessary for the development and validation of new pavement design procedures, collected from a full scale, destructive experiment. In many ways, road tests, the first of which had been conducted as early as the 1920s, were a product of the frustration of pavement engineering administrators with the mechanistic approach to design, which seemed to produce too little too slowly when it came to the development of practical advice about preventing the distresses that perennially seemed to plague pavements (Seely, 1991). The AASHO Road Test, however, owes much of its impact on two developments that had little to do with pavement engineering, namely, the development of modern computers, and the concomitant increasing effectiveness of statistical techniques in interpreting vast quantities of empirical data, especially in fields fraught with unpredictability and ignorance. The undesirable repercussions of the unbridled use of the latter in pavement engineering, generally recognized as a field not governed by accident or chance but by definite laws of nature and mechanics (Vesic and Saxena, 1970), became only too slowly apparent in the AASHTO design procedures developed between 1961 and 1993, and which are only currently in the process of being dislodged (Khazanovich et al., 2004). As far as pavement mechanics is concerned, however, the AASHO Road Test constitutes a turning point inasmuch as in the decades that followed it, there has been a proliferation of efforts for the development of numerical solutions to the differential equations governing the pavement problem. In the United States, as in other technologically advanced societies, engineers resorted to the computer, sometimes with undue haste and without due thought, compiling a series of codes to perform such basic operations as matrix manipulation and numerical integration. The products of this effort included various computer programs for the Burmister layered elastic theory (now applied to "n-layered" systems), as well as for the discrete element, the finite difference and the finite element methods, in that order. A review of these methods was presented in the PhD dissertation by Ioannides (1984), who indicated that the finite element method had become the approach of choice.

The prominent name in this area is that of Professor Yang Hsien Huang (b. 1927) of the University of Kentucky, who developed the first 2-D finite element (FE) code tailored for concrete pavements (Huang and Wang, 1973); this software was left unnamed by Huang, but was referred to as *KENWINK* by Ioannides (1984). It eliminated three of Westergaard's six original assumptions, namely those of the single slab panel (no load transfer), single-wheel load and infinite slab extent. A very similar code, also left unnamed by its developer, Arthur C. Eberhardt, was prepared at about the same time at the Construction Engineering Research Laboratory (CERL) in Champaign, Illinois (Eberhardt, 1973). This software did not accommodate curling or load transfer, but did provide for the presence of a second constructed layer: the formulation accounted for bonded, partially bonded and unbonded base layers (treated always as plates). The most widely known FE program for concrete pavements is the University of Illinois' *ILLI-SLAB*, first published by Tabatabaie and Barenberg (1978); this evidently combined the features of both its predecessors noted, and its formulation accommodated load transfer using aggregate interlock and/or dowel bars. Subsequent investigators elaborated on these original solutions, sometimes without adequate attribution. One of Huang's former co-workers, Chou (1981) published *WESLIQID*, retaining most of the original characteristics of *KENWINK*; Tayabji and Colley (1983) strove to improve *ILLI-SLAB*'s load transfer formulation in *J-SLAB*.

The Curling Problem

Just as Westergaard (1927) had returned to the slab-on-grade problem seeking to eliminate the uniform contact restriction, a series of investigators after him have tackled the four additional assumptions he had introduced in his curling analysis. Perhaps the first such attempt focused on the elimination of the limitation of a linear temperature distribution, and this should be rightfully credited to the British researcher J. Thomlinson. Like Westergaard, Thomlinson (1940) sought "a rational theory ... in a form that can be applied directly to practical problems," yet produced what at first sight might be perceived as an intimidating sequence of theoretical manipulations. While his solution to the problem of a nonlinear temperature gradient is complete, Thomlinson (1940) failed to present it as a step-by-step approach to the solution of practical problems, despite the fact that he even provided the tools for determining the diurnal variation of slab temperatures. Consequently, his truly pioneering effort remained largely unknown, with subsequent investigators either totally ignoring it or citing it inadequately.

The nonlinear gradient analysis by Thomlinson (1940) was reproduced more recently in the USA by Choubane and Tia (1992), who presented the methodology in a straight-forward, step-by-step manner for the simplest nonlinear temperature distribution, namely the quadratic. Consequently, Choubane and Tia were able to reach some important quantitative conclusions of interest to the practicing engineer, most notable among which is that Westergaard's linear gradient assumption leads to conservative (i.e., higher) estimates of combined stresses during the daytime than does a nonlinear (quadratic) distribution. The reverse is true under nighttime conditions. The same conclusion had been reached independently in Spain by Mirambell (1990). Thus, since night-time gradients are generally only about half the corresponding values during the day (Huang, 1993), Westergaard's linear temperature variation assumption may be belatedly justified in view of its conservative nature and its simplicity. The clear presentation of the solution to the nonlinear temperature distribution problem by Choubane and Tia (1992) was heavily indebted to the work of Bergström (1950) and of Richardson and Armaghani (1987), among others, as well as that of Thomlinson (1940). Essentially the same solution has been subsequently presented by Harik et al. (1994), by Mohamed and Hansen (1997), and possibly by others.

The next Westergaard curling assumption to be challenged was not surprisingly that of continuous contact between the slab and the subgrade, by explicitly accommodating the actual unit weight of the concrete slab. This enhancement was made possible through the use of an iterative scheme implemented in a FE code, developed at the University of Kentucky by Huang and Wang (1974). Naturally, this computerized implementation of curling analysis also eliminated the need for superposition and permitted the examination of day-time and night-time conditions independently. The Kentucky code, which eventually became known as *KENSLABS* (Huang, 1993), retained several Westergaard assumptions, including that of a linear temperature gradient. Nonetheless, its methodology became widely adopted, as evidenced by its subsequent incorporation - sometimes without adequate citation - in such codes as *WESLIQID* (Chou, 1981), *J-SLAB* (Tayabji and Colley, 1983), *FEACONS* (Tia et al., 1987), and *ILLI-SLAB* (Korovesis, 1990). Yet, despite the fact that explicit analysis of slab-on-grade pavements was now possible on a case-by-case basis, little effort was invested in interpreting FE results in a manner that could lead to general conclusions useful to practicing engineers, and even less in assessing the repercussions of Westergaard's assumptions.

The assumption of a single-placed layer (no base) in curling analysis was first tackled by Korovesis (1990), who implemented Huang's iterative approach into *ILLI-SLAB*, and extended it to the case of a two-layer slab-on-grade. In addition, Korovesis adapted multi-layered FE program *ROOF* (Barzegar, 1988), creating the *ILLI-LAYER* code, suitable for analyzing pavement systems incorporating more than two placed layers. Although Korovesis retained a linear temperature distribution through each placed layer, multiple layers now permitted the piecewise examination of a nonlinear distribution. In addition, the presence of adjacent slab panels and the provision of load transfer were also accommodated in Korovesis' curling analysis.

Regrettably, the work of Korovesis (1990) has remained in relative obscurity as it has not been widely published following the author's repatriation to Greece. Albeit quite preliminary in nature and incomplete in several respects, it spawned considerable interest in nonlinear curling analysis of multi-layered pavements, culminating in the work of Khazanovich (1994). The latter complemented the contributions of Thomlinson (1940), of Huang and Wang (1974) and of Korovesis (1990) with achievements promulgated in the former Soviet Union, notably those by Korenev and Chernigovskaya (1962). The result was the development of a robust and practical scheme for the nonlinear curling analysis of multi-layered slabs-on-grade, which Khazanovich (1994) implemented in a drastically expanded version of the *ILLI-SLAB* FE code, called *ILSL2*. This generalized FE formulation accommodates simultaneously a nonlinear temperature distribution, loss of support, multiple placed-layers, and load transfer, thereby effectively eliminating all of Westergaard's assumptions, except the one pertaining to the absence of a rigid subgrade bottom.

The most recent developments in 2-D FE analysis of concrete pavements were perhaps those presented by Ioannides and Khazanovich in a series of papers, following the submission of the latter's PhD thesis (Khazanovich, 1994). New formulations were presented so as to account for: the nonlinear shape of the temperature distribution through the thickness of the slab (Ioannides and Khazanovich, 1998); the interaction between slab and base (Ioannides et al., 1992); and the support provided by the subgrade (Khazanovich and Ioannides, 1993). The latter deserves a more detailed discussion since this was the first occasion that a foundation type other than those of the dense liquid and the elastic solid was introduced in concrete pavement finite element analysis.

Subgrade Models

The two subgrade models described above, namely those ascribed to Winkler and Boussinesq, can be considered as defining two extremes, with real soils probably lying somewhere in between (Ioannides, 1991b). The challenge presented in analyzing Porland cement concrete (PCC) pavement systems essentially consists of striking a balance between the requirement to reproduce closely in-situ behavior of real soils, while simultaneously ensuring that the mathematical formulation of the subgrade model adopted remains reasonably simple. In this effort, one of three distinct approaches have been employed by previous investigators (Kerr, 1964, 1993): (a) Adopt the elastic solid (continuum) model, but introduce simplifying assumptions regarding the internal stresses and displacements (Reissner, 1958, 1967; Vlasov and Leont'ev, 1960; Kerr, 1985a; etc.); (b) Adopt the Winkler model, but introduce a certain degree of interaction between adjacent foundation springs (Filonenko-Borodich, 1940; Hetenyi, 1950; Pasternak, 1954; Kerr, 1965; etc.); and, (c) Use a formal analytical procedure, as described by Kerr (1984). Kerr showed that in many cases it is possible to derive the same analytical expressions for pavement responses using any of these procedures (Kerr, 1985b, 1984). The expanded version of *ILLI-SLAB* (Khazanovich and Ioannides, 1993) incorporates options for the Pasternak, the Kerr, and the Zemochkin-Sinitsyn soil idealizations, reflecting the influx of knowledge to the West following the demise of the Soviet Union. A brief description of each of these models follows.

The Pasternak Foundation Model: Pasternak (1954) accounts for the existence of shear interactions between the subgrade spring elements by deriving the following relation between subgrade reaction, q, and deflection, w:

$$q = kw - G\nabla^2 w \tag{5}$$

where k is the vertical spring stiffness, G is a coefficient describing the shear interactions between adjacent springs (assumed to be equal in the x- and y-directions), and Λ^2 is the Laplace operator. The corresponding mechanical model consisting of a shear layer attached to the Winkler spring bed was proposed by Kerr (1964, 1985b). The governing equation given above is similar to an expression that had been presented earlier by Filonenko-Borodich (1940), who used a stretched membrane instead of the shear layer to connect the springs. Determination of the two model parameters has been discussed by several investigators, notably by Slivker (1981) in the East, and by Kerr (1985a) and by Girija Vallabhan et al. (1991) in the West. Note that if G is set to zero, the Pasternak foundation reduces to the Winkler model.

In general, the Pasternak model offers an attractive alternative to the elastic solid continuum, by providing a degree of shear interaction between adjacent soil elements, while remaining relatively simple to analyze. The predicted deflection profile vanishes much faster than the corresponding elastic solid basin, and may be a better approximation of the deflections observed in a real foundation of finite depth. Early analytical solutions in terms of Bessel functions for a plate attached to a Pasternak foundation and subjected to interior loading have been presented by Korenev and Chernigovskaya (1962). More recent solutions were presented by Kerr (1985a) and by Kerr and El-Sibaie (1989). Additional solutions may be obtained by finite element analyses. For the Pasternak foundation, these are much less tedious than for the elastic solid, since the subgrade stiffness matrix for the former is banded (Ioannides et al., 1985; Khazanovich and Ioannides, 1993).

The Zhemochkin-Sinitsyn-Shtaerman Foundation Model: Despite its advantages, Pasternak's model cannot be used for the estimation of subgrade stresses, because, like the elastic solid model, it predicts infinite stresses along the free edge of a slab resting on it. The subgrade idealization proposed by Zhemochkin and Sinitsyn (1947) and by Shtaerman (1949) overcomes this weakness, since it results from an in-series combination of the dense liquid and elastic solid models. Thus, it is related to both of these, allowing a limited degree of interaction between adjacent foundation elements, yet preventing the generation of infinite edge stresses. At the same time, the new model remains reasonably simple to formulate (Khazanovich and Ioannides, 1993). It can be noted that its stiffness matrix, like the one for the elastic solid, is symmetric but not banded, and, as a result, it requires a larger computer memory and more computational time than either the Winkler or the Pasternak idealizations.

The Kerr Foundation Model: To address the shortcomings of the Pasternak foundation, another three-parameter subgrade idealization was proposed by Kerr (1964, 1965). This model consists of a spring bed placed over a Pasternak foundation. From a finite element viewpoint, the Kerr model leads to a banded stiffness matrix, without predicting infinite soil pressures under the free edge of the slab. Kneifati (1985) and Kerr and El-Sibaie (1989) obtained several analytical solutions and discussed the characteristic features of this model. The finite element formulation for the Kerr model was presented by Khazanovich and Ioannides (1993). Although the Kerr model requires twice as many nodes as the Pasternak model, the fact that its stiffness matrix is banded allows considerable savings in computational effort. The disadvantage of this model is its requirement for three material parameters, k_U, G, and k_L, where k_U and k_L are the upper and lower spring stiffnesses, respectively. Kerr (1964, 1965) has shown, however, that these three parameters may be reduced to two, by setting $k_U = 3 k_L$. Furthermore, the ratio (k_U/k_L) can provide a basis for qualitative comparisons between the Kerr and the other models. Thus, as (k_U/k_L) tends to zero, the Kerr model reduces to the dense liquid, whereas as (k_U/k_L) tends to infinity (in practice, to a value between 10 and 100), the Kerr model yields the same results as Pasternak's.

Three-Dimensional Finite Element Analysis

More recently, pavement engineers have also turned to 3-D FE analysis (e.g., Ioannides and Donnelly, 1988; Ioannides and Peng, 2004), encouraged in part by the development of the BOEING-777 aircraft with its novel tridem axle main gear (Gervais, 2002), and by a devastating report by the U.S. General Accounting Office (GAO, 1997), which concluded that "the pavement design guide developed and updated by AASHTO over the years for designing and analyzing highway pavement structures is outdated."

Dimensional Analysis in Data Interpretation

It is apparent that the development of computerized analysis programs, notably those based on the FE method, enhance the engineer's ability to eliminate many of the constraints imposed by the original assumptions implicit in the closed-form solutions presented by Westergaard (1926, 1927). This ability in itself, however, is not adequate for extracting meaningful conclusions of a sufficiently general nature so that they become of value to practicing engineers and pavement designers. To address this limitation, Ioannides (1984) outlined a methodology founded on an application of the principles of dimensional analysis for the interpretation of analytical (e.g., from FE analyses) and field data. The methodology consists of identifying the dimensionless independent and dependent variables governing the problem, generating analytical or collecting field data from a relatively small experimental factorial, and establishing predictive algorithms through the use of statistical analysis. This methodology has been used successfully in a wide variety of applications (Ioannides, 1990; Ioannides et al., 1992), and has been incorporated into the mechanistic-empirical design procedure for PCC pavements developed under Project NCHRP 1-26 (NCHRP, 1992), as well as the proposed AASHTO 2002 Design Guide (Khazanovich et al., 2004).

In applying the dimensional analysis-FE-statistics methodology to the curling problem, for example, Ioannides and Salsilli-Murua (1989) identified the following governing independent variables: (a/P), (L/P), (W/P) and (∀)T), in which a: radius of tire print; P: radius of relative stiffness; L: slab length; W: slab width; ∀: coefficient of linear expansion of PCC;)T: linear temperature differential between top and bottom of slab. The fact that these variables are already evident in Westergaard's work testifies to the enduring value of the latter. The dimensionless form of the bending stress dependent variable had already been identified by Ioannides (1984) as $(\Phi h^2/P)$, with Φ: bending stress (assumed to be positive if tensile at the bottom of the slab); h: slab thickness; P: total applied load. These findings anticipated the definition of two additional independent variables governing the curling phenomenon by Lee (1993), as follows: $D_\gamma=((h^2 / kP^2)$; and $D_P=(Ph / kP^4)$, in which (: unit weight of concrete and k: modulus of subgrade reaction. The first of these is already encountered in the work of Korenev and Chernigovskaya (1962).

Future Directions: Fracture Mechanics and Similitude

The conventional approach to pavement design is commonly a two-stage one: first, a critical primary response is calculated, which is subsequently passed into a statistical/empirical algorithm that converts it into a measure of performance (Ioannides, 1992). The first stage involves the prediction of behavior from initial loading until shortly before failure, using methods based on the theory of linear elasticity. With respect to the prediction of performance, current methods resort to rather simple, mostly empirical and phenomenological concepts, such as Miner's cumulative linear fatigue hypothesis (Miner, 1945). Such concepts and practices have undergone little change or adaptation since their introduction several decades ago. A considerable number of these can be traced to investigations into the fatigue behavior of small specimens dating back to the early part of the twentieth century. There appears to be little doubt that both Portland cement and asphalt concrete specimens fail in fatigue when subjected to either constant or variable amplitude repeated loads. Moreover, test results exhibit a considerable scatter, lending credence to the suggestion that fatigue is a stochastic phenomenon. Even as pavement engineers employ Miner's fatigue hypothesis in making fatigue life predictions, however, most of them also acknowledge that this is a matter of expediency rather than acceptance of its validity. Studies seeking to verify Miner's hypothesis often provide adequate information that could justify its abandonment instead (Ioannides, 1997b). Data reported suggest that predictions of life using Miner can only be expected to approximate reality at best within an order of magnitude, and more commonly within two or even three orders of magnitude. Such predictive ability is considered entirely unsatisfactory, and calls for a Miner replacement.

Application of fracture mechanics concepts developed in various branches of engineering to the pavement problem can potentially address these limitations, thereby advancing considerably the state-of-the-art of pavement design. One of the most promising contributions to the development of rational, mechanistic pavement design procedures and in particular to the application of fracture mechanics to pavement engineering is a paper presented by Hans Henrik Bache and Ib Vinding at the 1990 Second International Workshop on the Design and Evaluation of Concrete Pavements, held in Sigüenza, Spain (Bache and Vinding, 1990). In this ground-breaking paper, the authors initiated a reconsideration of prevailing concrete pavement design concepts in the light of advances in fracture mechanics, particularly the introduction and use of Hillerborg's Fictitious Crack Model (FCM) (Hillerborg et al., 1976). They showed that accounting for the degree of ductility exhibited by concrete is an overriding consideration, but that this characteristic is not exclusively a material property, being considerably influenced by the entire structural system, as well. They pointed out that small-scale simulation of field behavior is only meaningful if rational scaling and similitude laws formulated on the basis of the physics and thermodynamics of the problem are adhered to. To this end, they provided some preliminary guidance, which can be enhanced by incorporating the experience gained through the introduction of the principles of dimensional analysis to concrete data interpretation (Ioannides, 1990). In addition, Bache and Vinding (1990), outlined the framework of an improved pavement design procedure, pending additional experimental and computational investigations.

Drawing from Swedish research into the application of the FCM, Bache and Vinding (1990) state that the size of the plastic zone ahead of (i.e., in the direction of) the crack tip is "of the order of magnitude" of a characteristic length of the material, P_c, defined as:

$$\ell_c = \frac{EG_f}{f_0^2} \tag{6}$$

where E is the modulus of elasticity, G_f is the fracture energy, and f_0 is the (direct) tensile strength. The relative size of P_c in comparison to the size of the structure determines which theory is applicable for the analysis of the cracking phenomenon. Thus, linear elastic fracture mechanics (LEFM) is valid only if this relative plastic zone size is very small, in which case the controlling parameters are E and G_f (recall that the critical stress intensity factor, $K_{IC} = [E\, G_f]$).

On the other hand, plasticity theory applies for very large relative values, with f_0 being the controlling variable. In the intermediate range, the contribution of all three parameters, E, G_f and f_0, must be accounted for. This is achieved by the introduction of brittleness number, B, defined by Bache (1989) as follows:

$$B = \frac{f_0^2 h}{EG_f} \tag{7}$$

in which h is "a relevant dimension, often the concrete slab thickness." The theoretical justification for this definition is provided by energy similitude considerations, as will be explained later. The usefulness of this definition is that it provides a measure of ductility reflecting both material and structural system properties. The brittleness number is particularly significant in efforts to simulate field behavior by small scale tests in the laboratory. Bache and Vinding (1990) argue that the equality of field and lab B-values must be assured if meaningful extrapolations are to result.

In applying the brittleness number concept, the implicit assumption is made that P_c is not specimen size dependent. The validity of this assumption needs further exploration and experimental verification. Furthermore, it can be postulated that in examining the behavior of pavement systems (as opposed to the behavior of the pavement material, i.e., of concrete), the appropriate structure length will no longer be the slab thickness, but probably the radius of relative stiffness of the pavement-subgrade system, with the radius of the applied load also having some degree of influence (Ioannides, 1997a).

Bache (1991) maintains that the lack of "simple, universal design principles" for cement-based materials can be attributed to the variety of behavior types exhibited by structures made of these materials, "spanning from brittle to extremely ductile behavior." Although his own interests focus exclusively on Portland cement-based materials, particularly those that are heavily reinforced, Bache's comments are relevant to most pavement materials, including those made of bituminous mixes. Consider, for example, the following comment, interpreted in the light of recent pavement engineering experiences: "It is hardly surprising that ... one is often forced to turn to full-scale tests [such as the AASHO Road Test and the Long Term Pavement Performance Initiative of the Strategic Highway Research Program] or to design principles with a doubtful basis in the physical reality" [such as the Equivalent Single Axle Load and Structural Number concepts and the use of Miner's hypothesis]. To address these shortcomings, Bache has made extensive use of the principles of similitude.

In general, Bache identifies four distinct types of material behavior:
  (a) Elastic;
  (b) Plastic (bulk);
  (c) Plastic (local); and
  (d) Brittle (local).

The distinction between bulk and local behavior is crucial in this context. Bulk deformations occur in the body as a whole, and increase as the applied load increases. In contrast, local deformations occur only after the applied load reaches a limit value, and affect only a narrow zone, leading to eventual physical separation. As local deformations increase, the sustained load decreases. Brittle and ductile behavior involve different relative components of bulk and local deformations. When the predominant mode of deformation before failure is bulk in character, the behavior is brittle, whereas ductile behavior is characterized by local crack zone deformations of a relatively large magnitude. Consequently, brittle failure is characterized by opening of cracks by peeling, with the material's resistance being concentrated in small zones at the crack tips. Conversely, ductile failure involves yielding over larger areas. In quantifying brittleness or ductility, Bache carefully distinguishes between "bulk ductility" and "fracture zone ductility" and demonstrates that only the latter is specimen size dependent.

The behavior classification proposed by Bache is particularly useful in addressing the specimen size effect, since it can be shown that only those involving "fracture zone (local) ductility", i.e., (c) and (d), are specimen-size dependent. A fundamental repercussion of the specimen size effect is that brittle behavior "cannot be scaled up from small specimens only on the basis of stresses and strains." A pavement engineer readily recognizes, of course, that this is precisely what is done in conventional applications of Miner's hypothesis, using (Φ/M_R) for concrete pavements, or the strain at the bottom of the asphalt concrete layer, $_{,AC}$, for bituminous pavements (Ioannides, 1997a).

Bache's application of similitude principles rests on the postulate that "geometrically similar objects exhibit similar behaviour if the same ratio exists between the significant forces or energies." Consideration of the forces and energies pertaining to the four behavior types identified above can lead to a number of interesting conclusions concerning the physics of failure in each instance. It is convenient to consider the four behavior types separately to begin with, recognizing, however, that real material and structural behavior more often consists of interacting components of all four types. Consideration of each of these behavior types individually can lead to a number of important observations. Thus, it can be noted that for elastic behavior, the relationship between stress and strain in independent of load history, whereas for plastic (bulk) behavior, load history influences the stress-strain relationship. Furthermore, LEFM is applicable only to brittle (local) behavior, whereas for plastic (local) behavior nonlinear fracture mechanics (or a suitably modified version of LEFM) needs to be applied.

More complex behaviors can be described in terms of interacting components of the fundamental behavior types. In general, when only two behavior types are predominant, consideration of the ratio of their respective force, energy or deformation relationships can be very instructive. Thus, for elastic-plastic (bulk) behavior, such ratios are independent of L, illustrating that "the degree of brittleness and ductility is independent of the object size when ductility is due to bulk flow" (Bache, 1991). Similarly, for elastic-brittle (local) behavior, the ratio of either deformations or energies leads to the brittleness number, $B = \Phi^2 L/EG_f$. The brittleness number, B, therefore, represents the ratio of elastic energy to fracture energy in the body. The stored elastic energy is proportional to the volume of the body ($\sim L^3 f_0^2/E$), whereas the fracture energy is proportional to the characteristic area ($\sim L^2 G_f$). This explains why the brittleness number B has dimension of a characteristic length, and illustrates that this behavior type is specimen size dependent.

On the basis of such considerations, a number of qualitative conclusions are possible at this time. These include the following (Bache, 1989; 1991):

(a)   Load capacity increases as the brittleness number decreases, and decreases as the brittleness number, B, increases.

(b)   For small values of the brittleness number, B, the behavior is ductile and plastic. The load capacity is high because all yield reserves are utilized; failure deformations are large and are dominated by crack zone deformations. The perfectly plastic solution, represented by B60, provides the asymptotic upper bound of load capacity. The behavior in this range is usually calculated in accordance to the theory of plasticity. Structures in this range are relatively insensitive to small cracks and local stress concentrations.

(c)   For large values of the brittleness number, B, the behavior is brittle and the load capacity is low because there are no yield reserves. This is true even for structures without cracks or stress concentrations. Load capacity and failure deformations are usually determined using the theory of elasticity of LEFM. Structures in this range are very sensitive to cracks and local stress concentrations. The perfectly elastic solution, represented by B64, leads to an asymptotic lower bound of load capacity, in the absence of initial cracks. For structures with large, sharp cracks, load capacity decreases even below the elastic lower bound, in accordance to LEFM.

(d)   The transitional range between plastic and brittle behavior is often quite broad. The prevailing conditions are also more complex. Here, more precise calculations of failure and crack behavior normally require considerable computational effort, usually involving finite element applications with special elements introduced in the region of crack zone deformations.

Determination of the value of the asymptotic bounds and delineation of load capacity for intermediate values of B remains the subject of current research. The tools used for this purpose are: plasticity theory for very low B-values; elastic theory or LEFM for very high B-values; and nonlinear fracture mechanics for the "transitional range." In addition, model tests can be conducted, in which changes in the failure mode can be observed in response to changes in the relative magnitudes of the pertinent material characteristics.

It is apparent from the preceding discussion, that Bache is concerned primarily with the determination of "load capacity," or "strength," i.e., with the determination of the maximum sustainable load before failure. This might at first sight be construed to refer only to one load repetition and to be inapplicable to fatigue loading considerations. Bache and Vinding (1990), however, re-interpret the conventional application of fatigue laws (Φ/M_R v. log Number of Load Repetitions, N) as involving the substitution of the static flexural strength, M_R, by a (lower) fatigue strength, M_f, which is a function of the number of repetitions, N. In conventional design, the allowable stress in the slab is set equal to M_f. This interpretation is akin to early use of safety factors in the PCA Airport Pavement Design (Packard, 1973). Bache and Vinding (1990) criticize the conventional application of Miner's hypothesis, involving testing of small unsupported beam specimens and applying the results to full-scale pavement sections. They stress that such an approach is pertinent only to crack initiation, and point out that "it tells us nothing about the consequences of local fracture—for example, whether this leads to total failure or only to the formation of harmless, small cracks." This shortcoming leads to the observation that "the carrying capacity of thick pavements is often found to be lower than expected from experience with thin pavements and small laboratory test specimens." A similar observation has been made by Darter (1990), who showed that for field data the number of expected repetitions could be overestimated by more than three orders of magnitude, especially at high N (low Φ/M_R) values. Darter (1990) indicates that "numerous theoretical explanations for these differences have been suggested, but have not at this time been adequately investigated." The similitude principles advocated by Bache and Vinding (1990) and the application of the FCM are very promising in this respect.

An illustration of how the FCM can be implemented using the finite element method to simulate analytically and to track the propagation of cracks in PCC structures, including pavements, has been presented in a series of papers by investigators at the University of Cincinnati. To begin with, a FORTRAN computer program, *CRACKIT*, was developed that can used in conjunction with a commercial finite element package, *GTSTRUDL*, to simulate crack propagation in simply supported beams (Ioannides and Sengupta, 2003). The robustness of the *GTSTRUDL/CRACKIT* combination was verified by reproducing results obtained by independent investigators and reported in the literature. Subsequently, commercial finite element code *ABAQUS* was used to verify these findings through a stand-alone procedure, and to track crack propagation in both simply supported beams and slabs resting on a Winkler foundation (Ioannides et al., 2003; Ioannides and Peng, 2004). The methods adopted for the analysis of slabs are an extension of those applied to beams, thereby affirming the suitability of a step-by-step approach, which can aid engineers in generating additional analytical data for various specimen sizes and fracture parameters. Such data can be interpreted using the principles of dimensional analysis in an effort to identify relationships that may ultimately lead to unraveling the specimen-size effect and its relationship to the fracture behavior of concrete pavement structures (Ioannides, 1997a).

Conclusions

It is a distinct honor for me to be included in this gathering honoring the career of Arnold D. Kerr, a man I have come to know as a colleague and have developed a close friendship with. It is also a great privilege to be able to have my contribution included alongside those of such a distinguished group of investigators, whose shoes I cannot possibly be expected to fill. Instead, confessing I am a pavement engineer, I have presented here a story that began with the Westergaard-Talbot romance in Illinois and has been evolving with particular fecundity ever since.

The history of concrete pavement analysis is inoxerably connected with the parallel evolution of concepts and techniques, not to mention personalities, responsible for admirable advances in the similarly budding discipline of geotechnical engineering. Both these civil engineering specialties have been graced with the lives and contributions of some of the most accomplished engineers of the twentieth century. Their shining examples of blending theory and practice, imaginative abstraction with practical solutions, mathematical elegance to sensitive aesthetics, can only motivate those of us who have the good fortune to be following in their pioneering footsteps, equipped with all that twenty-first-century electronic and computer technology has to offer.

The discussion presented was admittedly and of necessity selective, both in scope, as well as in depth of treatment. Nonetheless, the following broad conclusions may be reached:

1. The effort to overcome the limiting assumptions inherent in Westergaard's analysis has motivated and directed the efforts of pavement engineers for more than 85 years, but the profession can now boast that such restrictions need no longer plague modern pavement design.
2. Consequently, it is essential that the use of Miner's linear cumulative fatigue hypothesis be recognized widely as merely a matter of practical expediency. Its reliability is highly questionable.
3. Moreover, it must be admitted that early linear elastic fracture mechanics applications have yielded inconsistent predictions of pavement performance.
4. The Fictitious Crack Model (FCM) proposed by Hillerborg appears most promising for computerized application to pavements. Similitude concepts and the principles of dimensional analysis will be very useful in the pursuit of modern pavement design procedures.

This paper confirms both the desirability for and the scarcity of suitable candidates to replace Miner's cumulative linear fatigue hypothesis in conventional pavement design. Fracture mechanics is shown to be a very promising engineering discipline from which innovations could be transplanted to pavement activities. Nonetheless, rather slow progress characterizes fracture mechanics developments in general is pointed out. Pavement engineers clearly need to remain abreast of and involved in fracture mechanics activities. It is submitted that Portland cement concrete appears to be the most reasonable choice for a material to focus on at the beginning phases of an investigation into the behavior of composite pavement materials.

References

AASHTO, 1961. AASHTO Interim Guide for the Design of Flexible Pavement Structures. American Association of State Highway and Transportation Officials, AASHTO Committee on Design, October.

AASHTO, 1962. AASHTO Interim Guide for the Design of Rigid Pavement Structures. American Association of State Highway and Transportation Officials, AASHTO Committee on Design, April.

Bache, H.H., 1989. Brittleness/Ductility from Deformation and Ductility Points of View. Chapter 7.4 in *Fracture Mechanics of Concrete Structures: From Theory to Applications*, edited by L. Elfgren, RILEM Report, Chapman and Hall Ltd., London, England, 405.

Bache, H.H., 1991. Principles of Similitude in Design of Reinforced Brittle Matrix Composites. Presented at the International Workshop High Performance Fiber Reinforced Cement Composites, Mainz, Germany, June, 17. (Also, CBL Reprint No. 23, Aalborg Portland, Denmark).

Bache, H.H., and Vinding, I., 1990. Fracture Mechanics in Design of Concrete Pavements. Proceedings, 2nd International Workshop on the Design and Evaluation of Concrete Pavements, CROW/PIARC, Sigüenza, Spain, October 4-5, 139-165.

Baron, F.M., 1942. Variables in the Design of Concrete Runways of Airports. Proceedings, 22nd Annual Meeting, St. Louis, MO, Highway Research Board, National Research Council, Washington D.C., 225-239.

Barzegar, F., 1988. Layering of RC Membrane and Plate Elements in Nonlinear Analysis. *Journal of Structural Engineering*, ASCE, **114**(11), Nov., 2474-2492.

Bergström, S.G., 1950. Temperature Stresses in Concrete Pavements. Proceedings No. 14, Swedish Cement and Concrete Research Institute at the Royal Institute of Technology, Stockholm, Sweden.

Boussinesq, M.J., 1885. *Application des Potentiels*. (Application of Potential), Gauthier- Villars, Paris (in French).

Bradbury, R.D., 1938. *Reinforced Concrete Pavements*. Wire Reinforcement Institute, Washington, D.C.

Burmister, D.M., 1943. The Theory of Stresses and Displacements in Layered Systems and Application to the Design of Airport Runways. Proceedings, Highway Research Board, **23**, National Research Council, Washington D.C., 126-144.

Burmister, D.M., 1945. The General Theory of Stresses and Displacements in Layered Systems. *Journal of Applied Physics*, **16**(2), 89-96; No. 3, 126-127; No. 5, 296-302.

Burmister, D.M., 1956. Stresses and Displacement Characteristics of a Two-Layer, Rigid Base Soil System: Influence Diagrams and Practical Applications. Proceedings, Highway Research Board, **35**, National Research Council, Washington D.C., 773-814.

Childs, L.D., Colley, B.E., and Kapernick, J.W., 1957. Test to Evaluate Concrete Pavement Subbases. Proceedings of the American Society of Civil Engineers, Paper No. 1297, **83** (HW-3), July, 1-41.

Chou, Y.T., 1981. Structural Analysis Computer Programs for Rigid Multicomponent Pavement Structures with Discontinuities- *WESLIQID* and *WESLAYER*; Report 1: Program Development and Numerical Presentations; Report 2: Manual for the *WESLIQID* Finite Element Program; Report 3: Manual for the *WESLAYER* Finite Element Program. Technical Report GL-81-6, U.S. Army Engineer Waterways Experiment Station, May.

Choubane, B., and Tia, M., 1992. Nonlinear Temperature Gradient Effect on Maximum Warping Stresses in Rigid Pavements. Transportation Research Record No. 1370, Transportation Research Board, National Research Council, Washington D.C., 11-19.

Darter, M.I., 1990. Concrete Slab vs. Beam Fatigue Models. Proceedings, Second International Workshop on the Design and Evaluation of Concrete Pavements, CROW/PIARC, Sigüenza, Spain, 4-5 October, 472-481.

Eberhardt, A.C., 1973. Aircraft-Pavement Interaction Studies, Phase 1: A Finite Element Model of A Jointed Concrete Pavement on a Non-Linear Viscous Subgrade. USACE Construction Engineering Research Laboratory, Preliminary Report S-19, June, Champaign, IL, 23.

Filonenko-Borodich, M.M., 1940. Some Approximate Theories of the Elastic Foundation. *Uchenyie Zapiski Moskovskogo Gosudarstvennogo Universiteta (Mekhanika)*, No. 46, USSR (in Russian).

GAO, 1997. Transportation Infrastructure: Highway Pavement Design Guide is Outdated. Report to the Secretary of Transportation, GAO/RCED-98-9, United States General Accounting Office, Washington, D.C., 17.

Gervais, E.L., 2002. Flexible Pavement ACN/PCN Concepts: A Manufacturer's Point of View. Presented at the 6th International Conference On The Bearing Capacity of Roads, Railways And Airfields, Instituto Superior Tecnico, Lisbon, Portugal, June 21, *BOEING* Commercial Airplanes, Seattle, WA.

Girija, Vallabhan, C.V., Straughan, W.T., and Das, Y.C., 1991. Refined Model for Analysis of Plates on Elastic Foundations. *J. Engrg. Mech.*, ASCE, **117**(12), 2830-2844.

Goodman, R.E., 1999. *Karl Terzaghi: The Engineer as Artist*. ASCE Press, New York, NY, 112.

Harik, I.E., Jianping, P., Southgate, H., and Allen, D., 1994. Temperature Effects on Rigid Pavements.

Journal of Transportation Engineering, ASCE, **120**(1), New York, NY, 127-143.

Hertz, H., 1884. Uber das Gleichgewicht Schwimmender Elastischer Platten. (On the Equilibrium of Floating Elastic Plates), *Annalen der Physik und Chemie*, **22**, Wiedemann's (in German).

Hetenyi, M., 1950. A General Solution for the Bending of Beams on an Elastic Foundation of Arbitrary Continuity. *Journal of Applied Physics*, **21**.

Hillerborg, A., Modeer, M., and P.E., Petersson, 1976. Analysis of Crack Formation and Crack Growth in Concrete by Means of Fracture Mechanics and Finite Elements. *Cement and Concrete Research*, **6**(6), November, 773-782.

Hogg, A.H.A., 1938. Equilibrium of A Thin Plate, Symmetrically Loaded, Resting on an Elastic Foundation of Infinite Depth. *Philosophical Magazine*, Series 7, **25**, March, 576-582.

Holl, D.L., 1938. Equilibrium of A Thin Plate, Symmetrically Loaded, on a Flexible Subgrade. *Iowa State College Journal of Science,* **12**(4), July, Ames, IA, 455-459.

Huang, Y.H., 1993. *Pavement Analysis and Design.* Prentice-Hall, Englewood Cliffs, NJ, 805.

Huang, Y.H., and Wang, S.T., 1973. Finite-Element Analysis of Concrete Slabs and Its Implications for Rogid Pavement Design. Highway Research Record 466, Highway Research Board, Washington, D.C., 55-69.

Huang, Y.H., and Wang, S.T., 1974. Finite-Element Analysis of Rigid Pavements with Partial Subgrade Contact. Transportation Research Record No. 485, Transportation Research Board, National Research Council, Washington D.C., 39-54

Ioannides, A.M., 1984. *Analysis of Slabs-On-Grade for a Variety of Loading and Support Conditions.* Thesis presented to the University of Illinois, at Urbana, IL, in partial fulfillment of the requirements for the degree of Doctor of Philosophy.

Ioannides, A.M., 1988. The Problem of Slab on an Elastic Solid Foundation in the Light of the Finite Element Method. Proceedings, Sixth International Conference on Numerical Methods in Geomechanics, Innsbruck, Austria, 1059-1064.

Ioannides, A.M., 1990. Extension of Westergaard Solutions Using Dimensional Analysis. Proceedings, Second International Workshop on the Design and Evaluation of Concrete Pavements, CROW/PIARC, Sigüenza, Spain, 4-5 October, 357-388.

Ioannides, A.M., 1991a. Theoretical Implications of the AASHTO 86 Nondestructive Testing Method 2 for Pavement Evaluation. Transportation Research Record 1307, Transportation Research Board, National Research Council, Washington D.C.: 211-220. Reprinted in Proceedings, Second International Workshop on the Design and Rehabilitation of Concrete Pavements, Sigüenza, Spain (Oct., 1990), 494-530.

Ioannides, A.M., 1991b. Subgrade Characterization for Portland Cement Concrete Pavements. Proceedings, International Conference on Geotechnical Engineering for Coastal Development - Theory and Practice-, Port and Harbour Institute, Ministry of Transport, Yokohama 3-6, Japan, 3-6 (Sep., 1991), 809-814.

Ioannides, A.M., 1992. Mechanistic Performance Modeling: A Contradiction of Terms? Proceedings, Seventh International Conference on Asphalt Pavements Design, Construction and Performance, 2, Nottingham, U.K. (17-21 Aug.), 165-179.

Ioannides, A.M., 1997a. Fracture Mechanics in Pavement Engineering: The Specimen-Size Effect. Transportation Research Record 1568, Transportation Research Board, National Research Council, Washington, D.C., 10-16.

Ioannides, A.M., 1997b. Pavement Fatigue Concepts: A Historical Review. Proceedings, *Sixth International Conference on Concrete Pavement Design and Rehabilitation*, Purdue University, Indianapolis, IN (Nov.), 18-21.

Ioannides, A.M., and Donnelly, J.P., 1988. Three-Dimensional Analysis of Slab on Stress Dependent Foundation. Transportation Research Record 1196, Transportation Research Board, National Research Council, Washington D.C., 72-84.

Ioannides, A.M., and Hammons, M.I., 1996. A Westergaard-Type Solution for the Edge Load Transfer Problem. Transportation Research Record 1525, Transportation Research Board, National Research Council, Washington, D.C., 28-34.

Ioannides, A.M., and Khazanovich, L., 1994. Analytical and Numerical Methods for Multi- Layered Concrete Pavements. Proceedings, Third International Workshop on the Design and Rehabilitation of Concrete Pavements, Krumbach, Austria (Oct.).

Ioannides, A.M., and Khazanovich, L., 1998. Nonlinear Temperature Effects on Multi-Layered Concrete Pavements. *Journal of Transportation Engineering*, ASCE, **124**(2), (Mar./Apr.): 128-136.

Ioannides, A.M., Khazanovich L., and Becque, J.L., 1992. Structural Evaluation of Base Layers in Concrete Pavement Systems. Transportation Research Record 1370, Transportation Research Board, National Research Council, Washington D.C., 20-28.

Ioannides, A.M., and Korovesis, G.T., 1990. Aggregate Interlock: A Pure-Shear Load Transfer Mechanism. In *Transportation Research Record 1286*, Transportation Research Board, National Research Council, Washington, D.C., 1990, 14-24.

Ioannides, A.M., and Korovesis, G.T., 1992. Analysis and Design of Doweled Slab-On-Grade Pavement Systems. *Journal of Transportation Engineering*, ASCE, **118**,(6), New York, NY, November, 1992, 745-768.

Ioannides, A.M., and Peng, J., 2004. Finite Element Simulation of Crack Growth in Concrete Slabs: Implications for Pavement Design. Proceedings, Fifth International Workshop on the Design and Rehabilitation of Concrete Pavements, Istanbul, Turkey (Apr., 1-2), 56-68.

Ioannides, A.M., Peng, J., and Sengupta, S., 2003. Crack Propagation in Portland Cement Concrete: Combining Dimensional Analysis and Finite Elements. Proceedings, International Conference on Highway Pavement Data, Analysis and Mechanistic Design Applications, **1**, ORITE/FHWA, Columbus, Ohio, September 7-10, 271-279.

Ioannides, A.M., and Salsilli-Murua, R.A., 1989. Temperature Curling in Rigid Pavements: An Application of Dimensional Analysis. Transportation Research Record 1227, Transportation Research Board, National Research Council, Washington D.C., 1-11.

Ioannides, A.M., Salsilli, R.A., Vinding, I., and Packard, R.G.1992 "Super-Singles": Implications for Design. Proceedings, Third International Symposium on Heavy Vehicle Weights and Dimensions, Queens College Cambridge, U.K., Jun. 28-Jul. 2, 225-232.

Ioannides, A.M., and Sengupta, S., 2003. Crack Propagation in PCC Beams: Implications for Pavement Design. Transportation Research Record 1853, Transportation Research Board, National Research Council, Washington, D.C., 110-117.

Ioannides, A.M., Thompson, M.R., and Barenberg, E.J., 1985. Finite Element Analysis of Slabs-On-Grade Using a Variety of Support Models. Proc., Third Int. Conference on Concrete Pavement Design and Rehabilitation, Purdue University, Apr., 309-324.

Kerr, A.D., 1964. Elastic and Viscoelastic Foundation Models. *J. Appl. Mech. Trans.*, ASME, **31**(3), 491-498.

Kerr, A.D., 1965. A Study of a New Foundation Model. *Acta Mechanica*, **1**(2).

Kerr, A.D., 1984. On the Formal Development of Elastic Foundation Models. *Ingenieur-Archiv*, **54**, 455-464.

Kerr, A.D., 1985a. On the Determination of Foundation Model Parameters. *J. Geot. Engrg.*, ASCE, **111**(11), 1334-1340.

Kerr, A.D., 1985b. Application of Pasternak Model to Some Soil-Structure Interaction Problems: Vol. 1-Solutions for Plates Continuously Supported on a Pasternak Base. Technical Report K-85-1, U.S. Army Engineer Waterways Experiment Station, Vicksburg, MS.

Kerr, A.D., 1987. On the Vertical Modulus in the Standard Railway Track Analysis. *Rail International*, November, 37-45.

Kerr, A.D., 1993. Mathematical Modeling of Airport Pavements, in Airport Pavement Innovations - Theory to Practice. J.W. Hall, Jr. (Ed.), ASCE, New York, NY, 31-45.

Kerr, A.D., 1996. Bearing Capacity of Floating Ice Covers Subjected to Static, Moving, and Oscillatory Loads. *Applied Mechanics Review*, **49**(11), American Society of Mechanical Engineers, November, 463-476.

Kerr, A.D., and El-Sibaie, M.A., 1989. Green's Functions for Continuously Supported Plates. *J. Appl. Math. Phys.*, **40**, 15-38.

Kerr, A.D., and Shade, P.J., 1984. Analysis of Concrete Pavement Blowups. *Acta Mechanica*, **52**, Springer-Verlag, 201-224.

Khazanovich, L., 1994. *Structural Analysis of Multi-Layered Concrete Pavement Systems*. Thesis presented to the University of Illinois, at Urbana, IL, in partial fulfillment of the requirements for the degree of Doctor of Philosophy.

Khazanovich, L., and Ioannides, A.M., 1993, Finite Element Analysis of Slabs-On-Grade Using Improved Subgrade Soil Models, in *Airport Pavement Innovations—Theory to Practice*. Proceedings of a Specialty Conference sponsored by the Airfield Pavement Committee, Air Transport Division, ASCE, Jim W. Hall, Jr., Ed., New York, NY, 16-30.

Khazanovich, L., Yu, T.H., and Darter, M.I., 2004. Prediction of Critical JPCP Stresses in the Mechanistic-Empirical 2002 Design Guide. Proceedings, 9th International Symposium on Concrete Roads, April 4-7, Istanbul, Turkey.

Kneifati, M.C., 1985. Analysis of Plates on a Kerr Foundation Model. *J. Engrg. Mech.*, ASCE, **111**(11), 1325-1342.

Korenev, B.G., and Chernigovskaya, E.I., 1962. Analysis of Plates on Elastic Foundation. Gosstroiizdat, Moscow (in Russian).

Korovesis, G.T., 1990. *Analysis of Slab-On-Grade Pavement Systems Subjected to Wheel and Temperature Loadings*. Thesis presented to the University of Illinois, at Urbana, IL, in partial fulfillment of the requirements for the degree of Doctor of Philosophy.

Lee, Y.-H., 1993. *Development of Pavement Prediction Models*. Thesis presented to the University of Illinois, at Urbana, IL, in partial fulfillment of the requirements for the degree of Doctor of Philosophy.

Losberg, A., 1960. *Structurally Reinforced Concrete Pavements*. Doktorsavhandlingar Vid Chalmers Tekniska Högskola, Göteborg, Sweden.

Miner, M.A., 1945. Cumulative Damage in Fatigue. Transactions, American Society of Mechanical Engineers, **67**, *Journal of Applied Mechanics*, Sept., New York, NY, A-159 to A-164.

Mirambell, E., 1990. Temperature and Stress Distributions in Plain Concrete Pavements Under Thermal and Mechanical Loads. Proceedings, Second International Workshop on the Design and Evaluation of Concrete Pavements, CROW/PIARC, Sigüenza, Spain, 4-5 October, 121-135.

Mohamed, A.R., and Hansen, W., 1997. Effect of Nonlinear Temperature Gradient on Curling Stress in Concrete Pavements. Transportation Research Record No. 1568, Transportation Research Board, National Research Council, Washington D.C., 65-71.

Morton, J.O., 1936. The Application of Soil Mechanics to Highway Foundation Engineering. Proceedings, (First) International Conferences on Soil Mechanics and Foundation Engineering, **1**, June, Harvard University, Cambridge, MA, 243-247.

NCHRP, 1992. Calibrated Mechanistic Structural Analysis Procedures for Pavements, Phase 2. Final Report, Project 1-26: **1-2**, National Cooperative Highway Research Program, University of Illinois, Urbana, IL.

Newmark, N.M., 1942. Influence Charts for Computation of Stresses in Elastic Foundations. University of Illinois Experiment Station Bulletin 338, Urbana.

Odemark, N., 1949. Investigations as to the Elastic Properties of Soils and Design of Pavements According to the Theory of Elasticity. (In Swedish), Meddelande 77, Statens Väginstitut, Stockholm, Sweden. English translation by Michael A. Hibbs and Johan Silfwerbrand (A.M. Ioannides, Ed.), 1990.

Packard, R.G., 1973. Design of Concrete Airport Pavement. Engineering Bulletin EB050.03P, Portland Cement Association, Skokie, IL.

Pasternak, P.L., 1954. Fundamentals of a New Method of Analysis of Structures on Elastic Foundation by Means of Two Subgrade Coefficients. *Gosudarstvennoe Izdatel'stvo Literatury po Stroitel'stvu i Arkhitekture*, Moscow (in Russian).

Pickett, G., and Badaruddin, S., 1957. Influence Chart for Bending of a Semi-Infinite Pavement Slab. Proceedings, 9th Congress on Applied Mechanics, Vol. **6**, 1956, Université de Bruxelles, 396-402.

Pickett, G., Badaruddin, S., and Ganguli, S.C., 1955. Semi Infinite Pavement Slab Supported by an Elastic Solid Subgrade. Proceedings of the First Congress on Theoretical and Applied Mechanics, Indian Institute of Technology, Nov. 1-2, 51-60.

Pickett, G., and Ray, G.K., 1951. Influence Charts for Concrete Pavements. *Transactions*, ASCE, **116**, 49-73.

Reissner, E., 1958. A Note on Deflections of Plates on a Viscoelastic Foundation. *Journal of Applied Mechanics*, Transactions, ASME, **25**(1), March.

Reissner, E., 1967. Note on the Formulation of the Problem of the Plate on an Elastic Foundation. *Acta Mechanica*, **IV**(1), 88-91.

Richardson, J.M., and Armaghani, J.M., 1987. Stress Caused by Temperature Gradient in Portland Cement Concrete Pavements. Transportation Research Record No. 1121, Transportation Research Board, National Research Council, Washington D.C., 7-13.

Schiffman, R.L., 1957. The Numerical Solution for Stresses and Displacements in a Three-Layer Soil System. Proceedings, Fourth International Conference on Soil Mechanics and Foundation Engineering, London, **2**, 169-173.

Schiffman, R.L., 1962. General Analysis of Stresses and Displacements in Layered Elastic Systems. Proceedings, (First) International Conference on Structural Design of Asphalt Pavements, University of Michigan, Ann Arbor, 365-375.

Schiffman, R.L., 2000. *Theories of Consolidation*. University of Colorado Press, Boulder, CO.

Seely, B.E., 1991. The Scientific Mystique in Engineering: Highway Research at the Bureau of Public Roads, 1918-1940, In *The Engineer in America: A Historical Anthology from Technology and Culture.* (T.S. Reynolds, Ed.), The University of Chicago Press, Chicago, IL, 309-342.

Selvadurai, A.P.S., 1979. Elastic Analysis of Soil-Foundation Interaction. Developments in Geotechnical Engineering, **17**, Elsevier, Amsterdam, The Netherlands.

Shtaerman, I.Ya., 1949. *Contact Problems of the Theory of Elasticity.* Gostekhizdat, Moscow-Leningrad (in Russian).

Skarlatos, M.S., 1949. Deflections and Stresses in Concrete Pavements of Airfields With Continuous Elastic Joints. Report AD628501, Ohio River Division Laboratories, U.S. Army Corps of Engineers, Mariemont, OH. Also, D.Sc. Thesis, Harvard University, Cambridge, MA.

Skarlatos, M.S., and Ioannides, A.M., 1998. The Theory of Concrete Pavement Joints. Proceedings, Fourth International Workshop on the Design and Rehabilitation of Concrete Pavements, Bussaco, Portugal (Sept.).

Slivker, V.I., 1981. On Problem of Assignment of Characteristics for Two-Parametric Elastic Foundation. Stroitel'naya Mekhanika i Raschet Sooruzhenii, No. 1, Moscow, 36-39 (in Russian).

Tabatabaie, A.M., and Barenberg, E.J., 1978. Finite-Element Analysis of Jointed or Cracked Concrete Pavements. Transportation Research Record No. 671, Transportation Research Board, National Research Council, Washington D.C., 11-19.

Tayabji, S.D., and Colley, B.E., 1983. Improved Rigid Pavement Joints. Transportation Research Record No. 930, Transportation Research Board, National Research Council, Washington D.C., 69-78.

Teller, L.W., and Sutherland, E.C., 1943. The Structural Design of Concrete Pavements -- Part 5: An Experimental Study of the Westergaard Analysis of Stress Condition in Concrete Pavement Slabs of Uniform Thickness. *Public Roads*, **23**(8), Apr.-May-Jun., 167-212.

Terzaghi, K., 1943. *Theoretical Soil Mechanics.* John Wiley and Sons, Inc., New York, NY.

Terzaghi, K., 1955. Evaluation of Coefficients of Subgrade Reaction. *Géotechnique*, **5**(4), Dec., 297-326.

Thomlinson, J., 1940. Temperature Variations and Consequent Stresses Produced by Daily and Seasonal Temperature Cycles in Concrete Slabs. *Concrete Constructional Engineering*, **36**(6), 298-307; (7), 352-360.

Thompson, M.R., 1996. Mechanistic-Empirical Flexible Pavement Design: An Overview. Transportation Research Record 1539, Transportation Research Board, National Research Council, Washington D.C. 1996: 1-5.

Tia, M., Armaghani, J.M., Wu, C.-L., Lei, S., and Toye, K.L., 1987. FEACONS III Computer Program for An Analysis of Jointed Concrete Pavements. Transportation Research Record No. 1136, Transportation Research Board, National Research Council, Washington D.C., 12-22.

Van Cauwelaert, F., 1990. Westergaard's Equations for Thick Elastic Plates. Proceedings, Second International Workshop on the Design and Evaluation of Concrete Pavements, CROW/PIARC, Sigüenza, Spain, 4-5 October, 165-175.

Vesic, A.S., and Saxena, S.K., 1970. Analysis of Structural Behavior of AASHO Road Test Rigid Pavements. NCHRP Report 97, Highway Research Board, Washington, D.C., 35.

Vlasov, V.Z., and Leont'ev, N.N., 1960. Beams, Plates and Shells on Elastic Foundations. NASA-NSF, NASA TT F-357, TT 65-50135, Israel Program for Scientific Translations (translation date: 1966).

Westergaard, H.M., 1923. Om Beregning af Plader paa elastik Underlag med saerlight Henblik paa Sporgsmaalet om Spaendinger I Betonveje, (On the Design of Slabs on Elastic Foundation with Special

Reference to Stresses in Concrete Pavements), In *Ingeniøren* (in Danish), **32(42)**, Copenhagen, 1923, 513-524.

Westergaard, H.M., 1926. Stresses in Concrete Pavements Computed by Theoretical Analysis. In *Public Roads*, Vol. 7(2), April, 25-35. Also in *Proceedings*, Highway Research Board, 5th Annual Meeting 1925, published 1926, Part I, 90-112, under title Computation of Stresses in Concrete Roads.

Westergaard, H.M., 1927. Analysis of Stresses in Concrete Pavements Due To Variations of Temperature. Proceedings, Highway Research Board, **6**, National Research Council, Washington D.C., 201-217.

Westergaard, H.M., 1938. A Problem of Elasticity Suggested by a Problem in Soil Mechanics: Soft Material Reinforced by Numerous Strong Horizontal Sheets. In Contribution to the Mechanics of Solids, Stephen Timoshenko 60th Anniversary Volume, Macmillan, New York.

Westergaard, H.M., 1948. New Formulas for Stresses in Concrete Pavements of Airfields. In *Transactions*, ASCE, **113**, New York, NY, 1948, 425-439.

Winkler, E., 1867. *Die Lehre von der Elastizitt und Festigkeit.* (Theory of Elasticity and Strength), H. Dominicus, Prague (in German).

Woodhead, R.W., and Wortman, R.H., 1973. Proceedings, Allerton Park Conference on Systems Approach to Airfield Pavements: 23-26 March, 1970: Technical Report P-5. Construction Engineering Research Laboratory, Champaign, IL, June, 1973.

Yoder, E.J., and Witczak, M.W. 1975. *Principles of Pavement Design.* Second Ed., John Wiley and Sons, Inc., New York, NY, 711.

Zeevaert L., and Newmark, N.M., 1956. Aseismic Design of Latino Americana Tower in Mexico City. Proceedings, World Conference on Earthquake Engineering, Earthquake Engineering Research Institute, Berkeley, CA, 35-1 to 35-11.

Zhemochkin, B.N., and Sinitsyn, A.P., 1947. *Practical Method of Analysis of Plates and Beams on Elastic Foundation Without Winkler's Hypothesis.* Stroiizdat, Moscow (in Russian).

History of Sandwich Structures and Their Analyses

Jack R. Vinson, University of Delaware

Abstract

The use of sandwich structure is growing very rapidly around the world. It is used in many applications ranging from satellites, aircraft, ships, automobiles, rail cars, wind energy mills, and in bridge construction to mention only a few. Its many advantages, the development of new materials, and the need for high performance, low-weight structures insure that sandwich construction will continue to be in demand. The equations describing the behavior of sandwich structures are usually compatible with the equations developed for composite material thin-walled structures, simply by employing the appropriate in-plane, flexural, and transverse shear stiffness quantities. Only if a very flexible core is used, is a higher order theory needed.

Most often there are two faces, identical in material and thickness, which primarily resist the in-plane and lateral (bending) loads. However, in special cases the faces may differ in thickness, materials, or fiber orientation, or any combination of these three. This may be due to the fact that one face is the primary load carrying, low temperature portion of the structure while the other face may have to withstand an elevated temperature, corrosive environment, etc. Assuming a uniform core, the former sandwich is regarded as a mid-plane symmetric sandwich, the latter a mid-plane asymmetric sandwich.

Advantages of Sandwich Construction

One reason for the increased use of sandwich construction is because it has such a high ratio of flexural stiffness to weight. As a result, sandwich constructions result in lower lateral deformations, higher buckling resistance, and higher natural frequencies than do other constructions. Thus sandwich constructions quite often provide a lower structural weight than do other constructions for a given set of mechanical and environmental loads.

It is also interesting to compare an isotropic sandwich construction with a monocoque (thin walled) construction of the same facing weight. In Fig. 1, the sandwich construction employs two identical faces of thickness t_f and a core depth of h_c. The monocoque construction on the right is a flat plate construction of thickness $2t_f$, hence the same weight as the faces of the sandwich construction and using the same materials.

For an isotropic face material with a modulus of elasticity, E_f, the extensional stiffness per unit width for both the sandwich and the monocoque construction is

$$K = 2E_f t_f / (1 - v_f^2) \tag{1}$$

Thus, for in-plane tensile and compressive (not considering buckling) loads, the two constructions have the same extensional stiffness.

However, there is a marked difference in the flexural stiffness per unit width, D. For the panel construction of Fig. 1(b) below, the flexural stiffness is

(a) Sandwich Construction   (b) Monocoque Construction

Fig. 1. Cross-sections of sandwich and monocoque construction.

$$D_{mon.} = \frac{E_f(2t_f)^3}{12(1-v_f^2)} = \frac{2E_f t_f^3}{3(1-v_f^2)} \tag{2}$$

The flexural stiffness for the isotropic foam or honeycomb core sandwich construction is given by

$$D_{sand.} = 2E_f t_f (h_c/2)^2 \big/ (1-v_f^2) = \frac{E_f t_f h_c^2}{2(1-v_f^2)} \tag{3}$$

where it is assumed that the core does not contribute to the flexural stiffness, and $t_f/h_c \ll 1$.

From the above, it is seen that the ratio of the flexural modulus of foam or honeycomb core sandwich to a monocoque construction of the same face weight, using the same material is

$$\frac{D_{sand.}}{D_{mon.}} = \frac{3}{4}\left(\frac{h_c}{t_f}\right)^2 \tag{4}$$

For example, if $t_f/h_c = 20$, the flexural stiffness of the sandwich construction is 300 times the flexural stiffness of the monocoque construction. As a result, the use of sandwich construction results in a much lower lateral deflection, much higher overall buckling loads, and much higher flexural natural frequencies of vibration than does the monocoque construction of nearly the same weight. Rarely, does the core weight approach the face weight per unit area. However, in sandwich constructions subjected to in-plane compressive or shear loads, in addition to overall buckling, face wrinkling and monocell buckling must also be checked, which will be discussed later.

Now, looking for a comparison in face stresses, consider the same sandwich and monocoque constructions to be subjected to an in-plane load of *N*, a force per unit width of the structure, and a bending moment per unit width, *M*.

For the in-plane loads, in each case the resulting in-plane stress is given by

$$\sigma_f = \frac{N}{2t_f} \tag{5}$$

Thus, neither construction provides any advantage over the other regarding in-plane loads. However, for a bending moment, *M*, for the monocoque structure, the maximum stresses which occur at the top and bottom surfaces is

$$\sigma_{mon.} = \pm\frac{6M}{(2t_f)^2} = \frac{3}{2}\frac{M}{t_f^2} \tag{6}$$

Similarly, for the sandwich construction, the stress at the top and bottom faces is found to be

$$\sigma_{sand.} = \frac{M}{2t_f h_c} \tag{7}$$

Therefore the ratio of the maximum bending stress in a sandwich panel to the maximum bending stress in the monocoque plate in which the face weight of the sandwich equals the panel weight per unit planform area is

$$\frac{\sigma_{sand.}}{\sigma_{mon.}} = \frac{t_f}{3h_c} \tag{8}$$

To continue the example used previously in which the ratio of the face thickness to core depth is 1/20, the maximum bending stress in the sandwich face is 1/30 the maximum stress at the surfaces of the monocoque construction subjected to the same bending moment. Thus for many applications, even if the weight of the core causes the weight of the sandwich to be as much as twice the weight of the monocoque construction with the same weight as the sandwich face weight, the fact that the bending stiffness is 300 times while the maximum stresses are 1/30 that of the monocoque construction makes the sandwich construction very desirable.

Even with these advantages, it is important to develop the means by which to optimize the sandwich construction for minimum weight in order to:

(1) Determine the absolute minimum weight for a given geometry, loading, and material system.

(2) Rationally compare one type of sandwich construction with other types of sandwich construction.

(3) Rationally compare the best sandwich construction with alternative structural configurations (monocoque, rib-reinforced, etc.).

(4) Rationally select the best face and core materials.

(5) Select the best stacking sequences for faces that are composed of laminated composite materials.

(6) Rationally compare the weight of the optimized construction to weights required when there are some restrictions, i.e., cost, minimum gage, manufacturing limitations, material availability, etc.

Origins of Sandwich Construction

Noor, Burton, and Bert (1996) state that the concept of sandwich construction dates back to Fairbairn (1849). In England, sandwich construction was first used in the Mosquito night bomber of World War II. Feichtinger (1988) states that also during World War II, the concept of sandwich construction in the United States originated with the faces made of reinforced plastic and a lower density core. In 1943 the Wright Patterson Air Force Base designed and fabricated the Vultee BT-15 fuselage using fiberglass-reinforced polyester as the face material and using both glass-fabric honeycomb and balsa wood core (Rheinfrank and Norman, 1944).

The first research paper concerning sandwich construction was written by Marguerre (1944) dealing with sandwich panels subjected to in-plane compressive loads. Hoff (1950) derived the differential equations and boundary conditions for the bending and buckling of sandwich plates using the Principle of Virtual Displacements, but pursued only the buckling problem of the panel subjected to edgewise compression. Around the same time, Libove and Batdorf (1948) published a small deflection theory for sandwich plates. In 1949, Flugge published a paper on the structural optimization of sandwich panels in which he presented nomograms for the solution of several problems (Flugge, 1949). In all cases the materials studied were isotropic. He published another, related paper in 1952 (Flugge, 1952).

Also in the 1940s, two young World War II veterans formed Hexcel, perhaps the most prestigious firm associated with sandwich construction.

In a series of papers in 1951-1952, Bijlaard (1951a, 1951b, 1952a) studied sandwich optimization for the case of a given ratio between core depth and face thickness, as well as for a given total thickness of the isotropic sandwich plate. An abridgement of this research is in the proceedings of the 1st U.S. National Congress of Applied Mechanics in 1952 (Bijlaard, 1952b). Ericksen (1955) issued a U.S. Forest Products Laboratory (USFPL) Report accounting for the effects of transverse shear deformation on the deflections in isotropic sandwich plates, using a double Fourier series to represent the deflections in the simply-supported plate, i.e. a Navier solution. Also presented were general expressions for the strain components in sandwich panels with orthotropic faces and cores. In 1952, Eringen used the Theorem of Minimum Potential Energy to obtain four partial differential equations for the bending and buckling of rectangular isotropic sandwich plates under various loads and boundary conditions (Eringen, 1952). In the same year, March (1952) also published a USFPL report on sandwich panel behavior. Raville (1955) published a definitive study for the design and optimization of sandwich panels. Military handbook 23 (Anon, 1955) was published which largely involved the results of the many publication issued by the USFPL.

In 1956, Gerard discusses sandwich plate optimization in one chapter of his landmark book, "Minimum Weight Analysis of Compression Structures" (Gerard, 1956).

In 1957, Kaechele published a report on the minimum weight design of sandwich panels (Kaechele, 1957). Another series solution was presented by Cheng (1962). In 1960, Heath published a paper on the correlation among and an extension of the existing theories for flat sandwich panels subjected to lengthwise compression, including the treatment of optimum design (Heath, 1960).

This early theoretical work was all restricted to uniform lateral loads, and simply supported boundary conditions. During the early post World War II period, the USFPL was the primary group in the development of analysis and design methods for sandwich structures. The USFPL also led the efforts with MIL-HDBK-23 (Anon, 1955) referred to above and in the continual updating of that document.

By the mid 1960s efforts in sandwich structures had increased significantly. In 1966, Plantema published the first book on sandwich structures (Plantema, 1966), followed by another book on sandwich structures by Allen (1969). These books remained the "bibles" for sandwich structures until the mid 1990s.

Also in the mid 1960s, the Naval Air Engineering Center sponsored research to develop a fiberglass composite sandwich constructions to compete in weight with conventional aluminum aircraft construction. Much of this research effort was in the development of minimum weight optimization methods so that the fiberglass sandwich construction could compete with the established aluminum construction (Vinson and Shore, 1965b, 1965c, 1967a, 1967b; McCoy, Vinson and Shore, 1967). A complete bibliography of over 250 publications describing research before 1966 in the sandwich structures area was published (Vinson and Shore, 1965a).

In 1986, a text containing many of Nicholas J. Hoff's publications was published (Hoff, 1986). In 1989, Ha published an overview of finite element analysis applied to sandwich construction (Ha, 1989). Bert (1991) provided a review of sandwich plate analysis in 1991, while in 1996, a review by Noor, Burton, and Bert (1996) provides over 800 references, all discussed in the review, and another 559 references as a supplemental bibliography.

In 1995, a monograph by Zenkert (1995) supplemented much of the material contained earlier in the Plantema and Allen texts, which by this time were out of print. In 1999, another textbook on sandwich structures was published by Vinson (1999).

To date there have been only seven International Conferences on Sandwich Constructions: the first in Stockholm in 1989, the second in Gainesville in 1992, the third was in Southampton in

1995, the fourth in Stockholm in 1998 and the fifth Conference in Zurich in September of 2000. The sixth Conference was held in Ft. Lauderdale in March 2003. The seventh Conference was held in Aalborg, Denmark in 2005.

Uses of Sandwich Construction

In 1998, Bitzer gave an excellent overview of honeycomb core materials and their applications (Bitzer, 1998). In this publication Bitzer points out that one of every two (or more) aircraft in the western world utilizes some honeycomb core sandwich. He points out that while 8% of the wetted surface of the Boeing 707 is sandwich, that 46% of the wetted surface of the Boeing 757/767 is honeycomb sandwich. In the 747, the fuselage cylindrical shell is primarily Nomex honeycomb sandwich, and the floors, side-panels, overhead bins and ceiling are also of sandwich construction.

The Beech Starship uses Nomex honeycomb with graphite and Kevlar faces for the entire structure-the first all sandwich aircraft. A major portion of the space shuttle is a composite-faced honeycomb-core sandwich. Almost all satellite structures employ sandwich construction.

Europe leads the way in the use of sandwich for light weight railcars, while in the U.S. some of the rapid transit trains use honeycomb sandwich. The U.S. navy uses honeycomb sandwich construction for bulkheads, deck houses, and helicopter hangars to reduce weight above the waterline. Also recently, they have incorporated a complete hexagonally shaped mast of the USS *Radford* which is ninety three feet tall and weighs 90 tons. Not only is this a foam core sandwich but the use of exterior materials for stealth purposes make this an asymmetric sandwich. This is probably the largest sandwich structure in the United States.

Sailboats, racing craft, and auto race cars are all employing sandwich construction. Sandwich is also used for snow skis, water skis, kayaks, canoes, pool tables, and platform tennis rackets.

Honeycomb sandwich construction is excellent for absorbing mechanical and sound energy. It can also be used to transmit heat or to be an insulative barrier. In the former, a metallic honeycomb is used plus natural convection; for the latter, a non-metallic core is used with the cells filled with foam. For a sound barrier, the core can be filled with a fiberglass batting, and a thin porous Tedlar skin used for the interior surface, again a mid-plane asymmetric sandwich construction.

Boat hulls became a logical use for fiberglass sandwich, particularly in pleasure craft where the foam core increases the chance for flotation in emergencies. Ferry boats in the Scandinavian countries and the Pacific Rim countries use much fiberglass-sandwich construction.

The Royal Swedish navy has been using fiberglass and graphite composite sandwich construction for more than twenty years (Lonno and Hellbratt, 1996). The hulls were designed to withstand underwater explosion and debris strike requirements as steel hulls in the past. However, the fiberglass sandwich construction eliminates the attraction of the hulls to magnetic mines, as well as negating salt water corrosion. Furthermore, the state that their next generation of surface vessels, such as the YP2000 Visby, will be stealth-optimized graphite/epoxy composite vessels using sandwich construction primarily. Similarly, the Royal Australian Navy uses high performance foam composite sandwich for its inshore mine hunters (Robson, 1989).

In rail locomotives, since 1980 composite front cabs have been built for the XPT locomotives in Australia. the ETR 500 locomotives in Italy, the French TGV and the Swiss locomotive 2000. Interestingly, the major design criteria are the pressure waves occurring during the crossing of two high speed trains in a tunnel. Self supporting sandwich construction was selected with a weight saving of 1000 kilograms per locomotive. In Japan, the new Nozomi 500 bullet trains use honeycomb sandwich for their primary structure.

Also in 1996, Starlinger and Reif reported that sandwich construction is now being used in roof panels and intermediate floor panels in double-decker buses (Starlinger and Reif, 1996).

Kujala and Tuhkuri (1996) investigated the use of steel corrugated core sandwich panels for ship superstructures both analytically and experimentally. They found the sandwich construction to be 40-50% lighter than the conventional steel construction.

Most recently sandwich construction is being used increasingly in civil engineering infrastructure rehabilitation projects such as bridge decks. Karbhari (1997) provides an overview in the use of composite-sandwich usage in the twenty-first century. In 2000, Eckel describes in detail the analysis, design, fabrication and installation of a bridge deck, which comprised his Doctoral Dissertation at the University of Delaware (Eckel, 2000).

Davies (1977) states that sandwich-cladding panels composed of two metallic faces and a lightweight insulating core are finding increasing use as wall and roof cladding for a wide variety of buildings, where the primary attraction is their outstanding thermal performance.

Woldesenbet and Vinson (1996) have studied the use of sandwich construction for low cost and emergency housing.

It must also be remembered that boxes and packaging of all kinds comprise a multibillion dollar use each year of sandwich construction. This use often employs low cost Kraft paper construction, which quite often utilizes three face–two core construction.

Face Material Property Determination

For composite material used in sandwich faces there are several methods by which to calculate the composite elastic properties from the properties of the fibers and the matrix materials. These include methods developed by Halpin and Tsai (1967), Hashin (1972), Christensen (1979) and Hahn (1980). From their use the testing requirements for composite sandwich faces is reduced significantly.

Cores

The core of a sandwich structure can be of almost any material or architecture, but in general, cores fall into four types: (a) foam or solid core, (b) honeycomb core, (c) truss core, and (d) corrugated or truss core, as shown in Fig. 2. Since World War II, honeycomb cores have been widely used. The two most common types are the hexagonally-shaped cell structure (hexcell) and the square cell (egg-crate). Web core construction is analogous to a group of I-beams with their flanges welded together. The U.S. Navy refers to this web core construction as "double hull" construction which they plan to use in their future stainless steel ship hulls. Truss or triangulated core construction can vary from being very simple to being of a complex cross-section, such as in many box materials. In the truss core and web core constructions, the empty space in the core may be used for liquid storage, or as a heat exchanger. In all cases the primary loads, both in-plane and lateral, are carried by the faces, while the core resists the transverse shear loads. In most foam core and honeycomb core sandwich constructions, one can assume for all practical purposes that all of the in-plane loads and lateral bending loads are carried by the faces only. However in truss core and web core constructions, a portion of these loads are carried by the core.

Developments in new cores continue to be of primary interest, such as the new cores developed by Christensen and Kim (1999) and Czaplicki (1992). Christensen's core is a star shaped repeated pattern while Czaplicki suggests a cellular core formed by the progressive corrugation of a single continuous sheet of material. Recently Bitzer (1998) described several new honeycomb cores, including a thermoplastic CECORE, marine Nomex, Korex and non-metallic Tube-core.

Foam or solid cores are relatively inexpensive and can consist of balsa wood or an almost infinite selection of foam/plastic materials with a continuous variety of densities and shear moduli. Many of the cores that are commercially available today are of polyvinylchloride (PVC).

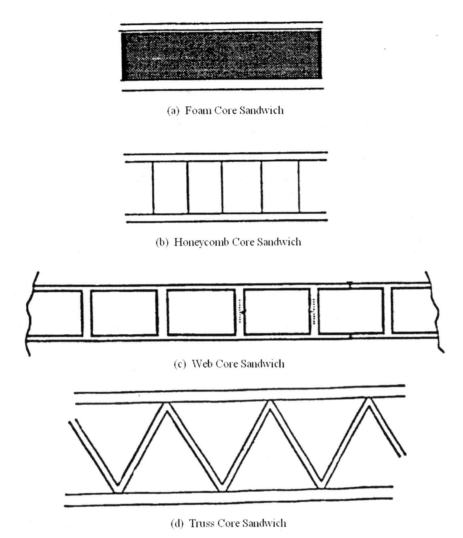

(a) Foam Core Sandwich

(b) Honeycomb Core Sandwich

(c) Web Core Sandwich

(d) Truss Core Sandwich

Fig. 2. Types of sandwich construction

Hygrothermal Effects

As in all composite constructions, thermal and hygrothermal considerations must be taken into account. Concerning the thermal effects, with increased temperature, there are three: thermal expansion, degradation of elastic properties; and an increase in non linear creep/viscoelastic effects. A fourth effect is that in many two dimensional boundary value problems the boundary conditions become non-homogeneous,, negating the use of separation of variables in partial differential equations, thus requiring a transformation of variables to solve the analytical problem.

For the "hygro" part of the effects, there is moisture expansion in all polymer materials mathematically analogous to the thermal expansion. The good news is that if one has the thermal effect solution then one also has the "hygro" solution and they can be superimposed. Moisture also affects the glass transition temperature, T_g. One major difference between the thermal and the moisture effects on a polymer matrix structure difference is in the time scales. It takes weeks or months to have a saturated specimen or structure. Weitsman (2000) points out that deleterious effects are found as late as eight years after composites have been soaking in salt water.

Noor, Starnes, and Peters (1997) studied the nonlinear effects of a temperature gradient through the thickness of a curved sandwich panel. More recently Noor, Starnes and Peters (2000) extended this study to include the effects of circular cutouts.

Long-Term Effects

Creep and viscoelastic effects can be important. Jurf and Vinson (1985) has shown that for some polymeric materials one can construct a master creep curve in which one can construct or define a temperature shift factor as well as a moisture shift factor, by which one can utilize short time creep data for predicting behavior at longer times in an hygrothermal environment.

Chevalier (1992) studied both creep and fatigue effects on sandwich core materials including end-grain balsa (EGB), hexagonal honeycomb made of aluminum as well as aramid fiber-paper (NOMEX), and PVC. He reports that little data is available, and is badly needed. Gellhorn and Starlinger (1992) studied creep effects on sandwich core foams, and emphasize the importance of including creep considerations in design for long-term loaded sandwich structures. Other studies have been conducted by Shenoi, Clark, Allen, and Hicks (1996); Feichtinger and Martensen (1996); Caprino, Teti, and Messa (1996). The latter team point out that although aluminum and Nomex honeycomb cores are satisfactory for heavily loaded structures, their high cost makes them less acceptable for many mass produced structural components. They state that polymeric foams provide a reasonable compromise between cost and mechanical properties, and in particular praise PVC because of its high stiffness and strength combined with good thermal insulation and low water absorption.

High Strain Rate Effects

Most materials have significantly different mechanical properties when subjected to dynamic loads that result in high strain rates. However most structural designs today are made using static properties, and much research is needed in this area to improve the designs and their reliability. Much can be learned from the publications of Lindholm (1964), Daniel, La Bedz, and Liber (1981), Nicholas (1981), Zukas (1982), Sierakowski (1997) and Feichtinger (1990).

Gellhorn and Reif (1992) investigated high strain rate effects on three different cross-linked PVC foams and one linear PVC foam up to strain rates of 500/s. They found that the linear PVC foam was the only one that did not suffer a significant loss in strain to failure. At the highest strain rate the linear PVC foam was 34% higher in compressive strength and 34% higher in strain to failure than the cross linked PVC foams. They give the perspective that a boat traveling at 10 knots combined with a panel length of 1.5 meters results in a shear strain rate of 6.9/s. they also state that materials which have high stiffness and strength at low strain rates may fail at high strain rates because of brittleness, while other materials such as linear PVC which has modest properties at low strain rates have superior properties at high strain rates.

A comprehensive program under ONR sponsorship of the high strain rate effects on many materials that are used in sandwich faces show that compared to static properties at high strain rates in the neighborhood of 1600/s (Powers et al., 1997, Woldesenbet and Vinson, 1999). Yield stresses may increase by factors of 3.6, yield strains may change by factors of 3.1, moduli of elasticity may increase by as much as 2.4, strains to failure may change as much as factors of 4.7, and equally important the elastic strain energy density functions may change by factors of 6 while the strain energy densities to failure may change by factors up to 8.1. It is clear that dynamic properties should be used rather than static properties for structures subjected to dynamic and shock loads.

Yet today there are no means by which the dynamic properties can be determined analytically if the static material properties are known. Experimental data must be generated at the strain rates and temperatures needed.

Sandwich Analysis

Keep in mind that most sandwich structures can be analyzed by using the laminate analysis methods of composite material structures by employing the A, B, and D stiffness matrices. Only in the case of sandwich constructions which use a very flexible core must a higher order sandwich theory be used. This will be discussed below. For an easy example consider that the lower face is lamina 1, the core is lamina 2, and the upper face is lamina 3. In that way you can include the effect of the core on the response to bending and in-plane loads or ignore it. One can easily handle sandwich constructions of mid-plane asymmetry as well as symmetric sandwich constructions. Also by involving the 4 and 5 terms, one can include the effects of transverse shear deformation. Likewise you can include thermal, hygrothermal and piezoelectric effects as well.

For a mid-plane symmetric sandwich construction in which the core stiffness can be neglected the A, B and D stiffness matrix quantities are given as shown below:

$$A_{ij} = \begin{bmatrix} \dfrac{2E_{11}t_f}{(1-v_{12}v_{21})} & \dfrac{2v_{21}E_{11}t_f}{(1-v_{12}v_{21})} & 0 \\[2ex] \dfrac{2v_{21}E_{11}t_f}{(1-v_{12}v_{21})} & \dfrac{2E_{22}t_f}{(1-v_{12}v_{21})} & 0 \\[2ex] 0 & 0 & 2G_{12}t_f \end{bmatrix} \tag{9}$$

$$D_{ij} = \begin{bmatrix} \dfrac{2E_{11}h_c^2 t_f}{2(1-v_{12}v_{21})} & \dfrac{v_{21}E_{11}h_c^2 t_f}{2(1-v_{12}v_{21})} & 0 \\[2ex] \dfrac{v_{21}E_{11}h_c^2 t_f}{2(1-v_{12}v_{21})} & \dfrac{E_{22}h_c^2 t_f}{2(1-v_{12}v_{21})} & 0 \\[2ex] 0 & 0 & \dfrac{G_{12}h_c^2 t_f}{2} \end{bmatrix} \tag{10}$$

$$B_{ij} = 0 \qquad\qquad (11)$$

It should be noted that for this sandwich

$$D_{ij} = A_{ij} h_c^2 / 4 \qquad\qquad (12)$$

And when for many cases in which the face thickness to core depth ratio is very small, then

$$A_{44} = A_{55} = G_c h_c + 2 t_f G_f \qquad\qquad (13)$$

Using the above, most solutions for isotropic and orthotropic laminated beam, plate and shell structures subjected to various loads, environments and boundary conditions can be employed to determine the behavior of the analogous problem associated with a sandwich construction.

Higher Order Sandwich Theory

Localized loads are one of the major causes of failure in sandwich constructions, because the faces are significantly thinner than the same material used in a monocoque construction to resist the same loads. These localized loads can cause the loaded face to deform significantly different from that of the unloaded face. The loaded face acts as a beam, plate or shell on an elastic foundation, i.e., the core. This causes the core to be subjected to significant deformations locally, which can cause high shear and normal stresses that can exceed the allowable stress for the flexible, weak core material. In addition because of the significant deformation of one face the stiffness matrix quantities shown above in Equations and, can be locally reduced significantly. This can cause a weak spot in the overall structure which can precipitate a premature failure.

Similarly, when there is a complex shape, such as the cross-section of a cylindrical shell which is a rectangular box with rounded corners, the results are similar to those described above, even under simple loads such as a constant internal pressure.

To account for these conditions in design and analysis, a higher order sandwich theory must be employed than that discussed above. So in the 1990s this problem was investigated, which accounted for the soft core, differing boundary conditions and behavior for the upper and lower faces. Frostig (1992, 1993a, 1993b, 1997a, 1997b, 1999), Frostig and Baruch (1990, 1993, 1995, 1996) and Thomsen and Frostig (1997) have developed a consistent rigorous closed-form, higher order theory for sandwich plates, curved panels and shells. The theory is valid for any loadings (localized or distributed). Accounts for discontinuities in load and geometry (ply drops) and includes transverse flexibility of the core. This theory has been used for buckling, vibrations, delaminations, tapered beam and stress concentration problems. It has also been used in comparisons with photoelasticity experimentally by Frostig (1999) and Thomsen and Frostig (1997) and found to be very accurate. Most recently this higher order theory has been used by Thomsen and Vinson (2000a, 2000b) to determine the behavior of non-circular boxy shells typical of a fuselage or truck tank, subjected to an internal pressure.

However, in some cases such as with localized loading where the core is compressed significantly, then according to Sikarskie and Mercado (1997), most of the common core materials behave linearly in shear initially, but then behave nonlinearly. PVC foams behave in an elastic-perfectly plastic manner, while end-grain balsa acts in a bi-linear fashion. They then studied the behavior of sandwich beams under four point bending, and studied plate behavior showing the growth of damage and the behavior with the nonlinear core materials.

Many of the modern sandwich structures involve foam cores which are compressible, and with some localized loadings, and discontinuous loads, the upper and lower faces undergo differing deformation patterns, which can lead to premature failures. Frostig (1997a) and several of his colleagues including Frostig and Baruch (1996), Patel and Frostig (1995), Frostig and Shenhar (1995) and Thomsen and Frostig (1997) have studied this type of problem extensively regarding buckling, vibrations and delaminations. Frostig points out that a stiffener edge support always causes stress concentrations that affect the faces as well as the skin-core interface for any type of loading.

Swanson and Kim (2000) recently compared solutions of beams using a higher order theory with solutions using finite elements and elasticity analyses.

Piezoelectric Considerations

Composite sandwich construction provides a unique opportunity to incorporate piezoelectric, optical and other materials for sensing, monitoring and advising regarding the "health" of the structure during manufacture and use. Thus, sandwich construction employing multifunctional materials are being developed.

Piezoelectric materials are being introduced into many structural applications today. Because of the brittleness of some of these piezoelectric materials, as well as the attendant wiring, soldered joints and other considerations, one promising use is with sandwich construction, where the piezoelectric materials and their wiring and joints can be safely contained in the core between the faces. The piezoelectric effects on a structure are analogous to thermal and hygrothermal effects analytically, and therefore can be treated analogously. Liebowitz (1993), Larson (1993), and Newill (1995) have studied these effects on elastic structures, and recently Yi et al. (1999) have studied these effects coupled with viscoelastic behavior.

Recently, Sun (1999) has studied the effects of piezoelectric (PZT) patch actuators on curved sandwich trim panels for improved acoustic control in vehicle interiors.

Most recently, a team involving Rabinovitch and Vinson (2002, 2003) and Rabinovitch, Weinacht, Bogetti, and Vinson (2004) investigated the use of piezoelectric materials in flight projectile fins to transform a ballistic projectile into a maneuvering vehicle.

Sandwich Plates

When using classical plate theory to solve problems involving sandwich plates, one may employ the usual methods used for isotropic an orthotropic monocoque plates, namely the Navier and the

Levy methods. In addition, for orthotropic sandwich plates, one may use the perturbation method (Vinson and Brull, 1962). When transverse shear deformation effects are to be included, again one can follow the approaches of Reissner and Mindlin, which have been used for fifty years. For the sandwich plate, the transverse shear stiffnesses to use should simply be the product of the transverse shear stiffness of the core in the respective directions times the core depth as shown in (14) above. This has been shown by Daniel et al. (1999) to match experimental results accurately.

In performing an analysis of a plate on an elastic foundation, such as an individual face on a core, one can consider a laminated plate on an elastic foundation, where the foundation modulus can be modeled as a Winkler foundation or a Pasternak foundation, the latter connecting the compressive-tensile springs with shear connections as well (Kerr, 1964).

Solutions to problems of isotropic and quasi isotropic sandwich plates are obtained from using either the Navier or Levy methods or from the Theorem of Minimum Potential Energy analogous to solutions of monocoque plates for the case of classical plate theory. The solutions have been catalogued for monocoque plates by Timoshenko and Woinowski-Krieger (1959) for various loads, boundary conditions, aspect ratios and material properties. The solutions have also been used for isotropic and quasi-isotropic sandwich plates (Vinson, 1998). However, for orthotropic material plates, monocoque or sandwich, such tables cannot be formulated because of the plethora of values of the flexural stiffnesses D with subscripts 1, 2, 12, and 6. However another technique can often be used, a perturbation method (Vinson and Brull, 1962), in which the series solution is expanded in terms of solutions of the isotropic plate. Only two terms of the solution are needed to obtain accurate values of stresses and deflections.

While on the subject of perturbation methods, recently a perturbation technique has been developed for mid-plane asymmetric structures (involving a B matrix) in terms of solutions of a mid-plane symmetric laminate or sandwich construction for the same problem.

The minimum weight optimization for sandwich panels subjected to lateral loads has been done for the isotropic or quasi-isotropic sandwich panel for a uniform lateral load, and various boundary conditions for classical plate theory, which is useful for preliminary design. Through this procedure a factor of merit is found for material selection. This factor of merit provides a means to compare weight savings with cost, manufacturability, etc. in making rational design decisions (Vinson and Dee, 1998).

Dobyns (1981) has provided solutions applicable to sandwich plates of composite materials including transverse shear deformation effects for simply supported panels subjected to a wide variety of dynamic lateral loads, which are most useful. These include explicit solutions for blast loads of various shapes, including, step, ramp, sinusoidal, linear decay, bilinear decay and exponential decay.

Pagano (1970) has provided an exact solution for the problem of a rectangular orthotropic sandwich plate subjected to a laterally distributed load. This solution has been used as a benchmark to which various numerical, finite difference, finite element and other approximate solutions are compared.

Sandwich Beams

All of the available beam solutions may be used to solve the problems of a sandwich beam with the same boundary conditions and loads. One must simply use the correct flexural stiffness and in-plane stiffness for the sandwich construction discussed in Eqs. (9)–(13) above. However, it must be remembered that after the deformations are solved for, one must use the correct expressions to determine the stresses.

To ascertain the natural overall frequencies and buckling loads of sandwich beams/columns, one may use all of the results from monocoque beams and columns, but must use the proper bending stiffness in the length direction as discussed above, and if including the effects of transverse shear deformation, include the transverse shear stiffness discussed above as well. The density per unit length must also be calculated since both the face and core material densities are involved. However it must be remembered that in some cases, the natural frequencies of the face, which is on an elastic foundation (the core) may be of importance along with the overall sandwich structure natural frequencies. Likewise, in buckling one must consider other failure modes, such as face wrinkling, core shear instability, and face dimpling (for honeycomb core) as well. These will be discussed later.

All of the valuable vibration results of Warburton (1968), Young and Felgar (1949) and Felgar (1959) are useful for sandwich beams, using the right stiffness and density quantities.

One very important dynamic loading unique to ship hulls is wave slamming. Allen and Shenoi (1992) have performed very extensive experimental research on sandwich beams involving two million cycles each lasting for one second, realistic of the loads on ship hull structures. Due to such slamming one important failure mode is cracking of the core due to shear stresses, generally followed by delamination. This is a rather unique, but important loading condition and requires special testing and core strength requirements.

Buene et al. (1992) has used four point bend tests of sandwich beams, following the ASTM C393 standard discussed below, to study the effects of slamming loads. They found that their core materials were both viscoelastic and strain rate dependent. Their results also agreed with those of Feichtinger (1990). Importantly, they found that they could create a master S/N fatigue curve for the foam core materials by normalizing the shear stresses to the static yield shear stress of the core material. Also in their experiments they noted no degradation of the sandwich beam properties until very shortly before failure.

Gordaninejad and Bert (1989) have used higher order theory to study the bending of sandwich beams. Hyer et al. (1976, 1989) have investigated the non-linear vibration of a sandwich beam with a viscoelastic core. Frostig, Baruch, Vilnay, and Sheinman (1991), and Satapathy and Vinson (1999) have investigated the behavior of mid-plane asymmetric sandwich beams with a flexible core.

Energy Solutions

When confronted with an analysis problem to solve involving sandwich constructions which is complicated by difficult loads, boundary conditions, changes in geometry, etc, (i.e., more complicated than a problem for which one could find the analytical solution for a monocoque isotropic structure for the same problem) it is advisable to obtain a solution utilizing the Theorem of Minimum Potential Energy. In that way one knows that a solution can be obtained, and the measure of accuracy for the solution obtained can be checked by several means, and one has retained algebraic generality. An alternative is to employ Reissner's Variational Theorm. In either case, a rudimentary knowledge of variational calculus is all that is needed, along with a little imagination to select good trial functions for the displacements. For dynamic problems The Theorem of Minimum Potential Energy can be used with Hamilton's Principle. Also thermal effects, hygrothermal effects, and piezoelectric material effects can be included. The potential energy expression to use for sandwich plates including general anisotropy, mid-plane asymmetry, thermal and hygrothermal effects, transverse shear deformation, in-plane and lateral loads is given in Eq. (6.15) of Vinson and Sierakowski (1986). It can be easily reduced for analyzing sandwich beams. For cylindrical sandwich shells, the expression to use is Eq's. 6.62 through 6.66 of the same reference.

For tapered sandwich plates with laminated composite faces and an orthotropic core, Hong and Jeon (1992) used Minimum Potential Energy and the Rayleigh-Ritz method to obtain solutions for various boundary conditions. Also Libove and Lu (1989) and Lu and Libove (1991) have studied the static and vibrational behavior of variable thickness sandwich plates to obtain solutions for aircraft empennage and control surfaces.

ASTM Standards

One useful and interesting experimental procedure is contained in ASTM Standard C393-62, in which a sandwich beam is subjected to 3-point and 4-point bending. From the load values and the mid-span deflection measurements, and the beam solution including transverse shear deformation, one can calculate the flexural stiffness D and the transverse shear modulus of the core, G_c. For instance this is very useful for assessing the quality of manufacturing, from the mean values obtained from several beams and the standard deviations obtained.

A rather complete list of all ASTM Standards for sandwich constructions is given in Appendix 3 of Vinson (1999).

Dynamic Effects—Vibrations

Seldom in real life is any structure subjected only to static loads. Most structures are subjected to many dynamic loads including vehicular, impact, crash, earthquake, handling and/or fabricating loads. In the linear elastic range dynamic effects can be divided into two categories: natural vibrations and forced vibrations. The latter can be divided into one time events or repetitive loading.

When a structure is excited repetitively at one of its natural frequencies , it takes very little input energy for the amplitude to grow until one or more of four things happens: 1) the amplitude grows until the ultimate strength of a brittle material is reached and the structure fails, 2) portions of the structure exceed the yield strength, plastically deform, and the behavior changes drastically, 3) the amplitude grows until nonlinear effects become significant, at which time there is no natural frequency, and 4) due to damping or some other mechanism the amplitude is limited, but if the structure continues to be excited, then fatigue failures will probably occur with time. In all structures, excitation of the natural frequencies is to be avoided if possible. In sandwich construction, there are overall in-plane natural frequencies, lateral (bending) natural frequencies, natural frequencies of each face (if the faces differ, as plates on an elastic foundation) and core natural frequencies. All of these frequencies should be avoided for the lower wave numbers.

Historically, Yu (1959, 1996) provided the linear equations of motion for the vibration of sandwich plates in 1959, and discussed both the linear vibration and nonlinear dynamics of sandwich plates in a text in 1996. Most recently, Yu (1997) included the effects of having piezoelectric layers in the faces of sandwich plates undergoing large deflections.

As to deflections associated with linear natural vibrations, Meyer-Piening (1992) states that for isotropic panels the maximum deflection shall not be greater than one half of the sandwich panel thickness for linear behavior to describe the motion. Experimentally, Bau-Madsen et al. (1992) found that a divergence from a linear behavior began to develop when the maximum deflection exceeded four tenths of the sandwich panel thickness.

A very important finding by Ramachandra and Meyer-Piening (1996) is the significant reduction in natural frequencies when a sandwich (or any other) panel is in contact with water on one surface. These equations should be used when designing or analyzing ship hulls.

Dobyns (1981) has provided the methods by which to calculate the natural flexural frequencies for orthotropic and isotropic sandwich plates, using the appropriate stiffness quantities, including the effects of transverse shear deformation. Over and above this he has provided forced vibration response solutions for these panels subjected to dynamic sine loads, step function loads, triangular loads, exponential decay (blast) loads, and stepped triangular loads typical of nuclear blasts. The solutions are given in easy to use Duhamel Integral formulations. Solutions for many static load problems are given as well.

Moyer et al. (1992) have analyzed the dynamic response of sandwich panels subjected to non contact underwater explosions (UNDEX). Williamson and Lagace (1993) performed an extensive study of damage to sandwich panels and compared indentations caused by static and impact loads. Sun has also studied the use of static loads to simulate the results of dynamic loading. Ferri and Sankar (1997) has also found that low velocity impacts can be approximated by considering a static indentation.

Dynamic Effects—Fatigue

The effects of fatigue in bending of sandwich panels with plastic foam cores include the research of Olsson and Lonno (1989) who found that the four point bend specimen was adequate for obtaining shear fatigue data for the core materials, Echtermeyer et al. (1991) and Buene et al. (1991) found that under constant amplitude loading the time for initiating fatigue damage was the major portion of the fatigue life, and that fatigue propagation was a rapid process. Also, they found that the specimen stiffness remained almost constant during the entire initiation process. Allen and Shenoi (1992) and Shenoi et al. (1995) also studied fatigue in bending. Burman and Zenkert (1997a, 1997b, 2000) studied fatigue in sandwich beams with cellular foam cores, and also in panels with a Nomex honeycomb core (Burman and Zenkert, 2000). In the latter they found that monitoring the stiffness variation during fatigue tests may not be a good measure of the health of the sandwich specimen.

Effects of Localized Loads

Localized loads on one face of sandwich panel can be extremely deleterious. The indentation resulting from such localized forces often negates the usual global approximation of constant thickness, constant stiffness and a fairly unstressed core. In this case the loaded face may locally undergo relatively high displacements and curvature, and the core therefore becomes compressed and sheared locally.

Thomsen (1992) states that sandwich panels are notoriously sensitive to failure by strongly localized point and line loads, resulting in premature failures. Ericsson and Sankar (1992) state that when a sandwich panel is subjected to a central concentrated load, compression of the core can be 60% of the deflection of the loaded face.

Swanson (2000) has studied the accuracy of using classical first order and higher order theories for analyzing sandwich plates subjected to a concentrated load.

Recent research in the area of localized dynamic loads includes that of Nemes and Simmonds (1992); Wu, Weeks and Sun (1995) and Wu and Sun (1996). Recently, Hasebe and Sun (2000) have found from the impact tests that they conducted that Rohacell foam reinforced web core construction shown in Fig. 4 decreases penetration damage compared to other sandwich constructions even though the surface damage may increase.

Linear and nonlinear large amplitude vibrations of composite plates with various boundary conditions including transverse shear deformation, which can be used for sandwich plates has been studied by Wu (Wu and Vinson, 1971), using Galerkin's method. As in the previous research of Yu (1962) and Ray and Bert (1969), it is assumed that the effect of coupling among individual vibration modes is not quantitatively significant, which has been confirmed by comparing analytical predictions with experimental results.

Concerning vibration damping, the texts by Nashif, Jones, and Henderson (1985) and Inman (1989) are highly recommended. Thamburaj and Sun (1999) have found that vibration and sound transmission losses can be increased significantly by using anisotropic facing and core materials compared to using isotropic materials in the same structures by inducing shear deformation and energy dissipation in the core leading to noise and vibration isolation.

Elastic Instability of Sandwich Panels

Historically there are four major textbooks dealing primarily with elastic instability or buckling. These are authored by Bleich (1952); Timoshenko and Gere (1961); Brush and Almroth (1975) and Simitses (1976). Although these texts deal with structures other than sandwich, the solutions can be applied to sandwich structures to investigate overall buckling by using the appropriate flexural stiffnesses given in Eqs. (9) through (13) above.

For the overall instability of honeycomb and solid core sandwich panels subjected to in-plane compressive loads or in-plane shear loads, the equations are given for all boundary conditions in Military Handbook 23, based upon research performed at the Forest Products Laboratory (Anon, 1955). Also, the equations to predict core shear instability are given as well, which occurs when the ratio of core shear stiffness to sandwich flexural stiffness is below a certain value. Core shear instability is only one of the local buckling phenomena that can occur.

Another local buckling that can occur is face wrinkling, there are two equations that have been given to express this type of buckling. One is by Heath (1960) which is in the form of a plate buckling type equation,: the other by Hoff and Mautner (1945) which involves the one third power of three stiffness quantities only, multiplied by a numerical coefficient of 0.5. There is still disagreement on which of these equation to use. For the Hoff-Mautner equation the numerical coefficient used by researchers, based upon various experiments (Smidt, 1996), varies between 0.5 and 0.8. Thus, it appears that the equations could be made to agree by simply adjusting the constant in the Hoff-Mautner equation. Recently, Thomsen and Mouritz (1999) have provided an equation by which to determine the numerical coefficient in the Hoff-Mautner equation. They even provide a knock-down factor to account for face-core debonding.

For honeycomb sandwich structures, either hexagonally shaped cells or the rectangular (egg-crate) type, monocell buckling or face dimpling can also occur wherein the face buckles over one cell of the honeycomb. This does not lead necessarily to panel failure but it can cause damage to the core, chip paint, trip boundary layers, in other words it should be avoided.

Unlike honeycomb core and solid or foam core sandwiches, in which one assumes that the in-plane and bending loads are resisted primarily by the faces and the core simply resists transverse shear loads (analogous to the web of an I-beam), for the truss core and web core sandwiches discussed below, the core carries a significant but measurable portion of the in-plane loads and bending loads as well. See Figs. 1.c and 1.d.

For truss core sandwich panels, the elastic and geometric constants were derived and presented by Libove and Hubka (1951) and by Anderson (1959). Overall buckling was treated by Seide (1952) and given in Mil HDBK 23, in a slightly different form, wherein the buckling coefficient is given in Figs. 2 and 4 of Seide (1952).

For a truss core sandwich, care must be taken to insure that the faces do not buckle locally, and that the core plates do not buckle as well, for both in-plane compressive loads and in-plane shear loads. This is discussed in detail in Vinson (1999).

For web core sandwich panels, the equation for overall buckling is given by Seide (1952). Again care must be taken to insure that the face plates do not buckle locally, and that the core plates do not buckle as well.

For the Rohacell type sandwich (Hasebe and Sun, 2000), both the face plates and the core plates are supported by the foam inserted in the core among the reinforcing webs, so that they behave and buckle locally as plates on an elastic foundation. Aimmanee (Aimmanee and Vinson, 2000) has performed the minimum weight optimization of this type of sandwich panel subjected to in-plane compressive loads.

Minimum weight optimization studies have been performed for the honeycomb core, the foam or solid core, the truss core and the web core sandwich panels subjected to in-plane compressive loads, and in-plane shear loads as well. In the process figures of merit were determined that are most helpful in material selection and comparison. In addition, these methods also provide the optimum stacking sequence for the face plates if a laminated construction is used. This research has appeared in many papers, but one succinct source to find all of the equations by which to analyze, design and optimize these sandwich panels is Vinson (1999).

Librescu, Hause, and Camarda (1997) developed methods of analysis for the static and dynamic behavior of curved sandwich panels subjected to complex mechanical and thermal loads in the prebuckled and postbuckled regimes, and have made comparisons with available experimental data. Smidt (1996) also investigated curved sandwich panels for high speed boats, containers, tanks and aircraft.

Recently, Niu and Talreja (1999) studied the buckling of face plates considering the core to be a Winkler foundation, and in particular investigated the effects of face-core debonding.

Laine and Rio (1992) have found that for foam core sandwich panels typical of those used in ship construction, loaded to 60% of their elastic buckling load, creep is significant (as defined by them to be a 10% influence on deformation), They therefore advocate that either a factor of safety of two on the critical buckling load, or a creep law be introduced into the structural analysis.

Sandwich Shells

Shell structures behave significantly different than plate and beam structures, in that under laterally distributed loads for example there exists a bending boundary layer in which bending stresses are superimposed upon the membrane type stresses over a small region close to any structural, load or material discontinuity. This is called the bending boundary layer. Further away from these discontinuities, the stresses are membrane only. The membrane stresses can quite often be calculated easily from equilibrium—a free body diagram. However, the bending stresses are sufficiently great that the total stress may be greater than twice the easily calculated membrane stresses. The bending layer length is on the order of four times the square root of the product of the local radius of curvature times the structural wall thickness. So for a sandwich shell the bending boundary layer is somewhat greater than it would be for the shell wall were the sum of the two face thicknesses (Vinson, 1999).

Regarding the buckling of shells, sandwich or not, the shells buckle usually at a fraction of the load predicted by standard methods of analysis, because shells are very imperfection sensitive. As a result, in spite of careful and inclusive methods of analysis developed over the decades. One must be very careful, and the predicted buckling load must be modified by an empirical factor developed from testing numerous shells.

NASA can be thanked for developing a very useful set of equations to use for the buckling of monocoque and sandwich shells, complete with equations and knockdown factors given in both graphical and equational form (Anon, 1968), Solutions are available for cylindrical shells subjected to axial compression, beam type bending, torsion, external lateral pressure, hydrostatic external pressure, transverse shear (beam type) loading and combinations of most of these leads.

Again for the buckling of sandwich shells, overall buckling solutions must also be supplemented by insuring that the sandwich structure will not buckle locally due to core shear instability, face wrinkling, or face dimpling in honeycomb core sandwich shells; the first two modes in foam or solid core sandwich shells; and face and web plate buckling in truss core and web core sandwich shells.

Again some minimum weight optimizations have been done, and some of these are given in Bitzer (1998), using the buckling analyses of Bert and his colleagues (Bert, Crisman, and Nordby, 1969; Reese and Bert, 1969).

Only recently has the behavior of shear-deformable elliptic sandwich shells been studied. The research of Birman and Simitses (2000) is part of an overall major effort to develop methods of analysis for non-circular cylindrical shells. In particular the research is concentrated on cylindrical shells in which the keel, crown and sides are flat panels joined by rounded corners which are portions of a circular cylindrical shell. The research is sponsored by the AFOSR and to date includes publications by Birman and Simitses (1998), and Thomsen and Vinson (2000a, b).

When dynamic loads are applied to the shell the analysis is complicated by inertial effects. Simitses has written numerous papers and books on this subject (Simitses, 1965, 1976; Huyan and Simitses, 1997).

Mid-Plane Asymmetric Sandwich Structures

Recently, considerable research is being conducted on the use of mid-plane asymmetric sandwich structures. Some of this research is being funded by the AFOSR in support of the Global Range Transport aircraft fuselage.

Through the use of mid-plane asymmetry considerations each face of the sandwich may be brought up to the allowable stress of that face, and the core depth can be selected to limit the maximum deflection to some predetermined value including the effects of transverse shear deformation. Thus one can utilize different materials and/or stacking sequences in each face, and can account for the difference in tensile and compressive strengths of the particular materials used. Also the materials for each face can be chosen for that particular environment to which that face is exposed. The only downside to this is that bending-stretching coupling associated with the asymmetry complicates the behavior, both in analysis and fabrication.

A mid-plane asymmetry parameter has been defined as

$$\phi = \frac{B_{11}}{\sqrt{A_{11}D_{11}}} = \frac{E_{x3}t_{f3} - E_{x1}t_{f1}}{E_{x3}t_{f3} + E_{x1}t_{f1}} = \left[\frac{-1 + \dfrac{E_{x3}t_{f3}}{E_{x1}t_{f1}}}{1 + \dfrac{E_{x3}t_{f3}}{E_{x1}t_{f1}}} \right] \tag{14}$$

where the A, B, and D are the stiffness matrix quantities defined by the following in the appropriate direction where here the inner (or lower) face is subscripted 1, the core subscripted 2 and the outer (or upper) face is subscripted 3.

It is seen that $-1 < \phi < 1$, so that this is a parameter that can be used to simplify the analysis of a mid-plane symmetric sandwich construction, and can be used to formulate a perturbation solution.

Conclusions

In the above, an attempt has been made to provide a review of several aspects of sandwich structures. In some ways, it is an update of Vinson (2001). Although not all significant research findings are included, with the references cited one can almost assuredly find all important literature on the subject. It is also remarked that there now exists a Journal of Sandwich Structures and Materials through which one may keep up to date on the latest research in this area.

The future for sandwich construction looks bright indeed. Sandwich construction will continue to be the primary structure for satellites. In aircraft, sandwich construction will be increasingly used particularly for large aircraft. Many countries are using composite sandwich constructions for their navy's ship hulls. However one of the largest uses may be for bridge constructions. Not only will it be used in those states whose Departments of Transportation (DOT) are or become knowledgeable, but there is a large international market in developing countries who may welcome the advantages, thus leapfrogging their bridge constructions into the twenty-first century without all of the conventional constructions used in the major countries today. Finally, with the growing need for alternative sources of energy, wind energy mills are being developed all of which rely heavily on sandwich constrictions.

References

Aimmanee, S., and Vinson, J.R., 2000. Analysis and Optimization of Foam Reinforced Web-Core Composite Sandwich Panels Under In-Plane Compressive Loads. Proceedings of the Fifteenth Annual Technical Conference of the American Society of Composites, September. Technomic Publishing Co. Inc. Lancaster, Pa., USA.

Allen, H.G., 1969. *Analysis and Design of Structural Sandwich Panels.* Pergamon Press, Oxford.

Allen, H.G., and Shenoi, R.A., 1992. Flexural Fatigue Tests on Sandwich Structures. Sandwich Constructions 2—Proceedings of the Second International Conference on Sandwich Construction, Gainesville, Florida. Editors D. Weissman-Berman and K-A Olsson, EMAS Publications. U.K., 499-517.

Anderson, M.S., 1959. Optimum Proportions of Truss Core and Web Core Sandwich Plates Loaded in Compression. NASA TN D-98, September.

Anon, 1955. Materials, Properties and Design Criteria, Part II, Sandwich Construction for Aircraft. *Military Handbook 23 (ANC 23),* Department of the Air Force Research and development Command, Department of the Navy, Bureau of Aeronautics, Department of Commerce, and the Civil Aeronautics Administration, Second Edition.

Anon, 1968. Buckling of Thin Walled Circular Cylinders. NASA SP-8007 Revised, August.

Bau-Madsen, N.K., Svendsen, K.H., and Kildegaard, A., 1992. Finite Deformation of Sandwich Plates. Sandwich Constructions 2—Proceedings of the Second International Conference on Sandwich Construction, Gainesville, Florida. Editors D. Weissman-Berman and K-A Olsson, EMAS Publications. U.K., 189-202.

Bert, C.W., 1991. Research on Dynamic Behavior of Composite and Sandwich Plates. Shock and Vibration Digest, **23** (6), Vibration Institute, Suite 212, 6262 S. Kingery Highway, Willowbrook, IL 60514, July, 9-21.

Bert, C.W., Crisman, W.C., and Nordby, G.M., 1969. Buckling of Cylindrical and Conical Sandwich Shells with Orthotropic Facings. *AIAA Journal,* **7,** (2), 250-257.

Bijlaard, P.P., 1951a. Analysis of Elastic and Plastic Stability of Sandwich Plates by the Method of Split Rigidities .1. *Journal of the Aeronautical Sciences,* **18** (5), 339-349.

Bijlaard, P.P., 1951b. Analysis of Elastic and Plastic Stability of Sandwich Plates by the Method of Split Rigidities .2. *Journal of the Aeronautical Sciences,* **18** (12), 790.

Bijlaard, P.P., 1952a. Analysis of Elastic and Plastic Stability of Sandwich Plates by the Method of Split Rigidities .3. *Journal of the Aeronautical Sciences,* **19** (7), 502-503.

Bijlaard, P.P., 1952b. On the Optimum Distribution of Material in Sandwich Plates Loaded in Their Plane. Proceedings of the First U.S/. Congress of Applied Mechanics, June, ASME, New York, 1952, 373-380.

Birman, V., and Simitses, G.J., 1998. Theory of Box-Type Sandwich Shells with Dissimilar facing Subjected to Thermomechanical Loads. *AIAA Journal,* **38,** 362-367.

Birman, V., and Simitses, G.J., 2000. Elliptical and Circular Cylindrical Sandwich Shells with Different Facings. *Journal of Sandwich Structures and Materials,* **2,** 152-176.

Bitzer, T.N., 1998. Recent Honeycomb Core Developments. *AIAA Journal,* **38,** 555-563.

Bleich, H.H., 1952. *Buckling of Metal Structures.* McGraw-Hill Book Co., Inc.

Brush, D.O., and Almroth, B., 1975. *Buckling of Bars, Plates and Shells.* McGraw Hill Book Co., New York.

Buene, L., Echtenmeyer, A.T., Hayman, B., and Engh, B., 1992. Shear properties of GRP Sandwich beams Subjected to Slamming Loads. Sandwich Constructions 2—Proceedings of the Second International Conference on Sandwich Construction, Gainesville, Florida. Editors D. Weissman-Berman and K-A Olsson, EMAS Publications. U.K., 737-756.

Buene, L., Echtermeyer, A.T., and Sund, O.E., 1991. Fatigue Properties of PVC Foam Core Materials. Det Norske Veritas Report, 91-2049.

Burman, M., and Zenkert, D., 1997. Fatigue of Foam Core Sandwich beams, Part 1: Undamaged Specimens. *International Journal of Fatigue*, **19**, (7), 551-561.

Burman, M., and Zenkert, D., 1997. Fatigue of Foam Core Sandwich Beams, Part II: Effect of Initial Damages. *International Journal of Fatigue*, **19**, (7), 563-578.

Burman, M., and Zenkert, D., 2000. Fatigue of Undamaged and Damaged Honeycomb Sandwich Beams. *Journal of Sandwich Structures and Materials*, **2**, 50-74.

Caprino, G., Teti, R., and Messa, M., 1996. Long Term Behavior of PVC Foam Cores for Structural Sandwich Constructions. Sandwich Construction 3Proceedings of the Third International Conference on Sandwich Construction, Southampton, Great Britain. Editor H.G.Allen, EMAS Publications, U.K., 813-824.

Cheng, S., 1962. On the Theory of Bending of Sandwich Plates, Proceedings of the Fourth U.S. National Congress of Applied Mechanics. ASME. New York, 511-518.

Chevalier, J.L., 1992. Creep and Fatigue Properties of End-Grain Balsa and Other Typical Sandwich Cores. Sandwich Constructions 2—Proceedings of the Second International Conference on Sandwich Construction, Gainesville, Florida. Editors D. Weissman-Berman and K-A Olsson, EMAS Publications. U.K., 519-539.

Christensen, R.M., 1979. *Mechanics of Composite Materials*, John Wiley and Sons. New York.

Christensen, R.M., and Kim, B., 1999. Properties of Core Materials for Sandwich Structures, Extended Abstracts. 12th International Conference on Composite Materials, Editors T. Massard and A. Vautrin, 195.

Czaplicki, R.M., 1992. Cellular Core Structures Providing Grid-like Bearing Surfaces on Opposing Parallel Planes of the Core. Sandwich Constructions 2—Proceedings of the Second International Conference on Sandwich Construction, Gainesville, Florida. Editors D. Weissman-Berman and K-A Olsson, EMAS Publications. U.K., 721-736.

Daniel, I.M., Abot, J.L., and Wang, K.A., 1999. Testing and Analysis of Composite Sandwich Beams. Extended Abstract Proceedings of the 12th international Conference on Composite Materials, EMAS Publication, U.K., 197.

Daniel, I.M., La Bedz, R.H., and Liber, T., 1981. New Method for Testing Composites at Very high Strain Rates. *Experimental Mechanics*, 71-72.

Davies, J.M., 1977. Design Criteria for Sandwich Panels for Building Construction. Proceedings of the ASME Aerospace Division, Structures and Materials Committee, ASME-AD **55**, New York, 273-284.

Dobyns, A.L., 1981. The Analysis of Simply Supported Orthotropic Plates Subjected to Static and Dynamic Loads. *AIAA Journal*, 642-650.

Echtermeyer, A.T., Buene, L., McGeorge, D., and Sund, O.E., 1991. Four Point Bend Testing of GRP Sandwich Beams-Part 1. Det Norske Veritas Report, VR-91-P0013.

Eckel, D.A. II, 2000. An All Fiber-Reinforced-Polymer-Composite Bridge: Design, Analysis, Fabrication, Full-Scale Experimental Structural Validation, Construction and Erection, PhD Dissertation, Civil Engineering. University of Delaware.

Ericksen, W. S., 1955. Effects of Shear Deformation in the Core of a Flat Rectangular Sandwich Panel: 1, Buckling Under Compressive End Load; 2, Deflection Under Uniform Transverse Load, Forest Products Laboratory Report 1583, August.

Ericsson, A., and Sankar, B.V., 1992. Contact Stiffness of Sandwich Plates and Applications to Impact Problems. Sandwich Constructions 2—Proceedings of the Second International conference on Sandwich Construction, Gainesville, Florida. Editors D. Weissman-Berman and K-A. Olsson, EMAS Publications, U.K., 139-159.

Eringen, A.C., 1952. Bending and Buckling of Rectangular Sandwich Plates. Proceedings of the First U.S. National Conference of Applied Mechanics, ASME, New York, 381-390.

Fairbairn, W., 1849. *An Account of the Construction of the Brittania and Conway Tubular Bridges.* John Weale, London.

Feichtinger, K.A., 1988. Test Methods and Performance of Structural Core Materials-1. Static Properties. 4th ASM International/ Engineering Society of Detroit Advanced Composites Conference and Exposition, September 13-15.

Feichtinger, K.A., 1990. Test Methods and Performance of Structural Core materials-IIA Strain Rate dependence of Shear properties. 5th Autumn INERN Conference, Composite Materials and Sandwich Structures, Lorient, October.

Feichtinger, K.A., and Martensen, S., 1996. A Comparison of the Creep Behavior of Aluminum and Nomex Honeycombs, PVC and Balsa. Sandwich Construction 3—Proceedings of the Third International Conference on Sandwich Construction, Southampton, Great Britain. Editor H.G. Allen, EMAS Publications. U.K., 801-802.

Felgar, R.F. Jr., 1959. Formulas for Integrals Containing Characteristic Functions of a Vibrating Beam. The University of Texas Bureau of Engineering Research Circular, **14**.

Ferri, R., and Sankar, B.V., 1997. Static Indentation and Low Velocity Impact Tests on Sandwich Plates. *ASME Aerospace Division Publication AD-55*, Analysis and Design Issues for Modern Aerospace Vehicles, 485-490.

Flugge, W., 1949. Determination of Optimum Dimensions of Sandwich Panels. *La Recherche Aeronautique*, **7**, Jan./Feb.

Flugge, W., 1952. The Optimum Problem of the Sandwich Plate. *Journal of Applied Mechanics*, **19**, (1) 104-108.

Frostig, Y., 1992. Behavior of Delaminated Sandwich Beams with Transversely Flexible Core—High Order Theory. *Composite Structures*, **20**, 1-16.

Frostig, Y., 1993a. Higher Order Behavior of Sandwich Beams with Flexible Core and Transverse Diaphragms. *Journal of the ASCE Engineering Mechanics Division*, **119**, (5), 955-972.

Frostig, Y., 1993b. On Stress Concentrations in the Bending of Sandwich Beams With Transversely Flexible Core. *Composite Structures*, **24**, 161-169.

Frostig, Y., 1997a. Hygrothermal (Environmental) Effects in High-Order Bending of Sandwich Beams With a Flexible Core and a Discontinuous Skin. *Composite Structures*, **37**, (2), 205-221

Frostig, Y., 1997b. Bending of Curved Sandwich Panels With Transversely Flexible Cores-Closed Form Higher Order Theory. *Composite Structures*, **37**, (2), 335-354.

Frostig, Y., 1999. Bending of Curved Sandwich Panels With Transversely Flexible Core-Closed Form High-Order Theory. *Journal of Sandwich Structures and Materials*, **1**, 4-41.

Frostig, Y., and Baruch, M., 1990. Bending of Sandwich Beams with Transversely Flexible Core. *AIAA Journal*, **28**, (11), 523-531.

Frostig, Y., and Baruch, M., 1993. High-Order Buckling Analysis of Sandwich Beams with Transversely Flexible Core. *Journal of the ASCE Engineering mechanics Division*, **119**, (3), 476-495.

Frostig, Y., and Baruch, M., 1994. Free Vibrations of Sandwich Beams with a Transversely Flexible Core: A High order Approach. *Journal of Sound and Vibration*, **176**, (2), 195-208.

Frostig, Y., and Baruch, M., 1996. Localized Load Effects in Higher Order Bending of Sandwich Panels With Transversely Flexible Core. *Journal of the ASCE Engineering Mechanics Division*, **122**, (11), 1069.

Frostig, Y., Baruch, M., Vilnay, O., and Sheinman, I., 1991, Sandwich Beams With Unsymmetrical Skins and A Flexible Core-Bending Behavior. *Journal of the ASCE Engineering Mechanics Division*, **117**, 1931-1952.

Frostig, Y., and Shenhar, I., 1995. High-Order Bending of Sandwich Beams With a Transversely Flexible Core and Unsymmetrical Composite Skins. *Composites Engineering*, **5**, (4), 405-414.

Gellhorn, E.V., and Reif, G., 1992. Sandwich Constructions 2—Proceedings of the Second International Conference on Sandwich Construction, Gainesville, Florida. Editors D. Weissman-Berman and K-A Olsson, EMAS Publications. U.K., 541-557.

Gellhorn, E.V., and Starlinger, A., 1992. An Efficient Extender Procedure for Creep Parameters of Structural Foams. Sandwich Constructions 2—Proceedings of the Second International Conference on Sandwich Construction, Gainesville, Florida. Editors D. Weissman-Berman and K-A. Olsson, EMAS Publications, U.K., 779-788.

Gerard, G., 1956. *Minimum Weight Analysis of Compressive Structures*. New York University Press, New York.

Gordaninejad, F., and Bert, C.W., 1989. A New Theory for Bending of Thick Sandwich Beams. *International Journal of Mechanical Science*, **31**, 925-934.

Ha, K.H., 1989. Finite Element and Sandwich Construction: A Critical Review. Sandwich Constructions 1—Proceedings of the First International Conference on Sandwich Constructions, Editors K-A Olsson and R.P. Reichard, EMAS Publication, U.K., 69-84.

Hahn, H.T., 1980. Simplified Formulas for Elastic Moduli of Unidirectional Continuous Fiber Composites. *Composite Technology Review*, **2**, (3), 5-7.

Halpin, J.C., and Tsai, S.W., 1967. Environmental Factors in Composite Materials Design. Air Force Materials Laboratory Technical Report, 67-423.

Hasebe, S.R., and Sun, C.T., 2000. Performance of Sandwich Structures with Composite Reinforced Core. *Journal of Sandwich Structures and Materials*, **2**, 75-100.

Hashin, Z., 1972. Theory of Fiber Reinforced Materials. NASA Contractor's Report, 1974.

Heath, W.G., 1960. Sandwich Construction, Part 1: The Strength of Flat Sandwich Panels. *Aircraft Engineering*, **32**, 186-191.

Heath, W.G., 1960. Sandwich Construction, Part 2: The Optimum Design of Flat Sandwich Panels. *Aircraft Engineering*, **32**, 230-235.

Hoff, N.J., 1950. Bending and Buckling of Rectangular Sandwich Plates. NACA Technical Note 2225, November.

Hoff, N.J., 1986. *Sandwich and Composite Aerospace Structures*. Technomic Publishing Company, Lancaster, Pa.

Hoff, N.J., and Mautner, E., 1945. Buckling of Sandwich Type Panels. *Journal of the Aeronautical Sciences*, **12**, (3), 285-297.

Hong, C.S., and Jeon, J.S., 1992. An Approximate Solution for Tapered Sandwich Plate With Composite Laminated faces. Sandwich Constructions 2—Proceedings of the Second International Conference on Sandwich Construction, Gainesville, Florida. Editors D. Weissman-Berman and K-A Olsson, EMAS Publications. U.K., 35-47.

Huyan, X., and Simitses, G.J., 1997. Dynamic Buckling of Imperfect Cylindrical Shells Under Axial Compression and Bending Moment. *AIAA Journal*, **35**, (8), 1404-1412.

Hyer, M.W., Anderson, W.J., and Scott, R.A., 1976. Non-Linear Vibrations of Three-Layer Beams with Viscoelastic Cores Part 1-Theory. *Journal of Sound and Vibration*, **46**, (1), 121-136.

Hyer, M.W., Anderson, W.J., and Scott, R.A., 1989. Non-Linear Vibrations of Three-Layer Beams with Viscoelastic Cores Part 2-Experiment. *Journal of Sound and Vibration*, **61**, (1), 25-30.

Inman, D.J., 1989. *Vibration and Control Measurement and Stability*. Prentice Hall, Englewood Cliffs, NJ.

Jurf, R.A., and Vinson, J.R., 1985. Effect of Moisture on the Static and Viscoelastic Shear Properties of Epoxy Adhesives. *Journal of Materials Science*, **20**, 2979-2989.

Kaechele, L.E., 1957. Minimum Weight Design of Sandwich Panels. USAF Project Rand Research memorandum, RM 1895, AD-133011, March.

Karbhari, V.M., 1997. Application of Composite Materials to the Renewal of Twenty First Century Infrastructure. Proceedings of the Eleventh International Conference of Composite Materials, July, Australian Composite Structures Society, RMIT, Melbourne.

Kerr, A.D., 1964. Elastic and Viscoelastic Foundation Models. *Journal of Applied Mechanics*, **31**, 491-498.

Kujala, K., and Tuhkari, J., 1996. All-Steel Corrugated-Core Sandwich Panels for Ship Structures. Sandwich Construction 3—Proceedings of the Third International Conference on Sandwich Construction, Southampton, Great Britain. Editor H.G. Allen, EMAS Publications, U.K., 411-422.

Laine, C., and Rio, G., 1992. Buckling of Sandwich Panels Used in Shipbuilding: Experimental and Theoretical Approach. Sandwich Constructions 2—Proceedings of the Second International Conference on Sandwich Construction, Gainesville, Florida. Editors D. Weissman-Berman and K-A Olsson, EMAS Publications. U.K., 243-252.

Larson, P.H., 1993. The Use of Piezoelectric Materials in Creating Adaptive Shell Structures. PhD Dissertation, Mechanical Engineering, University of Delaware.

Leibowitz, M., 1993. Studies on the Use of Piezoelectric Actuators in Composite Sandwich Constructions. PhD Dissertation, Mechanical Engineering, University of Delaware.

Libove, C., and Batdorf, S.B., 1948. A General Small Deflection Theory for Sandwich Plates. NACA Report 899, Washington, D.C.

Libove, C., and Hubka, R.E., 1951. Elastic Constants for Corrugated-Core Sandwich Plates. NACA TN 2289.

Libove, C., and Lu, C.H., 1989. Beam Type Bending of Variable Thickness Sandwich Plates. *AIAA Journal*, **27**, (4), 500-507.

Librescu, L., IIause, T., and Camarda, C.J., 1997. Geometrically Nonlinear Theory of initially Imperfect Sandwich Curved Panels Incorporating Non-classical Effects. *AIAA Journal*, **35**, (8), 1393-1403.

Lindholm, V.S., 1964. Some Experiments With the Split Hopkinson Pressure Bar. *Journal of Mechanics and Physics of Solids*, **12**, 813-824.

Lonno, A., and Hellbratt, S.A., 1996. Use of Carbon Fibre in a 63M High Speed Vessel, YS2000, for the Swedish Navy. Sandwich Construction 3—Proceedings of the Third International Conference on Sandwich Construction, Southampton, Great Britain. Editor H.G. Allen, EMAS Publications, U.K., 3-13.

Lu, C.H., and Libove, C., 1991. Beam Like Harmonic Vibration of Variable Thickness Sandwich Plates. *AIAA Journal*, **29**, (2), 299.

March, H.W., 1952. Behavior of a Rectangular Sandwich Panel Under Uniform Lateral Load and a Compressive Edge Load. Forest Products Laboratory Report 1834, September.

Marguerre, K., 1944. The Optimum Buckling Load of a Flexibly Supported Plate Composed of Two Sheets Joined by a Light Weight Filler When Under Longitudinal Compression. Deutsche Viertaljahrsschrist fur Literalurwissenschaft und Giests Geschichte, D.V.L. (ZWB UM 1360/2) 28 October.

McCoy, T.T., Vinson, J.R., and Shore, S., 1967. A Method for Weight Optimization of Flat Truss Core Sandwich Panels Under Lateral Loads. Naval Air Engineering Center Report NAEC-ASC.1111, July.

Meyer-Piening, H.-R., 1992. Nonlinear Investigation of Deflection, Stress and Frequencies of a Rectangular Thin Sandwich Plate. Sandwich Constructions 2—Proceedings of the Second International Conference on Sandwich Construction, Gainesville, Florida. Editors D. Weissman-Berman and K-A Olsson, EMAS Publications. U.K., 189-202.

Moyer, E.T. Jr., Amir, G.G., Olsson, K.-A., and Hellbratt, S.E., 1992. Response of GRP Sandwich Structures Subjected to Shock Loading. Sandwich Constructions 2—Proceedings of the Second International Conference on Sandwich Construction, Gainesville, Florida. Editors D. Weissman-Berman and K-A Olsson, EMAS Publications. U.K., 49-65.

Nashif, A.D., Jones, D.I.G., and Henderson, J.P., 1985. *Vibration Damping*. Wiley Interscience, New York.

Nemes, J.A., and Simmonds, K.E., 1992. Low Velocity Impact of Foam-Core Sandwich Composites. *Journal of Composite Materials*, **26**, (4), 500-519.

Newill, J.F., 1995. Composite Sandwich Structures Incorporating Piezoelectric Materials. PhD Dissertation, Mechanical Engineering, University of Delaware.

Nicholas, T., 1981. Tensile Testing of Materials at High Rates of Strain. *Experimental Mechanics*, 177-185.

Niu, K., and Talreja, R., 1999. Buckling of a Thin Face Layer on Winkler Foundation with Debonds. *Journal of Sandwich Structures and Materials*, **1**, 259-278.

Noor, A.K., Burton, W.S., and Bert, C.W., 1996. Computational Models for Sandwich Panels and Shells. *Applied Mechanics Reviews*, **49**, (3), 155-199.

Noor, A.K., Starnes, J.H. Jr., and Peters, J.M., 1997. Curved Sandwich Panels Subjected to Temperature Gradient and Mechanical Loads. *Journal of the Aerospace Division*, ASCE, **10**, (4), 143-161.

Noor, A.K., Starnes, J.H. Jr., and Peters, J.M., 2000. Recent Advances and Future Trends in Composite Materials and Structures. *Computer methods in Applied Mechanics and Engineering*, **185**, (2-4), 413-432.

Olsson, K.-A., and Lonno, A., 1989. Test Procedures for Foam Core Materials. Sandwich Constructions 1—Proceedings of the First International Conference on Sandwich Constructions, Eds. K-A Olsson and R.P. Reichard, EMAS Publication, U.K., 293-318.

Pagano, N.J., 1970. Exact Solutions for Rectangular Bi-directional Composite and Sandwich Plates. *Journal of Composite Materials*, **4**, 20-34.

Patel, D., and Frostig, Y., 1995. High Order Bending of Piecewise Uniform Sandwich Beams with a Tapered Transition Zone and a Transversely Flexible Core. *Composite Structures*, **31**, 151-162.

Plantema, F.J., 1966. *Sandwich Construction: The Bending and Buckling of Sandwich Beams, Plates and Shells*. John Wiley and Sons, New York.

Powers, B.M., Vinson, J.R., Hall, I.W., and Hubbard, R.F., 1997. High Strain Rate Properties of Cycom 5920/1582 Cloth Glass/Epoxy Composites. *AIAA Journal*, **35**, (3), 553-556.

Preissner, E.C., and Vinson, J.R., 1998. Circular Cylindrical Sandwich Shells With Mid-Plane Asymmetry Subjected to Axially Symmetric Loads. ASME-AD, **56**, Recent Advances in Mechanics of Aerospace Structures and Materials —1998, ed. B.V. Sankar, 245-252.

Rabinovitch, O., and Vinson, J.R., 2002. Analyses and Design Study of Piezoelectric Twist Actuation of Smart Fins. Proceedings of the Seventeenth Annual Technical Conference of the American Society for Composite, Purdue, October.

Rabinovitch, O., and Vinson, J.R., 2003. Analyses and Design of Sandwich Piezoelectric Actuators. Proceedings of the Sixth International Conference on Sandwich Structures, Ft. Lauderdale, March 31 – April 2, 817-826.

Rabinovitch, O., Weinacht, P., Bogetti, T., and Vinson, J.R., 2004. Analyses, Design and Structural – Aerodynamic Coupling of a Piezoelectric Actuator. Proceedings of the 45th AIAA/ASME/ASCE/AHS/ASC Structures, Structural Dynamics and Materials Conference, Palm Springs.

Ramachandra, L.S., and Meyer-Piening, H.-R., 1996. Dynamic Analysis of Sandwich Structures in Contact With Water. Sandwich Construction 3—Proceedings of the Third International Conference on Sandwich Construction, Southampton, Great Britain. Editor H.G. Allen, EMAS Publications, U.K., 363-375.

Ray, J.D., and Bert, C.W., 1969. Nonlinear Vibrations of a Beam with Pinned Ends. *Journal of Engineering for Industry*, 997.

Raville, M.E., 1955. Deflection and Stresses in a Uniformly Loaded Simply Supported, Rectangular Sandwich Plate. Forest products laboratory Report 1847.

Reese, C.D., and Bert, C.W., 1969. Simplified Design Equations for Buckling of Axially Compresses Sandwich Cylinders with Orthotropic Facings and Core. *AIAA Journal of Aircraft*, 515-519.

Rheinfrank, G.B., and Norman, W.A., 1944. Molded Glass Fiber Sandwich Fuselage for the BT-15 Airplane. Army Air Corps Technical Report No. 5159, November 8.

Robson, B.L., 1989. The Royal Australian NAVY Inshore Minehunter—Lessons Learned. Sandwich Constructions 1—Proceedings of the First International Conference on Sandwich Constructions, Eds. K-A Olsson and R.P. Reichard, EMAS Publication, U.K., 395-423.

Satapathy, N.R., and Vinson, J.R., 1999. Sandwich Beams With Mid-Plane Asymmetry Subjected to Lateral Loads-Analysis and Optimization. Extended Abstract Proceedings of the 12th International conference on Composite Materials, EMAS Publication, U.K, 91.

Seide, P., 1952. The Stability Under Longitudinal Compression of Flat Symmetric Corrugated-Core Sandwich Plates with Simply Supported Loaded Edges and Simply Supported or Clamped Unloaded Edges. NACA TN 2679.

Shenoi, R.A., Clark, S.D., and Allen, H.G., 1995. Fatigue Behavior of Polymer Composite Sandwich Beams. *Journal of Composite Materials*, **29**, (18), 2423-2445.

Shenoi, R.A., Clark, S.D., Allen, H.G., and Hicks, I.A., 1996. Steady State creep Behavior of FRP Foam Cored Sandwich Beams. Sandwich Construction 3—Proceedings of the Third International Conference on Sandwich Construction, Southampton, Great Britain. Editor H.G. Allen, EMAS Publications, U.K., 789-799.

Sierakowski, R.L., 1997. Strain Rate Effects in Composites. *Applied Mechanics Reviews*, **50**, (12, Pt 1), 741-761.

Sierakowski, R.L., and Chaturvedi, S.K., 1997. *Dynamic Loading and Characterization of Reinforced Composites*. Wiley Interscience.

Sikarskie, D., and Mercado, 1997. Analysis and Design Issues for Modern Aerospace Vehicles. Proceedings of the 1997 ASME IMECE, Aerospace Division, AD-**55**, 365-377.

Simitses, G.J., 1965. Dynamic Snap-Through Buckling of Low Arches and Shallow Spherical Caps. PhD Dissertation, Aeronautics and Astronautics, Stanford University, June.

Simitses, G., 1976. *Introduction to Elastic Stability of Structures*. Prentice Hall, Inc., Englewood Cliffs, N.J.

Simitses, G.J., 1987. Instability of Dynamically-Loaded Structures. *Applied Mechanics Reviews*, **40**, (10), 1403-1408.

Smidt, S., 1996. Testing of Curved Sandwich Panels and Comparison With Calculations Based on the Finite Element Method. Sandwich Construction 3—Proceedings of the Third International Conference on Sandwich Construction, Southampton, Great Britain. Editor H.G. Allen, EMAS Publications, U.K., 665-680.

Starlinger, A., and Reif, G., 1996. Sandwich Design of Lightweight Bus Structures-Engineering and Cost Saving Aspects. Sandwich Construction 3—Proceedings of the Third International Conference on Sandwich Construction, Southampton, Great Britain. Editor H.G. Allen, EMAS Publications, U.K., 103-115.

Sun, J.-Q., 1999. An Impedance Study of Curved Sandwich Trim Panels Driven by Piezoelectric Patch Actuators. *Journal of Sandwich Structures and Materials*, **2**, 128-146.

Swanson, S.R., 2000. Response of Orthotropic Sandwich Plates to Concentrated Loads. *Journal of Sandwich Structures and Materials*, **2**, 270-287.

Swanson, S.R., and Kim, J., 2000. Comparison of a Higher Order Theory for Sandwich beams With Finite Element and Elasticity Analyses. *Journal of Sandwich Structures and Materials*, **2**, (1), 33-49.

Thamburaj, P., and Sun, J.-Q., 1999. Effect of material Anisotropy on the Sound and Vibration Transmission Loss in Sandwich Aircraft Structures. *Journal of Sandwich Structures and Materials*, **1**, 76-92.

Thomsen, O.T., 1992. Analysis of Local Bending Effects in Sandwich Panels Subjected to Concentrated Loads. Sandwich Constructions 2—Proceedings of the Second International Conference on Sandwich Construction, Gainesville, Florida. Editors D. Weissman-Berman and K-A Olsson, EMAS Publications. U.K., 417-440.

Thomsen, O.T., and Frostig, Y., 1997. Localized Bending Effects in Sandwich Panels: Photoelastic Investigation vs. High-Order Sandwich Theory Results. *Composite Structures*, **37**, (1), 97-108.

Thomsen, R.S., and Mouritz, A.P., 1999. Skin Wrinkling of Impact Damaged Sandwich Composites. *Journal of Sandwich Structures and Materials*, **1**, 299-322.

Thomsen, O.T., and Vinson, J.R., 2000a. Design Study of a Non-Circular Pressurized Sandwich Fuselage Section Using a High-Order Sandwich Theory Formulation. Proceedings of the Fifth International Conference on Sandwich Constructions, Zurich, September, 2000, EMAS Publishing, Ltd. UK.

Thomsen, O.T., and Vinson, J.R., 2000b. Comparative Study of two Different Conceptual Design principles for Non-Circular Pressurized Fuselage Sections Using a High Order Sandwich Theory. ASME-IMECE, Orlando, November.

Timoshenko, S., and Gere, J., 1961. *Theory of Elastic Stability, 2nd Edition*. McGraw Hill Book Co., New York.

Timoshenko, S., and Woinowski-Krieger, A., 1959. *Theory of Plates and Shells, 2nd Edition*. McGraw-Hill Book Co., Inc. New York.

Vinson, J.R., 1998. Minimum Weight Optimization of Composite Sandwich Panels for Lateral Loading. ECCM-8 European Conference on Composite Materials Science, Technologies and Applications, Ed, I Crivelli-Visconti, Woodhouse publishing Limited, 1, 583-590.

Vinson, J.R., 1999. *The Behavior of Sandwich Structures of Isotropic and Composite Materials*. Technomic Publishing Company, Lancaster, Pa.

Vinson, J.R., 2001. Sandwich Structures. *Applied Mechanics Reviews*, 201-214.

Vinson, J.R., and Brull, M.A., 1962. New Techniques of Solution for Problems in the Theory of Orthotropic Plates. Transactions of the Fourth U.S. National Conference of Applied Mechanics, 2, 817-825.

Vinson, J.R., and Dee, A.T., 1998. Use of Asymmetric Sandwich Construction to Minimize Bending Stresses. Proceedings of the Fourth International Conference on Sandwich Constructions, Stockholm, June, 1, EMAS Publishing, Ltd. UK., 391-402

Vinson, J.R., and Shore, S., 1965a. Bibliography on Methods of Structural Optimization for Flat Sandwich Panels. Naval Air Engineering Center Report, NAEC-ASC-1082, April 15.

Vinson, J.R., and Shore, S., 1965b. Methods of Structural Optimization for Flat Sandwich Panels. Naval Air Engineering Center Report NAEC-ASC-1083, April 15.

Vinson, J.R., and Shore, S., 1965c. Design procedures for the Structural Optimization of Flat Sandwich Panels. Naval Air Engineering Center Report NAEC-ASC-1084, April 15.

Vinson, J.R., and Shore, S., 1967a. Structural Optimization of Corrugated and Web Core Sandwich panels Subjected to Uniaxial Compression. Naval Air Engineering Center Report NAEC-ASC 1109.

Vinson, J.R., and Shore, S., 1967b. Structural Optimization of Flat Corrugated Core Sandwich Panels under In-Plane Shear Loads and Combined Uniaxial Compression and In-Plane Shear Loads. Naval Air Engineering Center Report NAEC-ASC-1110, July.

Vinson, J.R., and Sierakowski, R.L., 1986. *The Behavior of Structures Composed of Composite Materials*. Martinus-Nijhoff Publishers (now Kluwer Academic Publishers), Dordrecht, 178-179.

Warburton, G., 1968. *The Vibration of Rectangular Plates, Proceedings of the Institute of Mechanical Engineers*. Institute of Mechanical Engineers, London, 371.

Weitsman, Y.J., 2000. Effects of Fluids on Polymeric Composites-A Review, Comprehensive Composite Materials. Eds. In Chief A. Kelly and C. Zweben, 2, *Polymeric Matrix Composites*, Eds R. Talreja and J.-A. E. Manson, Chapter 11, Elsevier Publishing, 369-401.

Williamson, J.E., and Lagace, P.A., 1993. Response Mechanisms in the Impact of Graphite/Epoxy Honeycomb Sandwich Panels. Proceedings of the Eighth Technical Conference of the American Society for Composites, Technomic Publishing Company, Lancaster, Pa., 287-297.

Woldesenbet, E., and Vinson, J.R., 1996. Sandwich Composite Structures for Low-Cost and Emergency Housing. Sandwich Construction 3—Proceedings of the Third International Conference on Sandwich Construction, Southampton, Great Britain. Editor H.G. Allen, EMAS Publications, U.K., 61-70.

Woldesenbet, E., and Vinson, J.R., 1999. Specimen Geometry Effects on High Strain Rate Testing of Graphite/Epoxy Composites. *AIAA Journal*, **37**, (9), 1102-1106.

Wu, C.L., and Sun, C.T., 1996. Low-Velocity Impact Damage in Composite Sandwich Beams. *Composite Structures*, **34**, 21-27.

Wu, C.L., and Vinson, J.R., 1971. Nonlinear Oscillations of Laminated Specially Orthotropic Plates With Clamped and Simply Supported Edges. *Journal of the Acoustical Society of America*, **49**, 1561-1567.

Wu, C.L., Weeks, C.A., and Sun, C.T., 1995. Improving Honeycomb Core Sandwich Structures for Impact Resistance. *Journal of Applied Materials*, 41-47.

Yi, S., Ling, S.F., Ying, M., Hilton, H.H., and Vinson, J.R., 1999. Finite Element Formulation for Anisotropic Coupled Piezo-hygro-thermo-viscoelastodynamic Problems. *International Journal for Numerical Methods in Engineering*, **45**, 1531-1546.

Young, D., and Felgar, R.F. Jr., 1949. Tables of Characteristic Functions Representing Normal Modes of Vibration for a Beam. The University of Texas Engineering research Series report, **44**, July 1.

Yu, Y.Y., 1962. Nonlinear Flexural Vibrations of Sandwich Plates. *Journal of the Acoustical Society of America*, **34**, 1176-1183.

Yu, Y.Y., 1959. A New Theory of Elastic Sandwich Plates-One Dimensional Case. *Journal of Applied Mechanics*, **26**, 415-421.

Yu, Y.Y., 1996. *Vibrations of Elastic Plates: Linear and Nonlinear Dynamic Modeling of Sandwiches, Laminated Composites, and Piezoelectric Layers*. Springer-Verlag, New York.

Yu, Y.Y., 1997. Dynamics for Large Deflections of a Sandwich Plate With Thin Piezoelectric Face Layers. ASME AD-**55,** *Analysis and Design Issues for Modern Aerospace vehicles*, Ed. G. J. Simitses, 285-292.

Zenkert, D., 1995. *An Introduction to Sandwich Construction*. EMAS Publication, West Midlands, UK.

Zukas, J.A., 1982. *Impact Dynamics*. John Wiley and Sons, New York.

The Evolution of the Concept of Green's Function and Its Use in the Mechanics of Solids

Ivar Stakgold, University of Delaware

Abstract

This paper is a somewhat expanded version of my talk at the Kerr Symposium in April 2004. Its title, provided some months before the Symposium, is more ambitious than its actual contents, as will soon become apparent to the reader.

Introduction

The paper is comprised of several sections. *Green's Function and Generalized Functions* reviews the principal approaches for finding a common mathematical framework to handle both concentrated and distributed sources. These efforts culminated in the Theory of Distributions (Laurent Schwartz, 1950-51), which is now the accepted standard.

In *Examples of Green's Function and Fundamental Solutions*, we formulate the *Dirichlet* problem on a bounded domain for the Green functions of the negative *Laplacian* and of the biharmonic operator (characterizing the transverse deflection of a clamped thin plate under a unit concentrated load). The solution of the associated boundary value problem (BVP) with distributed source and inhomogeneous boundary conditions can then be written explicitly as a sum of integrals involving Green's function and the given data. Since Green's function is usually difficult to calculate, we turn to the related concept of *fundamental solution* (or free-space Green's function) which is also the response to a concentrated unit source, but now without boundary conditions (except, perhaps, at infinity). The fundamental solution enables us to translate the associated inhomogeneous BVP into a boundary integral equation for which numerical methods are available. We give some examples of simple fundamental solutions including the one for the biharmonic operator in R^2. Kerr and various collaborators have studied more complicated problems such as thin plates on various types of foundations.

The Lamé System deals with the Lamé system of 3-dimensional elastostatics. We follow closely the recent paper of Hsiao and Wendland (2004). Since the applied force and the displacement are now vector functions, Green's function is replaced by Green's matrix and the fundamental solution will also be a matrix. The fundamental solution for the Lamé system can be calculated explicitly and permits us to reformulate the inhomogeneous BVP as a boundary integral equation (BIE). The BIE, in turn, is the basis for the Boundary Element Method (BEM) which is a powerful numerical method for solving the BIE and the original BVP. The reader is referred to Hsiao and Wendland (2004) for a systematic and rigorous treatment of the BEM.

In *Nonpositivity of Green's Function for the Clamped Plate*, we return to the clamped plate problem and show that, for a sufficiently eccentric ellipse, Green's function takes on negative values at some points within the ellipse. We use the remarkably simple proof of Shapiro and Tegmark (1994).

Green's Function and Generalized Functions

A Green's function represents the "response" to a concentrated unit source (which, depending on the physical context, might be a force, an impulse, a charge, a heat source, ….). It is a function of two (vector) variables; the point x where the response is measured and the point y where the source is located. As a simple example, consider steady heat conduction in a homogeneous solid occupying the domain Ω in R^3 whose boundary $\partial\Omega$ is maintained at zero temperature; Green's function $g(x,y)$ is then the temperature at $x\in\Omega$ due to a unit source of heat placed at $y\in\Omega$. One can use g to find the temperature $u(x)$ in Ω under the influence of sources distributed with a volume density $f(x)$, the boundary being kept at zero temperature. Energy conservation shows that $u(x)$ satisfies

$$-\nabla^2 u = f(x) \qquad x\in\Omega; \qquad u\,|_{\partial\Omega} = 0 \tag{1}$$

where ∇^2 is the Laplacian and the thermal conductivity has been set equal to one. Replacing the distributed source $f(x)$ by small concentrated sources, superposition leads to

$$u(x) = \int_\Omega g(x,y)f(y)dy$$

as the candidate for the solution of Eq. (1).

To make this argument rigorous, we need to write the BVP for g in the same format as Eq. (1). Intuitively, we may regard the unit concentrated source at y as a density $\delta(x-y)$, where $\delta(x)$ is Dirac's delta "function" which vanishes for $x\neq0$ and is infinite at the origin so that its integral over all space is one. Of course, no such function exists in the classical sense; to cope with this problem, we must generalize the concept of function. Such a generalization should also enable us to characterize other singular sources such as multipoles and surface layers.

Next, we describe four approaches to this question. Each of these has its own intuitive basis, but it is the last one – Schwartz's Theory of Distributions - which has been the most successful as it encompasses the others and is best suited to the study of differential equations and Fourier analysis.

1.   A generalized function is the limit of a sequence of well-behaved functions (Kirchhoff, Dirac, Lighthill, Temple.) For instance, the Dirac delta function $\delta(x)$ representing a unit source at the origin in R^1 is regarded as the limit of a sequence:

$$\delta(x) = \lim_{n\to\infty} \frac{n}{\pi(1+n^2x^2)} \tag{2}$$

This sequence of continuous functions is, for large n, highly peaked at the origin with unit integral on the real line. The rigorous interpretation of Eq. (2) is that for all bounded, continuous $\varphi(x)$

$$\lim_{n \to \infty} \frac{n}{\pi(1 + n^2 x^2)} \varphi(x) dx = \varphi(0),$$

so that, as n tends to ∞, the sequence $n/\pi (1+n^2x^2)$ has the sifting property expected of the delta function. Many other sequences have the same property and must therefore be regarded as equivalent. A useful reference to this approach is Lighthill (1959).

2. A generalized function is obtained through an extension of the notion of differentiation (Heaviside-Kirchhoff, Sebastião e Silva). Thus, $\delta(x)$ is regarded as the derivative of the Heaviside function H(x). A complete treatment of this approach can be found in Ferreira (1997).

3. A generalized function is identified as an element of the field of convolution quotients of continuous functions on the positive real line. This approach is due to Mikusinski (1959) and is presented in a fairly simple manner in Erdélyi (1962); it leads to an extension of the Heaviside operational calculus.

4. A generalized function, or distribution, is a continuous linear functional on a space of test functions. The theory of distributions is due to Laurent Schwartz, although there were earlier contributions by Sobolev, Leray and Wiener. In R^1, a suitable space of test functions is the set C_o^∞ of all infinitely differentiable functions with compact support. The delta distribution is then defined as

$$<\delta,\varphi> = \varphi(0), \qquad \forall \varphi \in C_o^\infty .$$

Note that any integrable function $f(x)$ defines a distribution through

$$<f,\varphi> = \int_{-\infty}^{\infty} f(x)\varphi(x)\, dx ,$$

where the integral always exists since φ has compact support.

Examples of Green's Functions and Fundamental Solutions

Using the standard terminology for the Dirac delta function, we see that Green's function $g(x,y)$ for the heat conduction problem described in the previous section satisfies

$$-\nabla^2 g = \delta(x\text{-}y), \qquad x,y \in \Omega ; \qquad g = 0 \quad x \in \partial\Omega \qquad (3)$$

We can use g to solve the e inhomogeneous Dirichlet problem

$$-\nabla^2 u = f(x), \qquad x \in \Omega ; \qquad u = h(x) \qquad x \in \partial\Omega . \qquad (4)$$

Applying Green's theorem, and appealing to the symmetry of g, we obtain

$$u(x) = \int_{\Omega} g(x,y)f(y)dy - \int_{\partial\Omega} h(y)\frac{\partial g(x,y)}{\partial n_y}ds_y \ , \qquad (5)$$

which gives the solution of Eq. (4) in terms of g and the data. Unfortunately, it is difficult to find g explicitly except for the simplest geometries. As an alternative, one can reduce Eq. (5) to a boundary integral equation (for which there are many good numerical methods) by using instead of g a so-called fundamental solution $e(x,y)$ satisfying the same differential equation as g but not the boundary condition. Such a fundamental solution is often called a *free space Green function*. Fundamental solutions for $-\nabla^2$ in R^2 and R^3 are, respectively,

$$e = -\frac{1}{2\pi}\ln r \ , \ e = \frac{1}{4\pi r} \quad \text{where } r = |x\text{-}y| \ .$$

Consider next a thin cylindrical plate whose central cross section occupies the bounded domain Ω in R^2. If the plate is subject to a distributed transverse load $f(x)$ and its boundary $\partial\Omega$ is clamped, then, according to Kirchhoff's theory, the transverse deflection $u(x)$ satisfies

$$\nabla^4 u = f(x) \qquad x \in \Omega \ ; \qquad u = \frac{\partial u}{\partial n} = 0 \qquad x \in \partial\Omega \ , \qquad (6)$$

where ∇^4 is the biharmonic operator (iterated Laplacian).

The corresponding Green function satisfies

$$\nabla^4 g = \delta(x-y) \qquad x,y \in \Omega ; \qquad g = \frac{\partial g}{\partial n_x} = 0 \qquad x \in \partial\Omega \ . \qquad (7)$$

The solution of Eq. (6) is then given by

$$u(x) = \int_{\Omega} g(x,y)f(y)dy \ . \qquad (8)$$

For the clamped circular plate, g can be obtained explicitly by conformal mapping, but otherwise Green's function can rarely be found in tractable form. On the other hand, it is easy to calculate

the fundamental solution for ∇^4 in R^2 : $e = \frac{1}{8\pi}r^2 \ln r$, where $r = |x\text{-}y|$.

In more complex cases, such as continuously supported plates, even fundamental solutions may be difficult to calculate. Here we mention a few of the cases studied in the literature:

  (a)   The uniformly stretched plate
         (Kerr, 1960)

(b) The plate on a liquid foundation
(Hertz, 1884 Wyman; 1950; Kerr, 1963)

(c) The continuously supported plate
(Kerr and El-Sibaie, 1989)

(d) The orthotropic plate
(Gilbert, Hsiao and Schneider, 1983)

(e) The "enhanced" theory of plates which takes into account transverse shear deformation and transverse normal strain
(Mitric and Schiavone, 2004; see also, Constanda, 1990)

(f) The anisotropic plate with cracks
(Cheng and Reddy, 2004)

Other examples of fundamental solutions can be found in the article by Dundurs and Santare in the present volume.

The Lamé System

The examples of Green's functions and fundamental solutions given so far are scalar functions. Looking at either Eq. (4) with $h=0$, or Eq. (6), Green's function is the kernel of an integral operator transforming the data $f(x)$, a scalar function, into the response $u(x)$, another scalar function. The reason for this simple characterization in the thin plate problem is that both the load and the deflection are unidirectional (transverse to Ω). In more general problems in elastostatics, both the applied force f and the resulting displacement u are vector functions, so that both the Green function and the fundamental solution will now be *matrices*. These ideas should become clearer as we deal with the Lamé system below, following closely the work of Hsiao and Wendland (2004).

Consider a homogeneous, isotropic, linearly elastic material occupying the bounded domain Ω in R^3. With f denoting the vector density of body forces per unit volume, the equilibrium equations are

$$\text{div } T + f = 0 \tag{9}$$

where T is the symmetric stress tensor. Thus Eq. (9) is a set of 3 linear PDEs in 6 unknowns. One can obtain a well-posed system for the stress components by requiring that the strains derived from T satisfy the compatibility conditions. Instead, we use Hooke's Law to express T in terms of the displacement u and then substitute into Eq. (9) to obtain the Lamé equations (also known by the names *Navier* and *Cauchy-Navier*)

$$-Lu = f, \tag{10}$$

with the Lamé operator L given by

$$L = \mu\nabla^2 + (\mu+\lambda)\text{ grad div}, \tag{11}$$

where μ is the shear modulus and $\lambda = 2\mu\sigma/(1\text{-}2\sigma)$, σ being Poisson's ratio. Note that Eq. (10) is a set of 3 simultaneous linear PDEs of the second order in the 3 Cartesian components of u.

It is sometimes useful to rewrite Eq. (11) as the sum of its transverse and longitudinal parts:

$$L = -\mu \text{ curl curl} + (2\mu+\lambda) \text{ grad div.} \qquad (12)$$

Using Eq. (12) in the homogeneous equation $Lu = 0$, we see by taking the divergence, that $\nabla^2 \text{div } u = 0$; then applying the Laplacian to $Lu = 0$ we find that $\nabla^4 u = 0$ so that the Cartesian components of u satisfy the biharmonic equation.

The principal BVPs associated with Eq. (10) are the *Displacement* problem in which u is given on the boundary and the *Traction* problem in which the stress is given on the boundary. In this article, we shall confine ourselves to the *displacement* problem

$$-Lu = f(x), \qquad x \in \Omega; \qquad u = \varphi(x) \qquad x \in \partial\Omega. \qquad (13)$$

BVP Eq. (13) is similar to the Dirichlet problem for Laplace's equation in that the boundary condition is *essential*, that is, it must be satisfied by admissible functions in the corresponding variational principle. By contrast, the boundary condition for the *Traction* problem is *natural*, and need not be satisfied by trial functions in the variational principle. Thus, the *Traction* problem resembles the Neumann problem for Laplace's equation.

It can be shown (Bergman and Schiffer, 1953) that the solution of the *Displacement* problem is unique, but this is not true for the *Traction* problem since a rigid body motion satisfies $Lu = 0$ and is stress-free on the boundary.

The Green matrix (or tensor) $G(x,y)$ associated with Eq. (13) satisfies

$$-LG = \delta(x-y)I \qquad x,y \in \Omega; \qquad G=0 \qquad x \in \partial\Omega, \qquad (14)$$

where I is the 3x3 identity matrix and $\delta(x)$ is the Dirac delta function representing a unit source at the origin in R^3. The solution of Eq. (13) can be written in terms of G; if $\varphi = 0$, then the solution takes the simple form

$$u = \int_\Omega G(x,y)f(y)dy.$$

Unfortunately, as in the scalar case, it is usually as difficult to find G as it is to solve Eq. (13) directly. So we turn to the fundamental matrix solution $E(x,y)$ of the differential equation in Eq. (14) in the whole space. Since L has constant coefficients, we have $E(x,y) = E(x-y,0)$ where $E(x,0)$ satisfies

$$-LE(x,0) = \delta(x)I \qquad\qquad x \in R^3. \qquad (15)$$

Using the spherical symmetry of *L*, it is relatively straightforward to calculate

$$E(x,0) = \frac{aI}{r} + \frac{b}{r^3}(x)(x)^T ,$$ (16)

where the superscript *T* stands for transpose, *r*=|*x*|, and *a*, *b* are explicit constants involving μ and λ. Note that Eq. (15) includes the case where the source is a unit force in the direction *n*. Since *In*=*n*, the solution of Eq. (15) corresponding to the source $\delta(x)n$ is

$$E(x,0)n = \frac{an}{r} + \frac{b(nx)}{r}x ,$$

a classical result due to Kelvin (Kanwal, 1997).

Let us now try to translate Eq. (13) into a boundary integral equation. Appealing to Betti's formula (Bergman and Schiffer, 1953), we have

$$u(x) = \int_{\Omega} E(x,y)f(y)dy + \int_{\partial\Omega} \left[E(x,y)^T [u(y)] - (TE)^T u(y) \right] ds_y ,$$ (17)

where we recall that *E*(*x*,*y*)=*E*(*x-y*,0) which is known from Eq. (16) by replacing *x* by *x-y*.

The first surface integral in Eq. (17) is a simple layer while the second is a double layer. The first term on the right side of Eq. (17) is an integral of known functions and is a solution of the inhomogeneous PDE; by subtracting it from *u*, the difference satisfies the homogeneous equation with a new known displacement on the boundary. It therefore suffices to consider the simpler problem

$$Lu = 0 \qquad x \in \Omega ; \qquad u = \varphi(x) \qquad x \in \partial\Omega ,$$ (18)

where, by an abuse of notation, we have used the symbols *u* and φ for new functions. Using Eq. (17) we can now write

$$u(x) = \int_{\partial\Omega} E(x,y)Tu \; ds_y - \int_{\partial\Omega} (TE)^T u(y)ds_y = V\sigma - W\varphi$$ (19)

where σ is the unknown value of the stress on $\partial\Omega$ and φ is the known value of the displacement on $\partial\Omega$. Now let $x \to \partial\Omega$ to obtain, in view of the jump in the double layer,

$$\frac{\varphi(x)}{2} = V\sigma - K\varphi ,$$ (20)

where $K\varphi = p.v. \int_{\partial\Omega} (T_y E)^T \varphi(y) ds_y$; the Cauchy principal value being needed because of the singularity of the kernel at $y=x$.

Equation (20) is an integral equation of the first kind for the unknown $\sigma = Tu$. Once σ has been found, Eq. (19) gives us "explicitly" the solution of Eq. (18). If we prefer to deal with an integral equation of the second kind, we can apply the traction operator to Eq. (19) as $x \to \partial\Omega$ to obtain

$$\frac{\sigma}{2} = D\varphi + K'\sigma \ , \tag{21}$$

where K' (like K) has a Cauchy singular kernel, whereas D is hypersingular. If φ is twice differentiable, then $D\varphi$ will be defined.

To discuss existence and uniqueness of solutions to Eq. (20) and Eq. (21) as well as to develop numerical methods based on these integral equations, it will be necessary to find an appropriate mathematical framework. For a careful examination of these questions we refer the reader to Hsiao and Wendland (2004).

A powerful numerical method for solving the integral Eqs. (20) and (21) is the boundary element method (BEM) in which the trial functions are finite elements on $\partial\Omega$, the so-called boundary elements. After an appropriate triangulation of the boundary, there are three principal schemes for treating the integral equations: collocation, Galerkin's method, and the least squares method. A rigorous analysis, including convergence results, error analysis, and stability is provided in Hsiao and Wendland (2004). We also want to mention other useful, related references: Ciarlet (1978), McLean (2000), Chen and Zhou (1992).

Nonpositivity of Green's Function for the Clamped Plate

The Green function for the clamped plate satisfies Eq. (7), repeated here for convenience.

$$\nabla^4 g = \delta(x - y) \quad x, y \in \Omega \ ; \qquad g = \frac{\partial g}{\partial n} = 0 \quad x \in \partial\Omega \ . \tag{22}$$

The great Hadamard conjectured that $g \geq 0$ so that it came as a surprise when Garabedian (1951) showed that it was false for a sufficiently thin ellipse. Some forty years later, Shapiro and Tegmark (1994) gave an elementary proof of this result, which we present here.

We recall that for the negative Laplacian, Green's function satisfies Eq. (3) and can be proved to be positive by appealing to the well-known result: if $u(\partial\Omega) = 0$ and $-\nabla^2 u \geq 0$ in Ω, then $u \geq 0$ in Ω. In the temperature problem this is obvious physically: with zero boundary temperature and positive sources, the interior temperature is positive.

For the clamped plate, we first show that the positivity of Green's function is equivalent to the statement:

$$\text{If } u = \frac{\partial u}{\partial n} = 0 \text{ on } \partial\Omega \text{ and } \nabla^4 u \geq 0 \text{ in } \Omega, \text{ then } u \geq 0 \text{ in } \Omega. \tag{23}$$

The proof is simple: a) if Eq. (23) holds, then g must be positive since it satisfies the B.C. and $\nabla^4 g \geq 0$. b) If $g < 0$ at (x_0, y_0) it must be negative for $x = x_0$ and y in a small neighborhood of y_0. Choose $f > 0$ in this neighborhood and 0 elsewhere; then $u = \int_\Omega g(x,y) f(y) dy < 0$ at $x = x_0$, yet $\nabla^4 u \geq 0$.

Having proved the desired equivalence, we now exhibit a domain Ω and a function u satisfying the boundary conditions and $\nabla^4 u \geq 0$ in Ω with $u < 0$ somewhere in Ω.

With x, y the usual Cartesian coordinates in the plane, let Ω be the ellipse $Q = x^2 + 25y^2 - 1 < 0$ and let $p(x) = (1-x)^2(4-3x)$. Then $Q^2 p$ is a polynomial of degree seven which vanishes together with its gradient on $\partial\Omega$. A straightforward but slightly tedious calculation shows $\nabla^4(Q^2 p) > 0$ in $\overline{\Omega}$. For ε sufficiently small, $u \doteq (p - \varepsilon)Q^2$ satisfies the boundary conditions and $\nabla^4 u > 0$ in Ω; yet u obviously takes on negative values for x in Ω near 1. Thus, for this eccentric ellipse, g must take on negative values in some part of Ω.

Acknowledgments

It is a pleasure to acknowledge useful conversations with my colleagues George Hsiao, Robert Gilbert, Peter Monk, and Dick Weinacht during the preparation of this paper.

References

Bergman, S., Schiffer, M., 1953. *Kernel Functions and Elliptic Differential Equations in Mathematical Physics*. Academic Press, New York.

Chen, G., Zhou, J., 1992. *Boundary Element Methods*. Academic Press, London.

Cheng, Z-Q., Reddy, J.N., 2004. Green's Function For An Anisotropic Thin Plate With A Crack Or An Anticrack. *International Journal of Engineering Science*, **42**, 271-289.

Ciarlet, P.G., 1978. *The Finite Element Method for Elliptic Problems*. North Holland, Amsterdam.

Constanda, C., 1990. *A Mathematical Analysis of Bending of Plates with Transverse Shear Deformation*. Longman, Harlow.

Erdélyi, A., 1962. *Operational Calculus and Generalized Functions*. Holt, Rinehart and Winston, New York.

Ferreira, J.D., 1997. *Introduction to the Theory of Distributions*. Longman, Harlow.

Garabedian, P.R., 1951. A Partial Differential Equation Arising in Conformal Mapping. *Pacific J. Math.* **1**, 485-524.

Gilbert, R.P., Hsiao, G.C., and Schneider, M., 1983. The Two-Dimensional, Linear, Orthotropic Plate. *Applicable Analysis*, **15**, 147-169.

Hertz, H., 1884. Uber Das Gleichgewicht Schwimmender Elastischer Platen. *Wiedemann's Annalen Phys., Chem.*, **22**, 449. Also *Gesammelte Werke*, **1**, 288. 1895.

Hsiao, G.C., and Wendland, W.L., 2004. Boundary Element Methods: Foundation and Error Analysis, *Encyclopedia of Computational Mechanics*, Wiley, New York.

Kanwal, R.P., 1998. *Generalized Functions*. Birkhäuser, Boston.

Kerr, A.D., 1960. Uniformly Stretched Plates Subjected to Concentrated Transverse Forces. *Quarterly Journal of Mechanics and Applied Mathematics*, **13**, (4), 462-472.

Kerr, A.D., 1963. Elastic Plates on a Liquid Foundation. *Journal of the Engineering Mechanics Division*, Proceedings of the American Society of Civil Engineers, **89**, EM3, 59-71.

Kerr, A.D., and El-Sibaie, M.A., 1989. Green's Functions for Continuously Supported Plates. *Journal of Applied Mathematics and Physics*, (ZAMP). **40**, 15-38.

Lighthill, M.J., 1959. *Introduction to Fourier analysis and generalized functions*. Cambridge University Press, Cambridge.

McLean, W., 2000. *Strongly Elliptic Systems and Boundary Integral Equations*. Cambridge University Press, Cambridge.

Mikusinski, J.G., 1959. *Operational calculus*. Pergamon Press, London.

Mitric, R., and Schiavone, P., 2004. Integral Solution of a Problem with Non-Standard Boundary Conditions in an Enhanced Theory of Bending of Elastic Plates. *Journal of Elasticity*, 75 (3), 267-289, 2004, Dordrecht.

Schwartz, L., 1950-51. *La Théorie des Distributions*, **1-2**, Hermann, Paris.

Shapiro, H.S., and Tegmark, M., 1994. An Elementary Proof that the Biharmonic Green Function of an Eccentric Ellipse Changes Sign. *SIAM Review*, **36**, 99-101.

Wyman, M., 1950. Deflections of an Infinite Plate. *Canadian Journal of Research*, May issue.

Dislocations as Green's Functions in Plane Elasticity

John Dundurs, Northwestern University and Michael H. Santare, University of Delaware

Abstract

After a brief discussion of the relationship between Green's functions and the stresses associated with dislocations, this paper presents a partial summary and critical review of the literature on the development and uses of edge dislocations as Green's functions for planar crack problems. In his book on dislocation based fracture mechanics, Weertman (1996) covers the mechanics of dislocations and their use in solving a variety of elastic and elastic-plastic fracture problems in simple geometries. This paper uses many of the same methods and mathematics discussed in that book and the references on which it is based. However, the focus here is on the use of dislocation solutions to study curved cracks and crack-inclusion interaction problems and so this paper expands upon and compliments the treatment in Weertman.

Introduction

Analytical solutions for elastic fracture mechanics problems invariably come down to the formulation and solution of a mixed boundary value problem. The usual conditions along the cracked surface are that the tractions are zero where the crack exists and is open, and relative displacement is zero where there is no crack, or the crack is closed. In three-dimensional problems, the cracked surface is a planar region of arbitrary shape and curvature, in two-dimensions, this surface is represented as a line segment or contour. At the edge of the cracked surface, commonly referred to as the crack tip, the governing conditions change abruptly from zero traction to zero displacement and consequently the elastic field at this point is singular. The mathematical formulation of the problem therefore involves writing one or more equations, which can be used to solve for the elastic fields that satisfy these mixed conditions. In some instances, the equations can be formulated directly in terms of displacements or tractions (or their potential functions) by recognizing the precise nature of the crack-surface conditions to be satisfied. This is true for many straight or circular arc cracks with simple interaction or no interaction effects. In other cases, such as interacting cracks, curved crack surfaces and or crack-inclusion interactions, the appropriate equations are less obvious. For these problems, the mathematical formulation of the equations follows naturally from a Green's function approach. Using the elastic field for a discrete dislocation as the Green's function, the solution is written in the form of a set of singular integral equations for a dislocation density distribution. Once this distribution is known, additional integrals can be used to solve for the resulting stresses and displacements. Although the Green's function approach generally gives an exact representation for the dislocation density, in all but the simplest cases, the integral equations must be solved numerically.

A computerized search of the literature reveals that there are hundreds of papers written that deal with the formulation and or use of dislocation solutions as Green's functions. Considering the first published solution for a dislocation interacting with a straight boundary (Head, 1953), there are well over two hundred papers written to date that reference this paper alone. Therefore, this survey does not attempt to reference all or even most of the papers on this subject. Instead, it is an attempt to summarize the basic concepts applied in this field and to point to a small fraction of the literature, chosen as examples to outline some of the major trends and events in this research over the past fifty years.

Dislocations as Green's Functions

In his book on boundary value problems, Stakgold (1967) defines a Green's function as the solution to a boundary value problem with a forcing function consisting of a concentrated unit of inhomogeneity

$$Lg(x,\xi) = \delta(x-\xi). \tag{1}$$

In Eq. (1), L is a linear differential operator on the Green's function $g(x,\xi)$, and δ is the Dirac delta function, which represents the concentrated loading at $x=\xi$. The mathematical properties of the Dirac delta function are that it has a value of zero everywhere except at $x=\xi$ where it becomes unbounded, such that its integral, over any domain including $x=\xi$ is one. As a way of identifying functions with these properties, Stakgold shows that for any nonnegative, locally integrable function $f(x)$, such that $\int_R f(x)dx = 1$, and $\beta > 0$,

$$f_\beta(x) = \frac{1}{\beta} f\left(\frac{x}{\beta}\right) \tag{2}$$

approaches the Dirac delta in the limit as $\beta \to 0$. Once evaluated, the Green's function can be used to calculate the solution to the general boundary value problem explicitly in terms of a continuous forcing function. In the case of elasticity, the boundary value problem consists of the elastic field equations and boundary conditions, the solution of which is usually in the form of the stress field in the given domain. The unit inhomogeneity can be a concentrated force or dislocation. Consequently, the elastic field solution due to a concentrated force or dislocation can be used as a Green's function in planar elasticity.

To show this more clearly, consider a discrete edge dislocation with a burger's vector b_y in the y-direction, acting at the point (ξ,η) of a two-dimensional elastic plane as shown in Fig. 1. For simplicity, consider the plane to be infinite in all directions and traction free on the boundaries. The elastic stress field due to this dislocation can be written as,

$$\sigma_{xx}(x,y,\xi,\eta) = \frac{2\mu b_y}{\pi(\kappa+1)} \frac{(x-\xi)\left[(x-\xi)^2 - (y-\eta)^2\right]}{r^4} \tag{3}$$

$$\sigma_{yy}(x,y,\xi,\eta) = \frac{2\mu b_y}{\pi(\kappa+1)} \frac{(x-\xi)\left[(x-\xi)^2 + 3(y-\eta)^2\right]}{r^4} \tag{4}$$

$$\sigma_{xy}(x,y,\xi,\eta) = \frac{2\mu b_y}{\pi(\kappa+1)} \frac{(y-\eta)\left[(x-\xi)^2-(y-\eta)^2\right]}{r^4}, \tag{5}$$

where σ_{ij} are the Cartesian stress components and $r^2 = (x-\xi)^2 + (y-\eta)^2$. The material constant μ is the shear modulus and κ is related to Poisson's ratio v such that $\kappa = 3-4v$ for plane strain and $\kappa = (3-v)/(1+v)$ for plane stress. Stresses for an edge dislocation in the x-direction can be found through a simple rotation of the coordinate axes. For displacement jumps out of the plane, one uses the elastic field due to a screw dislocation. The solution for an arbitrarily oriented dislocation is then just the sum of the stresses for each of its three vector components.

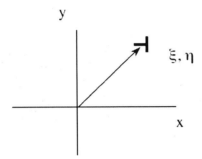

Fig. 1. Discrete, y-oriented, edge dislocation at ξ, η

Lets consider what happens to σ_{yy} when the dislocation is at the origin. Ignoring the constant term out front in Eq. (4), we have

$$\sigma_{yy} \propto \frac{x\left(x^2+3y^2\right)}{\left(x^2+y^2\right)^2} = \frac{x}{\left(x^2+y^2\right)}\left(1+\frac{2y^2}{\left(x^2+y^2\right)}\right) = \frac{1/x}{\left(1+\left(y/x\right)^2\right)}\left(1+\frac{2y^2}{\left(x^2+y^2\right)}\right). \tag{6}$$

From this expression, it is clear that $\lim\limits_{y\to 0}\sigma_{yy} \propto 1/x$. Now, consider the limit of this stress component for $x = \beta$, a parametric value, as $\beta \to 0$,

$$\lim_{\beta\to 0}\sigma_{yy}\Big|_{x=\beta} \propto \lim_{\beta\to 0}\frac{3/\beta}{\left(1+\left(y/\beta\right)^2\right)}. \tag{7}$$

Noticing that $\int_{-\infty}^{\infty} 1/\left(1+y^2\right)dy = \pi$, we can conclude from Eq. (2), that

$$\sigma_{yy}\Big|_{x\to 0^+} \propto \delta(y). \tag{8}$$

Following the same argument, we can show that the other stress components behave in a similar manner as the dislocation is approached. Thus, the stress state due to an edge dislocation is the elastic field response to a point singularity in an otherwise undisturbed domain and consequently serves as a Green's function in planar elasticity.

Crack Solutions Using Dislocations as Green's Functions

The similarity between a series, or pile-up, of dislocations and a crack was noticed early in the development of crystal dislocation theory. A detailed explanation of this relationship can be found in (Bilby and Eshelby, 1968) and serves as the basis for the following description. In order to find the σ_{yy}^D stress due to a continuous distribution of dislocation $B(x, y)$, in the y-direction, we simply apply the integral

$$\sigma_{yy}^D(x,y) = \int B(\xi,\eta)\frac{1}{b}\sigma_{yy}(x,y,\xi,\eta)d\xi d\eta \qquad (9)$$

over the entire region where $B(x, y)$ is nonzero. Similar expressions apply for the other stress components.

As a simple illustrative example, consider a linear elastic medium containing a continuous distribution of dislocations along the x-axis, $B(x)$, $-a < x < a$ with Burgers vector in the y direction. The region of dislocation, $-a < x < a$, $y = 0$ will be designated as ℓ. The net dislocation change between x and $x + dx$ is given as $db = B(x)dx$. In this case, the stress along the crack line can be simplified, so that Eq. (9) can be written

$$\sigma_{yy}^D(x) = \frac{2\mu}{\pi(\kappa+1)}\int_{-a}^{a}\frac{B(\xi)}{x-\xi}d\xi \qquad -a < x < a. \qquad (10)$$

Similar discussions hold for the other stress components and the other crack opening modes, Mode II and Mode III. The difference being that the dislocation distribution and corresponding Burgers vector would have a different orientation. One way to visualize a pile-up of dislocations is shown in Fig. 2. The figure depicts material corresponding to three matching pairs of edge dislocations, inserted into a region ℓ between the two points $-a < x < a$. Each dislocation corresponds to an additional plane of material of thickness b. In the limit, as we introduce more dislocations, with smaller magnitudes, we approach the case of a continuous distribution of dislocation density $B(x)$.

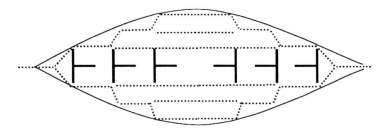

Fig. 2. Schematic interpretation of crack dislocations.

Inserting this material causes a compressive traction on $y = 0$ along ℓ, $-a < x < a$, and a tensile traction for $|x| > a$. By applying an external tensile load, σ_{yy}^A, such that $\sigma_{yy}^D + \sigma_{yy}^A = 0$ over ℓ (the region of the crack is traction free), one can then cut along ℓ and remove the inserted material without disturbing the surrounding field. Now the dislocation distribution, $B(x)$, represents a discontinuity in displacements across ℓ, free from traction in the applied field σ_{yy}^A, and thus, describes a thin crack.

The way the question is normally posed in fracture mechanics is, "What distribution of dislocations will produce a given traction along the specific crack path?" Referring to Fig. 3, consider an infinite plate loaded by a uniaxial traction, σ_{yy}^A in the y direction. The traction across the x-axis in this problem is clearly equal to the applied traction. To introduce an open crack along the x-axis from $-a < x < a$, this surface must be made traction free. So the problem is to find a distribution of dislocations along the segment $-a < x < a$ that produces a traction, that exactly cancels the applied traction, σ_{yy}^A. This is equivalent to finding a function $B(x,0)$, which satisfies the following integral equation,

$$\frac{2\mu}{\pi(\kappa + 1)} \int_{-a}^{a} \frac{B(\xi)}{(x - \xi)} d\xi = -\sigma_{yy}^A \qquad\qquad -a < x < a, \qquad\qquad (11)$$

subject to the following additional constraint, known as a crack closure condition,

$$\int_{-a}^{a} B(\xi) d\xi = 0. \qquad\qquad -a < x < a \qquad\qquad (12)$$

The solution to this set of singular integral equations is well know and can be found in numerous sources on integral equations or fracture mechanics

$$B(x) = \frac{-\sigma_{yy}^A(\kappa + 1)}{2\mu} \frac{x}{\sqrt{a^2 - x^2}} \qquad\qquad -a < x < a \qquad\qquad (13)$$

$$B(x) = 0. \qquad\qquad |x| > a. \qquad\qquad (14)$$

When we add the traction due to this distribution of dislocations, to the traction from the applied load, the result is a traction free surface on the interval $-a < x < a$ and the stresses due to a crack elsewhere.

$$\sigma_{yy}(x,0) = 0 \qquad\qquad -a < x < a \qquad\qquad (15)$$

$$\sigma_{yy}(x,0) = \frac{\sigma_{yy}^A |x|}{\sqrt{x^2 - a^2}} \qquad\qquad |x| > a. \qquad\qquad (16)$$

Additionally, the crack face displacements can be found directly from the dislocation distribution, B,

$$u_y(x,0^+) - u_y(x,0^-) = \int_x B(\xi)d\xi \qquad\qquad -a < x < a. \qquad\qquad (17)$$

Perfectly analogous solutions can be found for the Mode II and Mode III cracks in an unbounded medium. In the case of Mode II, the edge dislocations are oriented in the x-direction and the Green's function (The kernel in Eq. (9)) is the shear traction σ_{xy}, across the crack faces. In Mode III, the burger's vector and loading are out-of-plane, requiring the solution for the stress component σ_{yz} due to a screw dislocation. By considering arbitrarily oriented dislocations, this technique can be used to find the stress field due to an open crack in a planar elasticity, with any mixed-mode, far-field loading imposed. Additionally, by taking the integral (Eq. (10)) along an arbitrary path, one can, in principle, solve the basic problem for any crack configuration in an unbounded medium. This will be described in the next section on curved cracks. However, as soon as the problem deviates from the simple straight crack in an infinite medium, the solution to the integral equation, analogous to Eq. (10), must be found numerically.

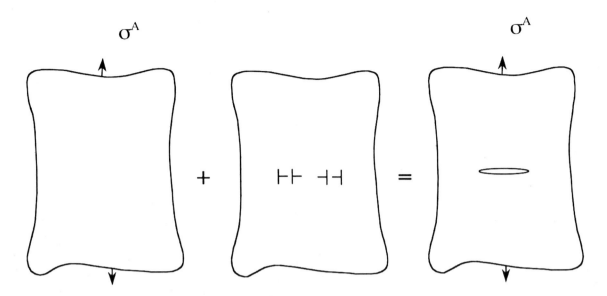

Fig. 3. Superposition of dislocation distribution to solve crack problem

In order to use this technique for the solution of cracks near boundaries or interfaces, one must have the solution for the dislocation in the particular domain of interest. For example, to study the interaction of a crack with a material inhomogeneity or inclusion, the basic solution is the dislocation interacting with the inclusion. This solution is then used as the Green's function for the interacting crack problem. Since the solution for the single dislocation is required to satisfy the boundary condition at the inclusion interface, the superposition solution for a distribution of dislocations must also satisfy it. In a latter section, we will discuss some of the dislocation-inclusion interactions solutions that appear in the literature.

Curved Crack Solutions

The curved crack problem was first studied by Muskhelishvili (1963), who analyzed a circular arc cut using complex stress potentials. In this case, he formulated the problem directly, without using the Green's function approach. The problem is solved in the same manner as the straight crack problem, using polar coordinates. Sih et al. (1962) and Sih (1965) revisited the problem, solving for different far-field loadings. However, Cotterell and Rice (1980) note that the solution of Sih et al. (1962) contained an error in the transcription of the solution and the correct expressions for the stress intensity factors of the circular arc crack can be found in Cotterell and Rice (1980).

Banichuk (1970), and Goldstein and Salganik (1974) studied slightly curved cracks using polynomial approximations for the shape of the crack path. These solutions are based on a Green's function approach to formulating the integrals, but in this case, an additional integral is added to account for the curvature effect in an approximate manner. Cotterell and Rice (1980) reported that these approximations are accurate to within 5% of the exact solutions when the local tangent angle at the tips of the curved crack differed by up to 15 degrees from that of the main crack. Lo (1978) solved the kinked crack problem by mapping the crack path conformally, onto a circle, solving the boundary value problem in this mapped domain, and then mapping the solution back to the physical domain. Although these papers take several different creative approaches to the solution of curved crack problems, the most general and commonly used approach, by far, is the Green's function approach.

This method was used in several papers by Ioakimidis and Theocaris who looked at periodic arrays of curvilinear cracks in an infinite medium (Ioakimidis and Theocaris, 1977) and in an elastic half-plane (Ioakimidis and Theocaris, 1979). Assuming a polynomial approximation to the dislocation density, Chen, Gross and Huang (1991) used a collocation technique to solve the singular integral equations for several different curved crack geometries. The results for a parabolic crack loaded in biaxial tension were compared with those reported in Savruk (1981) and good agreement was obtained. In a series of papers, Chen has also approached the curved crack problem in a variety different manners: A review of these different approaches is given in Chen (1995).

The simplest way to formulate the curved crack problem is through the use of complex potentials. The application of complex potentials to theoretical elasticity has its roots with Goursat's work with the biharmonic equation in the late 1890s (Goursat, 1898). However, it was Kolosov (1909) who applied the theory of complex potentials to problems in planar elasticity. An interesting fact noted by Muskhelishvili is that S. A. Chaplygin derived equations similar to some of those of Kolosov around 1900. Chaplygin never published them and they were discovered in his manuscripts only after his death (Muskhelishvili, 1963). Kolosov's work was extended to a variety of problems by a number of Soviet researchers, perhaps the best known being the Georgian mathematician N. I. Muskhelishvili. To show the basic formulation, consider the complex potentials for the planar elastic field. Let σ_{xx}, σ_{yy}, and σ_{xy} denote the components of stress for an isotropic, homogeneous body in a state of plane stress or plane strain with no body forces or thermal stresses. It can be shown that the stresses and displacements can be expressed in terms of two analytic functions of the complex variable z, $\phi(z)$ and $\psi(z)$ (Muskhelishvili, 1963)

$$\sigma_{xx} + \sigma_{yy} = 2\left\{\phi'(z) + \overline{\phi'(z)}\right\} \tag{18}$$

$$\sigma_{yy} - \sigma_{xx} + 2i\sigma_{xy} = 2\left\{\overline{z}\phi''(z) + \psi'(z)\right\} \tag{19}$$

$$2\mu(u + iv) = \kappa\phi(z) - z\overline{\phi'(z)} - \overline{\psi(z)} \tag{20}$$

where the prime denotes the derivative, the over bar denotes the conjugate, μ is the shear modulus and κ is related to Poisson's ratio v such that $\kappa = 3 - 4v$ for plane strain and $\kappa = (3 - v)/(1 + v)$ for plane stress. For an edge dislocation with burgers vector $\mathbf{b} = b_x + ib_y$, at the point $z = \zeta$, the appropriate potentials are

$$\phi(z) = \gamma \ln(z - \zeta) \tag{21}$$

$$\psi(z) = \bar{\gamma}\ln(z-\zeta) - \gamma\frac{\bar{\zeta}}{z-\zeta} \tag{22}$$

where

$$\gamma = \frac{\mu\mathbf{b}}{i\pi(\kappa+1)}. \tag{23}$$

Another useful formula, which aids in the determination of the elastic field is the resultant traction on the contour ℓ is expressed as

$$X + iY = -i\left[\phi(z) + z\overline{\phi'(z)} + \overline{\psi(z)}\right]_{\ell} \tag{24}$$

where X and Y are the traction components in the x and y direction, respectively. In addition, if we let n and t denote the normal and tangent to ℓ, respectively, then the normal stress, σ_{nn}, and the tangential stress, σ_{nt}, along the contour are given as

$$\phi'(z) + \overline{\phi'(z)} - e^{2i\alpha}\left\{\bar{z}\phi''(z) + \psi'(z)\right\} = \sigma_{nn} - i\sigma_{nt} \tag{25}$$

where α is the angle between the normal and the x-axis as shown in Fig. 4.

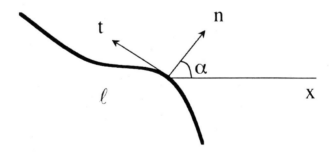

Fig. 4. Defining the normal and tangent to a contour.

Consider a planar, curved crack in an infinite elastic medium as shown in Fig. 4. Recall that the problem of a crack can be considered as the sum of an uncracked medium subjected to some loading at infinity, and that of a distribution of dislocations, the tractions for which, exactly cancel the tractions along these faces in the uncracked problem. Along the crack path, ℓ, this can be expressed as

$$\frac{\mu}{i\pi(\kappa+1)}\left\{\int_\ell \frac{\mathbf{b}}{z-\zeta}d\zeta + e^{i2\alpha}\int_\ell \frac{\overline{(z-\zeta)}\,\mathbf{b}}{(z-\zeta)^2}d\zeta\right.$$

$$\left. + \int_\ell \frac{\overline{\mathbf{b}}}{\overline{z-\zeta}}\overline{d\zeta} - e^{i2\alpha}\int_\ell \frac{\overline{\mathbf{b}}}{(z-\zeta)}\overline{d\zeta}\right\} = \sigma_{nn} - i\sigma_{nt}$$

$$z\in\ell,\ \zeta\in\ell, \tag{26}$$

where σ_{nn} and σ_{nt} can be found by substituting the potential functions for the external loading into Eq. (25). The terms on the left hand side of Eq. (26) are determined by substituting the potentials for a dislocation (Eq. (21) and (22)) into Eq. (25) and integrating along the crack contour ℓ. Eq. (26) is a Cauchy singular integral equation for the unknown dislocation density, $\mathbf{b} = b_x + ib_y$. In order to find a unique solution to Eq. (26), we must also satisfy the crack closure condition,

$$\int_\ell \mathbf{b}\,d\zeta = 0. \tag{27}$$

Except for a few special cases of curved cracks, Eq. (26) and (27) must always be solved numerically.

Dislocation-Inclusion Interaction Solutions

Interaction between dislocations and interfaces has been the topic of considerable research for over 50 years. The earliest work looks at the interaction of a dislocation with a straight boundary between dissimilar elastic half-spaces (Head, 1953). These solutions were found by the method of images. In an image solution, one considers the elastic field due to the dislocation in an infinite body to be modified by the stresses and displacements due to another, fictitious, dislocation on the other side of the boundary. By adding the effects of several such images, strategically placed, the prescribed boundary conditions and point singularities can be satisfied simultaneously. Numerous permutations of this solution have been developed over the years. By examining different geometries, interface conditions and material properties, researchers have all but exhausted the seemingly endless possibilities. In the following, we will describe a few of these studies that we feel laid the groundwork for most of the others.

In 1964, Dundurs and Mura (1964) solved the elastic interaction for an arbitrarily oriented edge dislocation located outside a circular inclusion as shown in Fig. 5. The solution is in the form of the Airy stress potential and was found using a semi-inverse method, where the basic structure of the solution was taken from a known, related solution and the specific terms and coefficients were evaluated through a combination of intuition and trial and error. It is easy to see that this solution includes the straight-boundary solution as a special case when the radius of the inclusion is much larger than the distance to the dislocation. In fact, this solution can equally well be formulated using the method of images. In the paper, Dundurs and Mura pay particular attention to the Peach-Koehler force on the dislocation and the effects of orientation and material mismatch. Notice that the dislocation resides on the x-axis, but its orientation is arbitrary. Since there is no external loading, this geometry covers all possible configurations for an edge dislocation outside of a circular inclusion. This paper laid the groundwork for much of the research in this field. Dundurs and Sendeckyj (1965) followed up on this work by considering the edge dislocation inside the circular inclusion. Over the next 50 years, researchers have developed many solutions, which represent variations on this theme. For a detailed discussion of the research on dislocation-inclusion interaction up to 1969, please refer to the review chapter by Dundurs (1969).

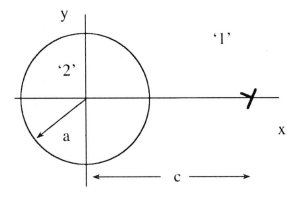

Fig. 5. Schematic of arbitrary dislocation outside a circular inclusion

Dislocation interaction with elliptical inclusions was a logical extension of the circular inclusion work. These solutions are usually derived through the use of conformal mapping and complex potentials. One starts with the Muskhelishvili complex potentials for the elastic field due to a dislocation in an otherwise undisturbed infinite medium, and adds to these a set of undetermined potentials which represent the effects of the inclusion interface. At the same time, the elastic field on the other side of the interface is expressed in terms of a different set of undetermined complex potentials.

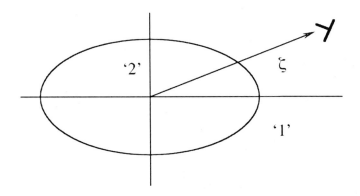

Fig. 6. Schematic of arbitrary dislocation outside an elliptical inclusion

By matching (or prescribing) tractions and/or displacements along the interface, and adhering to the conditions at infinity, the undetermined potentials are evaluated. Using this technique, Warren (1983) solved the edge dislocation inside of an elliptical inclusion in the form of a Laurent series, and Stagni and Lizzio (1983) employed a similar technique to solve the related problem where the dislocation is outside the ellipse. The importance of the rigid body rotation term, not accounted for in Stagni and Lizzio, was pointed out by Santare and Keer (1986) in their study of a rigid elliptical inclusion. The Warren and Stagni and Lizzio solutions are analogous to the earlier circular inclusion solutions referenced in the previous paragraph. They are perfectly general in terms of the orientation and location of the dislocation and the relative elastic properties of the inclusion and the surrounding medium. In order to cover general dislocation geometry in the elliptical case, orientation and angular location must be treated separately, as depicted in Fig. 6. The problem was generalized to an anisotropic elliptical inclusion in an anisotropic matrix for a dislocation outside the inclusion in Hwu and Yen (1993) and (1994) and inside or on the interface of the elliptic inclusion in Yen, Hwu, and Liang (1995). Ting (1996) generalized the previous cases to include generalized plane strain for dislocations outside, inside or on the interface of an anisotropic elliptic inclusion in an anisotropic matrix.

One, often-cited advantage of the complex potential technique, is that in principle, it can be used to solve the interaction problem with any arbitrarily shaped inclusion. All one needs to find, is the proper conformal mapping function $z = \omega(\sigma)$ to map the inclusion boundary onto the unit circle and substitute this function into the governing equations. This reduces the problem to the same form as the circular inclusion solution. It has been shown that such a function exists for a wide variety of possible inclusion shapes (Muskhelishvili, 1963). However, solutions of this type are rarely found in the literature due to the difficulty of identifying and then applying the proper conformal mapping function. A notable exception is the work by Hasebe et al. (2003). In this work, the authors apply a rational mapping function to the complex equations to find the interaction of a dislocation and a rhombic hole or rigid inclusion.

To show the basic methodology, outlined in Stagni (1982), consider an edge dislocation at point ζ interacting with an elastic inclusion in an otherwise unbounded medium as depicted in Fig. 6. Denote the region outside the inclusion as region 1, and the region inside as region 2. Assume that the known rational function $z = \omega(\sigma)$ conformally maps region 1 into the region outside the unit circle and region 2 to the inside. By substitution, the potentials in Eq. (13), (14) and (15) can be written as functions of the mapped-plane variable ζ. For the sake of this demonstration, assume the dislocation is in region 1. In general the potentials for both regions can be expressed in terms of Laurent series expansions. For the matrix, region 1, the potentials can be written as

$$\phi_1(\sigma) = \sum_{k=-\infty}^{\infty} A_k \sigma^k + \sum_{k=-\infty}^{\infty} c_k \sigma^k \qquad (28)$$

$$\psi_1(\sigma) = \sum_{k=-\infty}^{\infty} B_k \sigma^k + \sum_{k=-\infty}^{\infty} d_k \sigma^k . \qquad (29)$$

The first term in each expression, is the series expansion for the potential function for the elastic field due to a dislocation in an undisturbed, unbounded medium (Eqs. (21), (22) and (23)).

$$\sum_{k=-\infty}^{\infty} A_k \sigma^k = \left[\gamma \ln(z-\zeta) \right]_{z=\omega(\sigma)} \tag{30}$$

$$\sum_{k=1}^{\infty} B_k \sigma^k = \left[\overline{\gamma} \ln(z-\zeta) - \gamma \frac{\overline{\zeta}}{z-\zeta} \right]_{z=\omega(\sigma)} \tag{31}$$

The second terms in Eq. (28) and (29) represent the additional elastic field due to the presence of the inclusion, the coefficients of which are undetermined at this point. These additional series must be bounded everywhere in region '1' and on the boundary between regions 1 and 2. Therefore only the k-negative terms will be considered in the series. This is due to the fact that the potentials must also be bounded at large values of z.

The potentials for the inclusion, region 2, can be written as series with undetermined coefficients, bounded everywhere in region 2 and on the boundary.

$$\phi_2(\sigma) = \sum_{k=-\infty}^{\infty} a_k \sigma^k \tag{32}$$

$$\psi_2(\sigma) = \sum_{k=-\infty}^{\infty} b_k \sigma^k \tag{33}$$

Given the forms of the potentials, the undetermined coefficients can now be determined from the conditions of displacement and traction continuity at the interface of the inclusion and the matrix. From Eq. (20), the displacement on the interface, as determined separately from regions 1 and 2 must match

$$\frac{1}{\mu_1}\left\{ \kappa_1\phi_1(\sigma) - \frac{\omega(\sigma)}{\omega'(\sigma)}\overline{\phi_1'(\sigma)} - \overline{\psi_1(\sigma)} \right\} =$$
$$\left. \frac{1}{\mu_2}\left\{ \kappa_2\phi_2(\sigma) - \frac{\omega(\sigma)}{\omega'(\sigma)}\overline{\phi_2'(\sigma)} - \overline{\psi_2(\sigma)} \right\} \right|_{\sigma \text{ on interface}} \tag{34}$$

where the material constants are subscripted for regions 1 and 2. From Eq. (24), the tractions must also be made to match on the interface

$$\left. \phi_1(\sigma) + \frac{\omega(\sigma)}{\omega'(\sigma)}\overline{\phi_1'(\sigma)} + \overline{\psi_1(\sigma)} = \phi_2(\sigma) + \frac{\omega(\sigma)}{\omega'(\sigma)}\overline{\phi_2'(\sigma)} + \overline{\psi_2(\sigma)} \right|_{\sigma \text{ on interface}} \tag{35}$$

By substituting the series expressions for the potentials (Eq. (28)-(33)) into the continuity conditions (Eqs. (34) and (35)), one derives a set of equations in the unknown coefficients, a_k, b_k, c_k, d_k. Since these conditions are imposed on the unit circle, $\sigma = e^{i\theta}$, each value for $\sigma^k = e^{ik\theta}$, gives a linearly independent set of equations. These, together with boundedness and continuity conditions at infinity and near the origin, provide a sufficient set of equations to solve for each of the unknowns.

In the dislocation-inclusion literature, mention is often made to the pioneering work of Eshelby (1961). In this famous paper on inclusions, Eshelby derives general expressions that could, in theory, be used to solve any elastic inclusion problem. For example, any arbitrarily shaped inclusion surrounded by an infinite medium, with any arbitrary loading conditions in the inclusion, surrounding medium or along the interface. And, indeed, this formulation has been used to solve numerous two and three dimensional elastic inclusion problems. However, it was not until 2002 (Li and Shi, 2002; Shi and Li, 2003), that the Eshelby method was actually used to solve an inclusion-dislocation interaction problem. In the first paper, the authors show how the Eshelby method can be used to solve the screw dislocation interaction with an arbitrarily shaped, two-dimensional inclusion. In the second they solve the same problem for the edge dislocation.

These solutions have lead to a greater understanding of plasticity, material defects and the phenomenon of strain hardening. But perhaps more importantly, they have been used as Green's functions to solve a wide class of crack-inclusion interactions. Without attempting to review this extensive body of literature, we will point out some of the more well known studies, and a few of the most recent studies and their implications for the literature in general.

Crack-Inclusion Solutions

The first published, mathematical solution of a crack-inclusion interaction problem was due to Tamate (1968) who investigated the effect of a circular elastic inclusion on an external radial crack using a direct complex stress potentials approach. In this solution, the infinite domain was subjected to uniaxial loading perpendicular to the crack and a Laurent series expansion was used to determine the complex stress potentials required for solving the resulting dual Hilbert problem. Among other things, this paper showed that a relatively rigid inclusion decreases the stress intensity factor of the crack whereas a more compliant inclusion increases the stress intensity factor. A few years later, Atkinson (1972) noted that convergence of this solution was assured only for certain combinations of crack lengths and distances. To resolve this problem, he reformulated the problem as a continuous distribution of dislocations interacting with the inclusion. The Green's function for this solution was taken from the Dundurs and Mura (1964) solution. Atkinson gives values of the stress intensity factor for radial cracks under both uniaxial and biaxial tension at different distances from the inclusion for various ratios of material properties. Atkinson also makes the point that the implicit solution of the interaction equations differs significantly from the solution where the traction from the boundary conditions is applied directly to the crack path and the dislocation density is solved for separately.

Another paper written around the same time also uses the Dundurs and Mura dislocation solution as a Green's function for crack problems. Erdogan, Gupta, and Ratwani (1974) solved the arbitrary, straight crack outside the circular inclusion. Solutions for various straight cracks interacting with a hole and with a metallic inclusion embedded in an epoxy-type matrix are given for several different geometries under a uniaxial loading. The tabulated solutions in this paper are often cited for verification for the accuracy of new solutions. However, these tabular results were generated on 1974 computers, so some of the solutions are off, and some others are wrong (see for example, the discussion in Helsing and Jonsson, 2002). This solution was extended to an antiplane shear loading case by Theocaris and Demakos (1985). The related questions of a crack inside an inclusion and a crack terminating at the interface or crossing into the inclusion were later solved by Erdogan and Gupta (1975), again using the Dundurs and Mura solution as the Green's function.

Other techniques have been applied to the solution of crack-circular inclusion problem. Kunin and Gommerstadt (1985) applied a projection integral equation method to investigate the interaction energy of the crack and a small inclusion to determine the translational and expansional energy release rates for the crack-inclusion system. Using Eshelby's point force method along with the dislocation method, Wang (1995) gives the exact analytical expressions for the stress intensity factors of the crack-inclusion problem; although, unfortunately, no numerical examples are given. However, the solution techniques detailed in Erdogan, Gupta and Ratwani (1974) and Erdogan and Gupta (1975) have become the most widely accepted and many related crack-inclusion solutions have since been formulated using this method.

Muller has applied these techniques in his investigations of the stress induced transformation toughening of zirconia containing ceramics (Muller, 1989) and of radial cracks in fiber reinforced composites under a thermal and a mechanical loading (Muller and Schmauder, 1992 and 1993). The latter contains an exhaustive study of the effect of material constants, in the form of the Dundurs parameters, for both thermal and mechanical loadings on crack-inclusion interactions. The Dundurs parameters, α and β, can be written,

$$\alpha = \frac{\mu_2(\kappa_1+1)-\mu_1(\kappa_2+1)}{\mu_2(\kappa_1+1)+\mu_1(\kappa_2+1)}, \qquad \beta = \frac{\mu_2(\kappa_1-1)-\mu_1(\kappa_2-1)}{\mu_2(\kappa_1+1)+\mu_1(\kappa_2+1)} \qquad (36), (37)$$

where the subscript '1' denotes the matrix and '2' denotes the inclusion. Their findings include that for thermal loads, the stress intensity factors increase with increasing Dundurs parameters; the β parameter having a larger influence than the α parameter. For mechanical loads, the α parameter causes more significant changes in the stress intensity factor and that crack stability is a function of the sign and magnitude of the α parameter. And, stiff inclusions or those with higher thermal expansion coefficients, promote increased toughness of the matrices (Muller and Schmauder, 1993).

This approach was extended to study two equal length cracks interacting with a circular inclusion in Lam and Wen (1993) and to multiple radial cracks interacting with a circular inclusion in Hu, Chandra and Huang (1993a). Lam and Wen (1993) studied the effects of the crack position and orientation, the size of the inclusion and the material constants of the inclusion and matrix on the stress intensity factors of the cracks. Hu, Chandra and Huang (1993a) found that for the case of a remote external loading, a hard inclusion decreased the stress intensity factor for a radial crack system but a transformation loading (where the inclusion produces a dilatational strain) of a hard inclusion produces an amplification of the stress intensity factor. The opposite effect occurred for a soft inclusion. In addition, their investigation revealed that the stress intensity factor typically decreased as the number of radial cracks increased due to the shielding effect produced by the crack-crack interactions (Hu, Chandra, and Huang, 1993a). The case of multiple cracks, not necessarily radial, interacting with a circular inclusion was examined by Hu and Kemeny (1994) in their investigation of backfill on discontinuous rock masses and the interaction of multiple cracks with multiple inclusions or voids was considered in Hu, Chandra and Huang (1993b). They determined that the arrangement of the holes and cracks has a significant effect on the increase or decrease of the stress intensity factors. In particular, it was shown that a strong reduction in the stress intensity factors was obtained between voids and cracks if the spacing between them was reduced (Hu, Chandra, and Huang, 1993b).

The case of a semi-infinite crack penetrating the circular inclusion has been analyzed separately by Steif (1987) and by Wu (1988) using approximate solution techniques. Muller, Gao and Chiu (1996) investigated the case of a semi-infinite crack in front of a circular, thermally mismatched inclusion. Using the different quadrature techniques for solving the singular integral equations, they found the convergence of these numerical schemes to be of importance. Specifically, the number of collocation points had to be increased to ensure convergence as the crack tip approaches the inclusion.

The problem of a crack interacting with an elliptical hole was addressed by Ishida et al. (1985) using a force doublet solution as their Green's function. Patton and Santare (1990) solved the related crack-rigid ellipse problem using the distributed dislocation density approach. Anlas and Santare extended this solution to arbitrarily oriented straight cracks inside (Anlas and Santare, 1993a) and outside (Anlas and Santare, 1993b) an elastic, elliptical inclusion by using the dislocation solutions of Stagni and Lizzio (1983) and Warren (1983) as the Green's functions. Hwu, Liang and Yen (1995) published a series of work concerning anisotropic, elliptical inclusions. They investigated cracks inside, outside and penetrating the anisotropic, elliptic inclusion. The antiplane shear case for anisotropic inclusions in an anisotropic matrix was considered in Chao and Chiang (1996).

Curved Crack-Inclusion Interaction

Recently, Chen and Chen (1997) have investigated the interaction of a curved crack and an elastic inclusion. These researchers first solved the fundamental problem of a point dislocation interacting with an inclusion in an infinite plane. Using this solution, the crack was then represented as a distribution of dislocations and the resulting singular integral equations were formulated with a weaker logarithmic singularity. Numerical examples of a parabolic crack interacting with an inclusion in an infinite plain under a uniaxial loading were given. The authors note that the interaction between a curved matrix crack and an elastic inclusion is complicated since inclusions that are more rigid than the matrix and less rigid than the matrix can lower the stress intensity factors of the curved crack.

Muller and Kemmer (1995) have also studied the interaction of a curved crack interacting with elastic inclusion. These researchers have shown the mathematical equivalence of the dislocation and the crack opening displacement approaches to the problem for the case of semi-circular cracks. (The mathematical equivalence for arbitrarily shaped cracks has been shown by Kemmer (1995)). Results are given for the case of a semi-circular crack interacting with a circular inclusion where the crack faces are loaded by a pressure. Cheeseman and Santare (2000) revisited the problem, looking at other crack geometries. They formulated the integrals using the same methodology as Muller and Kemmer, but solved them using a different numerical scheme.

Crack-Inclusion Formulation

The development of the singular integral equations for a crack interacting with a circular inclusion follows that of a crack in an infinite medium derived in a previous section. Referring to Fig. 7, the total problem of a crack interacting with an inclusion can be found as the sum of an uncracked medium subjected to some loading at infinity and that of a medium with dislocations, where the tractions from the dislocations exactly cancel those which occur along these faces in the uncracked problem. Along the crack this can be expressed as

$$\sigma_{ij}^{D} + \sigma_{ij}^{A} = 0 \tag{38}$$

where, again, σ_{ij}^{A} are the tractions due to the applied load and σ_{ij}^{D} are the tractions due to the distribution of dislocations. This time, however, σ_{ij}^{A} are the tractions, along the crack path, for an uncracked medium with an inclusion. The elastic field for this type of problem has been the subject of uncountable solutions in the literature (for a review of this literature see Mura, 1988). Similarly, σ_{ij}^{D} are the crack path tractions due to a distribution of dislocations, all of which are interacting with the same inclusion. These solutions are briefly discussed in the previous section of this paper. Since each of the solutions must satisfy the boundary conditions along the inclusion interface, the sum of the solutions will also satisfy these boundary conditions.

For example, if we have the solution for the uncracked problem, in terms of complex potentials, $\phi_A(z)$ and $\psi_A(z)$, we can write the following expression for the stresses anywhere in the uncracked plane.

$$\phi_A'(z) + \overline{\phi_A'(z)} - e^{2i\alpha}\left\{\overline{z}\phi_A''(z) + \psi_A'(z)\right\} = \sigma_{nn}^{A} - i\sigma_{nt}^{A} \tag{39}$$

Remembering that a crack is represented by a distribution of dislocations, substituting Eq. (28) through (33) into the left hand side of Eq. (39) and integrating along ℓ gives equations of the form

$$\frac{\mu}{i\pi(\kappa+1)}\left\{\int_\ell \frac{\mathbf{b}}{z-\zeta}d\zeta + e^{i2\alpha}\int_\ell \mathbf{b}K(z,\zeta)d\zeta \right.$$

$$\left. + \int_\ell \frac{\overline{\mathbf{b}}}{\overline{z}-\overline{\zeta}}\overline{d\zeta} - e^{i2\alpha}\int_\ell \overline{\mathbf{b}K(z,\zeta)d\zeta}\right\} = \sigma_{nn}^{A} - i\sigma_{nt}^{A}$$

$$z\in\ell,\ \zeta\in\ell. \tag{40}$$

The Cauchy singular terms, are identical to the ones formulated in the section on curved cracks. They represent the singular stress field due to the dislocation. The K terms in this equation are determined from the dislocation-inclusion interaction terms. In a complex potential formulation, these terms can be derived from the interaction potentials (the second series terms in Eq. (28) and (29)) and are generally not singular. As before, a crack closure condition such as Eq. (27), is needed to find the appropriate, unique solution for the integral equations. Together, Eqs. (27) and (40) are a set of singular integral equations, which need to be solved for the unknown dislocation density distributions.

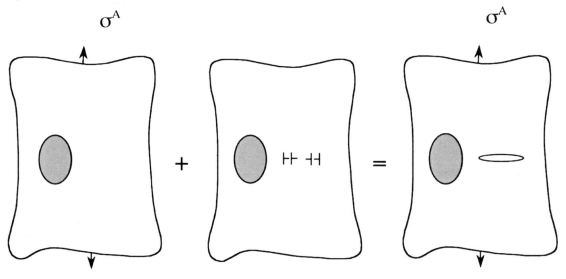

Fig. 7. Superposition of dislocation distribution to solve crack-inclusion problem.

Summary

This paper represents an abridged historical survey and user's guide to the development and use of dislocation solutions as Green's functions in planar elasticity. As mentioned in the introduction, it would be impossible to review all the literature written on this subject in the past fifty years in a single manuscript. Instead, we have tried to identify the most important published work upon which the rest of literature is based. We have then traced some of the general developments that have followed from this early work up to the present. Along the way, we have found that a few important developments have led to the proliferation of a large body of related literature. And, as expected, the rate of growth seems to be increasing as each new problem solved opens the door to the solution of another larger class of problems. For example, the edge dislocation-circular inclusion solution of Dundurs and Mura (1964) has led to body of published solutions examining general crack-inclusion boundary value problems, as outlined in a previous section. Each of these solutions in turn, has lead to publications outlining the application of these general boundary value problem to specific materials and fracture geometries of interest. (These publications were intentionally overlooked in the preparation of this paper, but there are many hundreds of such papers in the literature.)

Several interesting developments have spun-off from this line of intellectual pursuit. One worth mentioning appears in the micromechanics of damage literature. In 1987, Aboudi and Benveniste (1987) used the cracked circular inclusion solution, originally developed by Erdogan and Gupta (1975), as the basis for a generalized self-consistent method for predicting the modulus of a material with distributed microcrack damage. The crack-inclusion solution allowed them to improve upon the solution for the problem developed earlier by Budianski and O'Connell (1976). As other crack-inclusion solutions became available, it was possible to improve the Aboudi and Benveniste solution to examine higher crack densities and damage-induced anisotropy (Santare et al., 1995).

For the most part, however, the boundary value problems are solved and resolved, examining all possible combinations of parameters and numerical solution procedures. (The body of literature on the numerical solution of singular integral equations is at least as large as the body on dislocation-inclusion interaction.) Then it seems, just as the possibilities are nearly exhausted, a new dislocation solution, or numerical technique is discovered, and the process repeats itself. The most recent examples of this are the new approach by Hasebe et al. (2003) for solving the interaction problem for a more generally shaped inclusion and the papers by (Li and Shi, 2002; and Shi and Li, 2003) which revisit the use of the Eshelby equivalent inclusion method. Each of these discoveries is already beginning to generate a new rash of journal publications. So, while the use of dislocations as Green's functions in plane elasticity is already a mature field of study in engineering mechanics, this new research will in turn, lead to new mathematical discoveries and new physical insights.

References

Aboudi, J., and Benveniste, Y., 1987. The Effective Moduli of Cracked Bodies in Plane Deformations. *Engineering Fracture Mechanics*, **26**. 97.

Anlas, G., and Santare, M.H., 1993a. A Model for Matrix Cracks in Short Fiber Composites. *International Journal of Solids and Structures*, **30**, 1701-13.

Anlas, G., and Santare, M.H., 1993b. Arbitrarily Oriented Crack Inside an Elliptical Inclusion. *Transactions of the ASME Journal of Applied Mechanics*, **60**, 589.

Atkinson, C., 1972. The Interaction Between a Crack and an Inclusion. *International Journal of Engineering Science*, **10**, 127.

Banichuk, R.V., 1970. Determination of the Form of a Curvilinear Crack by Small Parameter Technique. *Izv. An SSSR*, **7**, 130.

Bilby, B.A., and Eshelby, J.D., 1968. Dislocations and the Theory of Fracture. *Fracture: An Advanced Treatise. Volume 1: Microscopic and Macroscopic Fundamentals*, ed. H. Liebowitz. Academic Press.

Budianski, B., and O'Connell, R.J., 1976. Elastic Moduli of a Cracked Solid. *International Journal of Solids and Structures*, **12**, 81.

Chao, C.K., and Chiang, T. F., 1996. Antiplane Interaction of an Anisotropic Elliptic Inclusion with an Arbitrarily Oriented Crack. *International Journal of Fracture*, **75**, 229.

Cheeseman, B.A., and Santare, M.H., 2000. The Interaction of a Curved Crack With a Circular Elastic Inclusion. *International Journal of Fracture*, **103**, 259.

Chen, Y.Z., Gross, D., and Huang, Y.J., 1991. Numerical Solution of the Curved Crack Problem by Means of Polynomial Approximation of the Dislocation Distribution. *Engineering Fracture Mechanics*, **39**, 791.

Chen, Y.Z., and Chen, R.S., 1997. Interaction Between Curved Crack and Elastic Inclusion in an Infinite Plane. *Archive of Applied Mechanics*, **67**, 566-575.

Chen, Y.Z., 1995. A Survey of New Integral Equations in Plane Elasticity Crack Problem. *Engineering Fracture Mechanics*, **51**, 97.

Cotterell, B., and Rice, J.R., 1980. Slightly Curved or Kinked Cracks. *International Journal of Fracture*, **16**, 155.

Dundurs, J., 1969. Elastic Interactions of Dislocations with Inhomogeneities. *In Mathematical Theory of Dislocations*. T. Mura, ed. The American Society of Mechanical Engineers New York, 70.

Dundurs, J., and Mura., T., 1964. Interaction Between an Edge Dislocation and a Cirular Inclusion. *Journal of the Mechancis and Physics of Solids*, **12**, 177.

Dundurs, J., and Sendeckyj, G.P., 1965. Edge Dislocation Inside a Cirular Inclusion. *Journal of the Mechancis and Physics of Solids*, **13**, 141.

Erdogan, F., Gupta, G.D., and Ratwani, M., 1974. Interaction Between a Circular Inclusion and an Arbitrarily Oriented Crack. *Transactions of the ASME Journal of Applied Mechanics*, **41**, 1007.

Erdogan, F., and Gupta, G.D., 1975. The Inclusion Problem with a Crack Crossing the Bounary. *International Journal of Fracture*, **11**, 13.

Eshelby, J.D., 1961. Elastic Inclusions and Inhomogeneities. In *Progress in Solid Mechanics 2*. I. N. Sneddon and R. Hill, eds. North Holland, Amsterdam, 89.

Goldstein, R.V., and Salganik, R.L., 1974. Brittle Fracture of Solids with Arbitrary Cracks. *International Journal of Fracture*. **10**, 507.

Goursat, E., 1898. Sur l'equation $\Delta\delta u=0$. *Bulltin de ls Soc. Math. de France*, **26**, 236.

Hasebe, N., Wang, X., and Kondo, M., 2003. Interaction Between Crack and Arbitrarily Shaped Hole with Stress and Displacement Boundaries. *International Journal of Fracture*, **119**, 83.

Head, A.K., 1953. The Interaction of Dislocations and Boundaries. *Philosophical Magazin,* **44**, 92.

Helsing, J., and Jonsson, A., 2002. On the Accuracy of Benchmark Tables and Graphical Results in the Applied Mechanics Literature. *Transactions of the ASME Journal of Applied Mechanics*, **69**, 88.

Hu, K.X., Chandra, A., and Huang, Y., 1993a. Fundamental Solutions for Dilute Distributions of Inclusions Embedded in Microcracked Solids. *Mechanics of Materials*, **16**, 281.

Hu, K.X., Chandra, A., and Huang, Y. 1993b. Multiple Void-Crack Interaction. *International Journal of Solids and Structures*, **30**, 1473.

Hu, K.X., and Kemeny, J., 1994. A Fracture Mechanics Analysis of the Effect of Backfill on the Stability of Cut and Fill Mine Workings. *International Journal of Rock Mechancis, Mining Science and Geomechanical Abstracts*, **31**, 231.

Hwu, C., and Yen, W. J., 1993. On the Anisotropic Elastic Inclusions in Plane Elastostatics. *Transactions of the ASME Journal of Applied Mechanics*, **60**, 626.

Hwu, C., Liang Y.K., and Yen, W.J., 1995. Interactions Between Inclusions and Various Types of Cracks. *International Journal of Fracture*, **73**, 301.

Ioakimidis, N.I., and Theocaris, P.S., 1977. Array of Periodic Curvilinear Cracks in an Infinite Isotropic Medium. *Acta Mechanica*, **28**, 239.

Ioakimidis, N.I., and Theocaris, P.S., 1979. A System of Curvilinear Cracks in an Isotropic Elastic Half-Plane. *International Journal of Fracture,* **15**, 299.

Isida, M., Chen, D.H., and Nisitani, H., 1985. Plane Problems of an Arbitrary Array of Cracks Emanating From the Edge of an Elliptical Hole. *Engineering Fracture Mechanics*, **21**, 983.

Kemmer, G., 1995. *Uber die Anwendung Singularer Integralgleichungen zur Charakterisierung Gerkrummter Risse in der Ebenen Elastizitatstheorie.* Studienarbeit, Universitat Paderborn.

Kolosov, G.V., 1909. *On an Application of Complex Function Theory to a Plane Problem of the Mathematical Theory of Elasticity.* Dissertation at Dorpat University.

Kunin, I., and Gommerstadt, B., 1985. On the Elastic Crack-Inclusion Interaction. *International Journal of Solids and Structures*, **21**, 757.

Lam, K.Y., and Wen, C., 1993. Enhancement/Shielding Effects of Inclusion on Arbitrarily Oriented Located Cracks. *Engineering Fracture Mechanics*, **46**, 443.

Li, Z.G., and Shi, J.Y., 2002. The Interaction of a Screw Dislocation with Inclusion Analyzed by Eshelby Equivalent Inclusion Method. *Scripta Materialia*, **47**, 371.

Lo, K.K., 1978. Analysis of Branched Cracks. *Transactions of the ASME Journal of Applied Mechanics*, **45**, 797.

Muller, W.H., 1989. The Exact Calculation of Stress Intensity Factors in Transformation Toughened Ceramics by Means of Integral Equations. *International Journal of Fracture*, **41**, 1.

Muller, W.H., and Kemmer, G., 1995. On the Mathematical Description of Curved Cracks in Composite Materials. *Deutscher Verband fur Materialforschung und –prufung e. V.* DVM, Berlin.

Muller, W.H., and Schmauder, S., 1992. On the Behavior of r- and θ-Cracks in Composite Materials Under Thermal and Mechanical Loading. *International Journal of Solids and Structures*, **29**, 1907.

Muller, W.H., and Schmauder, S., 1993. Stress-intensity Factors or r-cracks in Fiber Reinforced Composites Under Thermal and Mechanical Loading. *International Journal of Fracture*, **59**, 307.

Muller, W.H., Gao, H., and Chiu, C.H., 1996. A Semi-infinite crack in Front of a Circular, Thermally Mismatched Heterogeneity. *International Journal of Solids and Structures*, **33**, 731.

Mura, T., 1988. Inclusion Problems. *Applied Mechanics Review*, **41**, 15.

Muskhelishvili, N.I., 1963. *Some Basic Problems in the Mathematical Theory of Elasticity.* P. Noordhoff, Ltd.

Patton, E.M., and Santare, M.H., 1990. The Effect of Rigid Elliptical Inclusion on a Straight Crack. *Engineering Fracture Mechanics*, **44**, 195.

Santare, M.H., Crocombe, A.D., and Anlas, G., 1995. Anisotropic Effective Moduli of Materials with Microcracks. *Engineering Fracture Mechanics*, **52**, 833.

Santare, M.H., and Keer, L.M., 1986. Interaction Between an Edge Dislocation and a Rigid Elliptical Inclusion. *Transactions of the ASME Journal of Applied Mechanics*, **53**, 382.

Savruk, M.P., 1981. *Two-Dimensional Problems of Elasticity for Body with Cracks.* Naukova Dumka. (in Russian).

Shi, J.Y., and Li, Z. H., 2003. The Interaction of an Edge Dislocation with an Inclusion of Arbitrary Shape Analyzed by the Eshelby Inclusion Method. *Acta Mechanica*, **161**, 31.

Sih, G.C., Paris, P.C., and Erdogan, F., 1962. Crack-Tip, Stress-Intensity Factors for Plane Extension and Plate Bending Problems. *Transactions of the ASME Journal of Applied Mechanics*, **29**, 306.

Sih, G.C., 1965. Stress Distribution Near Internal Crack Tips for Longitudinal Shear Problems. *Transactions of the ASME Journal of Applied Mechanics*, **32**, 51.

Stagni, L., 1982. On the Elastic Field Perturbation by Inhomogeneities in Plane Elasticity. *ZAMP*, **33**, 315.

Stagni, L., and Lizzio, R., 1983. Shape Effects in the Interaction Between an Edge Dislocation and an Elliptical Inhomgeneity. *Applied Physics*, **A30**, 217.

Stakgold, I., 1967. *Boundary Value Problems of Mathematical Physics*. Macmillan Series in Advanced Mathematics and Theoretical Physics, Macmillan Company London.

Steif, P.S., 1987. A Semi-infinite Crack Partially Penetrating a Circular Inclusion. *Transactions of the ASME Journal of Applied Mechanics*, **54**, 87.

Tamate, O., 1968. The Effect of a Circular Inclusion on the Stresses Around a Line Crack in a Sheet Under Tension. *International Journal of Fracture Mechanics*, **4**, 257.

Theocaris, P.S., and Demakos, C.B., 1985. Antiplane Shear Crack in an Infinite Plate with a Circular Inclusion. *Ing. Arch.*, **55**, 296.

Ting, T.C.T., 1996. Green's Functions for an Anisotropic Elliptic Inclusion under Generalized Plane Strain Deformations. *Quarterly Journal of Mechanics and Applied Mathematics*, **49**, 1.

Wang, R., 1995. A New Method for Calculating the Stress Intensity Factor of a Crack with a Circular Inclusion. *Acta Mechanica*, **108**, 77.

Warren, W.E., 1983. The Edge Dislocation Inside an Elliptical Inclusion. *Mechanics of Materials*, **2**, 319.

Weertman, J., 1996. *Dislocation Based Fracture Mechanics*. World Scientific Publishing Co. Pte. Ltd. Singapore

Wu, C.H., 1988. A Semi-infinite Crack Penetrating an Inclusion. *Transactinos of the ASME Journal of Applied Mechanics*, **55**, 736.

Yen, W.J. and Hwu, C., 1994. Interactions Between Dislocations and Anisotropic Elastic Elliptical Inclusions. *Transactions of the ASME Journal of Applied Mechanics*. **61**, 548.

Yen, W.J., Hwu, C., and Liang, Y.K., 1995. Dislocation Inside, Outside, or on the Interface of an Anisotropic Elliptic Inclusion. *Transactions of the ASME Journal of Applied Mechanics*, **62**, 306.

On the Mechanical Behavior of Carbon Nanotubes in Hexagonal Arrays

R. Byron Pipes, The University of Akron; S. J. V. Frankland, NASA Langley Research Center; Pascal Hubert, McGill University; Erik Saether, NASA Langley Research Center

Abstract

The development of nano-reinforcements for composite materials applications requires both the understanding of the physical properties of the reinforcements and the mixing rules that relate weight fraction used to measure mixing fractions to volume fraction that determines the composite physical properties. Given the mono-atomic thickness of the single-walled carbon nanotube, the reinforcing element can be thought of as a hollow cylindrical geometry and as such, all physical properties must be related to the same volume element. In the following the geometry of the carbon nanotube is described and utilized in the definition of effective properties. The properties of both single-walled and multi-walled carbon nanotubes of arbitrary chirality are described.

Single-walled carbon nanotubes are self assembled into arrays or crystals with hexagonal symmetry in order to satisfy the minimum potential energy configuration. The arrays consist of large numbers of nanotubes that possess cohesion that resulting from the van der Waals interactions. To assess the level of the van der Waals interactions the strain energy density function of the array is examined to calculate the effective transverse properties. Plane strain bulk modulus and transverse shearing modulus are taken as measures of the magnitude of the cohesive forces in the array.

Single Walled carbon Nanotubes and Their Arrays

The use of single-walled carbon nanotubes (SWNT) for reinforcement of polymeric materials has been given as one of the primary applications for this new material form since its discovery more than a decade ago (Iijima, 1991). Yet no unified basis for reporting the physical properties of SWNT or arrays of SWNT as reinforcements has emerged. For example, the reference volume, required to measure volume dependent properties, is taken as the entire cylindrical volume enclosed by the carbon atom crystal lattice when reporting density (Collins and Avouris, 2000), while for the effective Young's modulus, the monoatomic layer of carbon atoms is used (Majumdar, 2001; Krishnan et al., 1998). These different geometry definitions have led to reported axial moduli ranging from 1.25 Tpa (Krishnan et al., 1998), for SWNT to 67 GPa (Salvetat et al., 1999), in arrays. The need for a self-consistent set of properties is evident. Another author (Salvetat-Delmotte and Rubio, 2002), has also stated this need, but stopped short of offering a set of relationships that unify these data. In order to account for the contribution of the SWNT to the overall properties of the polymeric composite wherein the SWNT serves the role of the reinforcement, it is necessary to view its effective properties so that they correspond to a specified volume within the composite. In this work, the representative volume element for all physical properties is defined as that of a cylindrical volume with diameter equal to twice the effective SWNT radius, where the effective SWNT radius includes one-half the van der Waals equilibrium separation distance.

The van der Waals equilibrium distance for SWNT can be defined as the average center-to-center distance between the carbon atom of the SWNT and the nearest atom of the adjacent medium. Therefore, the van der Waals standoff distance is dependent upon the properties of the adjacent medium. In the case of the array of SWNT of equal diameter, the individual SWNT are surrounded by other SWNT of like properties and the equilibrium distance is designated as λ. Suspension of SWNT in a polymeric phase yields a different separation distance, ν. When employing the equations developed in the following, it is necessary to establish the value of ν corresponding to the suspending medium which can vary widely. In this paper, the equilibrium separation distance in polymers is arbitrarily taken as 0.342 nm, the equilibrium separation distance for graphene sheet (Dresselhaus and Saito, 1995). The SWNT and SWNT array physical property predictions presented in the following include density, principal Young's modulus (longitudinal direction) and specific modulus. It is common practice when mixing SWNT with a polymeric second phase to do so using weight fraction in determining proportions in the mixture. Yet the physical properties of the mixture depend upon the volume fractions of the constituents. In the following, relationships are developed not only to predict the physical properties of the SWNT and its arrays, but also those necessary to convert SWNT weight fraction to volume fraction in mixtures. These property relationships then serve to provide precursor data for computational methods in the prediction of effective properties of SWNT composites.

The literature is filled with reported values for the Young's modulus of carbon nanotubes and their arrays (Salvetat et al., 1999; Lourie and Wagner, 1998; Sanchez-Portal et al., 1999; Hernandez and Rubio, 1998; Cornwell and Wille, 1997). These reported values differ by almost two orders of magnitude. The SWNT density has also been reported in Chesnokov et al. (1999) and Gao et al. (1998); however, no clear methodology for density calculation has been published in the literature.

It is the objective of this paper to develop a self-consistent set of properties for the SWNT and its hexagonal arrays, as well as, to provide the mixing rules for conversion of weight fraction to volume fraction for mixtures of SWNT and SWNT arrays with polymers. Properties predicted include effective radius, density, principal Young's modulus, and specific Young's modulus.

SWNT Geometry

The SWNT has been described as a single graphene sheet rolled up with varying degrees of twist as described by its chiral vector, $\mathbf{C_h}$ (Dresselhaus and Saito, 1995):

$$\mathbf{C_h} = n\mathbf{a_1} + m\mathbf{a_2} \tag{1}$$

where $\mathbf{a_1}$ and $\mathbf{a_2}$ are unit vectors in the two-dimensional hexagonal lattice and the chiral vector, $\mathbf{C_h}$ is also referred to by its indices, n and m. Nanotubes with chiral vectors of (n,n) and (n,0) have no twist and are classified as achiral nanotubes. These two special cases are sometimes denoted "armchair" and "zig zag," respectively, referring to the pattern of the carbon atoms around the nanotube circumference (Dresselhaus and Saito, 1995). Geometric representations of SWNT structures of three different chiralities are shown in Fig. 1.

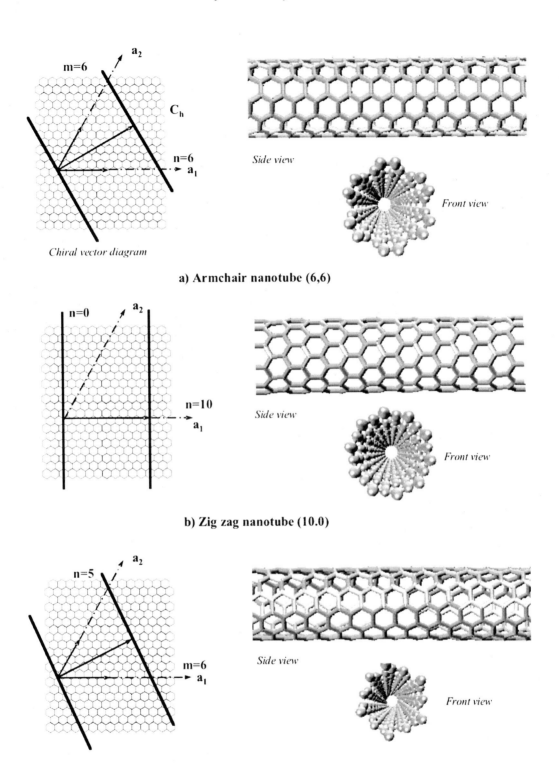

a) Armchair nanotube (6,6)

b) Zig zag nanotube (10.0)

c) Chiral nanotube (6,5)

Fig. 1. SWNT structure and example of nanotubes.

These images illustrate the dependence of the SWNT upon the components of the chiral vector, n,m. The SWNT radius, R_n as shown in Fig. 2 is given in the following relation as a function of the integer pair and the C-C bond length, b (Dresselhaus and Saito, 1995):

$$R_n = \frac{b}{2\pi}\sqrt{3}\Lambda \tag{2}$$

where

$$\Lambda = \sqrt{\left(n^2 + m^2 + mn\right)} \tag{3}$$

For the SWNT, the C-C bond length, b is equal to 0.142 nm (Dresselhaus and Saito, 1995). The carbon nanotube structure as shown in 2a is replaced by the effective reinforcing element in 2b.

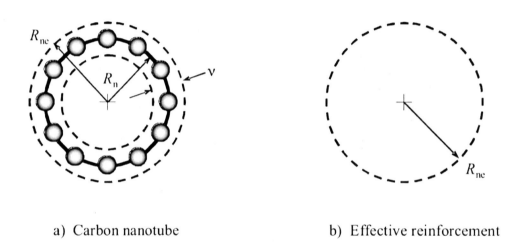

a) Carbon nanotube   b) Effective reinforcement

Fig. 2. SWNT nomenclature.

The diameter of the SWNT can vary over a wide range as shown in Table 1. However, it is the smaller diameters of SWNT that are of most interest as a polymeric reinforcement since the transverse properties of the larger tube diameters can lead to their collapse (Yakobson et al., 1996). The remaining geometric descriptors of the SWNT are the cross-sectional area and length. As stated earlier, several investigators have taken the SWNT equilibrium separation distance to be equal to that of the equilibrium separation distance of graphene sheets of 0.342 nm (Yakobson et al., 1996). This issue requires more discussion since, as stated earlier, the cross-sectional area occupied by the SWNT is influenced by the character of the adjacent medium.

The length of the SWNT is likely to be a function of both carbon nanotube process technology, as well as, handling and mixing of the SWNT with the polymeric phase just as it is for discontinuous fiber systems. Bundle lengths of up to 20,000 nm have been measured (Arepalli et al., 2001; Thess et al., 1996).

Table 1 SWNT Diameter for Various Chiral Integers.

n	m	$2R_n$ (nm)
5	5	0.68
4	6	0.68
3	7	0.70
2	8	0.72
10	10	1.36
50	50	6.78
6	0	0.48
8	0	0.63
10	0	0.78
12	0	0.94
18	0	1.41
24	0	1.88
50	0	3.91
96	0	7.52

The effective radius of the SWNT, R_{ne}, is defined as the radius given in Eq. (2) plus one-half the equilibrium separation distance between the SWNT and the polymer. By defining SWNT radius in this way, the total volume of the heterogeneous mixture is accounted for.

$$R_{ne} = \frac{b}{2\pi}\sqrt{3}\Lambda + \frac{v}{2} \qquad (4)$$

SWNT Hexagonal Array

The synthesis of SWNT typically results in the generation of collimated arrays of SWNT with hexagonal cross-sectional arrangement (Thess et al., 1996). In the present study we examined the equilibrium separation distance of such a hexagonal array, λ calculated using the methods discussed in Saether et al. (2002), wherein the van der Waals interactions of the SWNT are modeled with the Lennard-Jones potential (ε =34.0 K and σ = 0.3406 nm) (Saether et al., 2002).

The results for (n,0) SWNT were calculated in static molecular simulations are presented in Table 2.

Table 2 SWNT (n,0) Hexagonal Array Separation Distance.

n	Diameter (nm)	Separation Distance, λ (nm)
6	0.48	0.316
12	0.94	0.317
18	1.41	0.317
24	1.88	0.318
54	4.23	0.318
96	7.51	0.319

These results suggest that the separation distance is not a function of the SWNT diameter, but rather can be taken as a constant equal to 0.318 nm in agreement with the equilibrium distance of 0.313 nm determined in Girifalco et al. (2000) and 0.315 nm in Popov et al. (2000) for a similar range of SWNT diameters. Thus, the SWNT effective radius, R_{na}, in an array configuration (See Fig. 3) is expressed as:

$$R_{na} = \frac{b}{2\pi}\sqrt{3}\Lambda + \frac{\lambda}{2} \tag{5}$$

The carbon nanotube array shown in Fig. 3a is replaced by the effective reinforcement array shown in Fig. 3b. For a constant separation distance, λ, the SWNT volume fraction of the hexagonal array, V_a is a constant and is equal to the volume packing fraction of 0.906:

$$V_a = \frac{\pi}{2\sqrt{3}} = 0.906 \tag{6}$$

It should be pointed out that the hexagonal SWNT array could be thought of as a reinforcement form in like fashion to the SWNT. As such, the descriptions presented are for the effective properties of the array in the absence of a second phase.

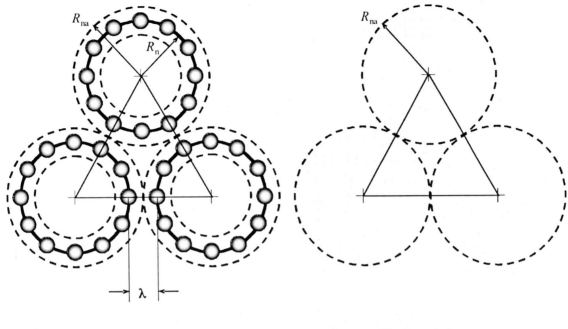

a) Carbon nanotube array   b) Effective reinforcement array

Fig. 3. SWNT array nomenclature.

Density

The density of the SWNT in the present work is defined as the total mass of the carbon atoms in the enclosed volume as defined in Eq. (5). To calculate the mass of the carbon atoms, it is necessary to express their number per unit length, N in the nanotube representative volume element:

$$N = \frac{4\Lambda}{3b} \tag{7}$$

Then the nanotube density, ρ_n can be expressed in terms of the chiral vector by combining Eqs. (4) and (7) as follows:

$$\rho_n = \frac{N M_w}{\pi N_a R_{ne}^2} = \frac{16\pi M_w \Lambda}{3 N_a b \left(3b^2\Lambda^2 + 2\sqrt{3}b\pi v \Lambda + \pi^2 v^2 \right)} \tag{8}$$

where v, measured in nm, is the equilibrium standoff distance between the SWNT and the adjacent medium, M_w is carbon atomic weight and N_a is Avogadro's number. SWNT density as a function of diameter (Eq. 8) is presented in Fig. 4 and for specific SWNT in Table 3. These results show that SWNT density decreases with increase in diameter by an order of magnitude over a range in diameter between 1 and 14 nanometers.

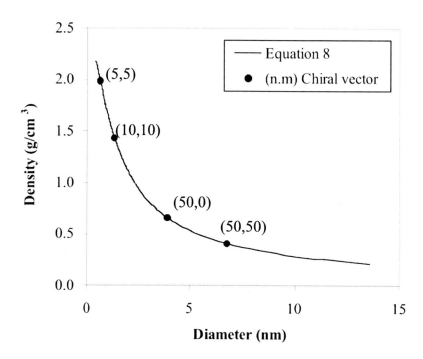

Fig. 4. SWNT density versus diameter.

The density results presented in Table 3 show some sensitivity to the choice of the van der Waals distance, v as shown for a (10,10) SWNT in Table 4.

The density of the hexagonal array of SWNT is the product of the SWNT density in an array configuration (Eqs. 5 and 7) and the volume fraction of the hexagonal array (Eq. 6):

$$\rho_a = V_a \rho_{na} = V_a \frac{NM_w}{\pi N_a R_{na}^2} = \frac{\pi}{2\sqrt{3}} \left[\frac{16\pi M_w \Lambda}{3N_a b \left(3b^2 \Lambda^2 + 2\sqrt{3}b\pi\lambda\Lambda + \pi^2\lambda^2 \right)} \right] \qquad (9)$$

Like the density of the SWNT, the array density shows a significant reduction with SWNT diameter (Eq. 9). It is noteworthy that density of the SWNT array differs from that of the SWNT only by the product of the maximum volume fraction (0.906) and the difference in separation distances, v and λ. Since the separation distances for the array (0.318 nm) and SWNT (0.342 nm) are similar, it is likely that the array density will not differ significantly from those of the SWNT. Fig. 5 shows a comparison between density values for SWNT and arrays for several nanotube chiralities in tabular form.

Table 3 SWNT Chiral Integers Versus Density

n	m	Density, g/cm^3
5	5	1.99
4	6	1.99
3	7	1.98
2	8	1.95
10	10	1.44
50	50	0.41
6	0	2.18
8	0	2.04
10	0	1.89
12	0	1.75
18	0	1.40
24	0	1.16
50	0	0.66
96	0	0.37

Table 4 van der Waals Distance and Density for (10,10) SWNT

ν, nm	Density, g/cm^3
0.32	1.471
0.33	1.453
0.34	1.436
0.35	1.419
0.36	1.403

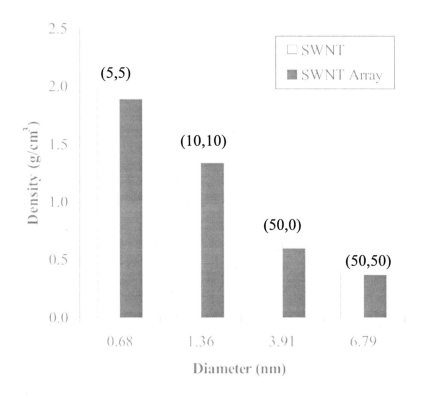

Fig. 5. Comparison of SWNT and array densities.

Principal Young's Modulus

Here the principal modulus is taken as the modulus of the SWNT and its arrays in the direction parallel to the SWNT longitudinal axis. In the following development it is assumed that the SWNT are continuous and that the array consists of SWNT of identical diameter. The present model is a simplified approach wherein the stiffness of a graphene sheet, rolled into the SWNT configuration, is mapped onto the enclosed volume defined by the effective SWNT radius.

SWNT Young's Modulus

The present approach to estimate the Young's modulus, E_n, of the SWNT is to represent the SWNT as a thin-walled cylinder of outer radius equal to that predicted by Eq. (4). It is assumed that the stiffness of the thin wall of the SWNT is equal to that of the Young's modulus of the graphene sheet, Y and that the SWNT occupies the enclosed cylindrical volume.

$$E_n = \frac{8YR_n v}{4R_n^2 + 4R_n v + v^2}$$

(10)

$$Y = \frac{\left(C_{11}^2 - C_{12}^2\right)}{C_{11}}$$

The C_{11} and C_{12} are the stiffness constants of graphene (Popov et al., 2000).

Substituting Eq. (2) into (10) yields:

$$E_n = \frac{4\sqrt{3}\pi b Y \Lambda v}{3b^2\Lambda^2 + 2\sqrt{3}b\pi\Lambda v + \pi^2 v^2}$$

(11)

Eq. (11) shows that the modulus of the SWNT decreases with increase in radius. Earlier work described in Salvetat et al. (1999); Cornwell and Wille (1997) and Ruoff and Lorents (1995) has shown a similar behavior. Here we have chosen to illustrate the work described in Cornwell and Wille (1997) and show a comparison with the present work prediction in Eq. (11).

$$E_n = \frac{A}{R_n} + B$$

(12)

Where the constants in Eq. (12) are defined as A = 429.6 GPa-nm and B = 8.42 GPa. The choice of the equilibrium separation distance, v and Young's modulus of the graphene sheet in Eq. (11) uniquely determine the quantitative values of the SWNT Young's modulus for all radii. In the present study, we take the experimental value of the Young's modulus of the graphene sheet as reported in Popov and Van Doren (2000) as 1029 GPa and the separation distance as that for the graphene sheet of 0.342 nm. Predictions of SWNT modulus as a function of diameter using Eq. (11) are shown in Fig. 6. Table 5 illustrates the utility of Eq. (11) in determining Young's modulus for five specific sets of chiral integers, (n,m). In addition, a comparison with the results of Salvetat et al. (1999) and Ruoff and Lorents (1995) show that all four models provide very similar results.

Table 5 SWNT Young's modulus for specific chiral integers.

n	m	E_n, GPa Eq. 11	E_n, GPa Eq. 12	E_n, GPa Salvetat et al. (1999)	E_n, GPa Ruoff and Lorents (1995)
10	10	662	642	608	680
18	0	647	618	594	664
24	0	536	465	492	550
50	0	304	228	279	311
96	0	171	123	157	175

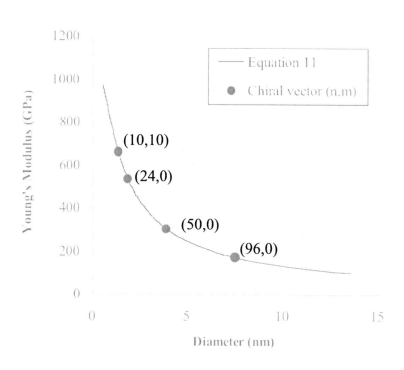

Fig. 6. SWNT Young's modulus versus diameter.

SWNT Hexagonal Array Young's Modulus

Utilizing the Young's modulus, E_a of a SWNT array containing SWNT with modulus predicted by Eq. (11) we obtain the following array modulus:

$$E_a = V_a E_{na} = \frac{2\pi^2 b Y \Lambda \lambda}{3b^2 \Lambda^2 + 2\sqrt{3}b\pi\Lambda\lambda + \pi^2\lambda^2} \tag{13}$$

Results for the principal modulus of the SWNT array shown in Fig. 7 predicted by Eq. (13) show that the array modulus follows identical trends as for the individual SWNT. This is true because the two predictions differ only by the product of the maximum volume fraction (0.906) and the difference in van der Waals distances ($\lambda = 0.318$ for the array). Predictions of Eq. (13) also show excellent agreement with those presented in Popov and Van Doren (2000) obtained from lattice dynamics calculations for (3n,3n) "armchair" and (3n,0) "zigzag" SWNT as shown in Fig. 7.

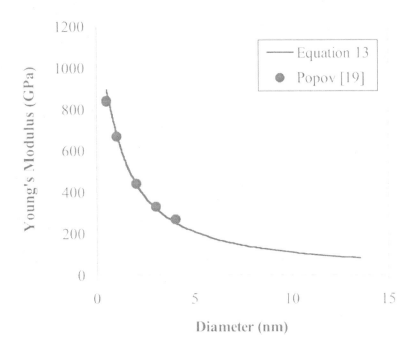

Fig. 7. SWNT array Young's modulus versus diameter.

The data taken from Popov and Van Doren (2000) were for three data points and one interpolated value. To further clarify the differences in moduli between the SWNT and the array, results are presented in tabular form in Fig. 8 for several nanotube chiralities.

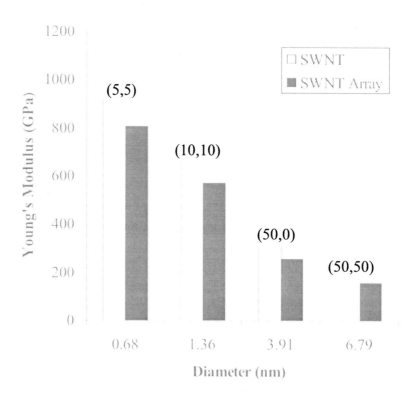

Fig. 8. Comparison of SWNT and array Young's modulus.

Specific Axial Modulus

The specific modulus, $E_{\rho n}$ of the SWNT is defined as the ratio of the principal Young's modulus, E_n divided by a quantity analogous to specific gravity of the SWNT, $\bar{\rho}_n$. Combining Eqs. (8) and (10) yields the expression:

$$E_{\rho n} = \frac{E_n}{\bar{\rho}_n} = \frac{3N_a\sqrt{3}b^2 Y v}{4M_w} \qquad (14)$$

It is not surprising to note that Eq. (14) yields results independent of the SWNT diameter since the same SWNT volume is utilized in the calculation of both modulus and density.

In a similar manner the specific modulus for the SWNT array, $E_{\rho a}$, is obtained by combining Eqs. (9) and (13):

$$E_{\rho a} = \frac{E_a}{\bar{\rho}_a} = \frac{3N_a\sqrt{3}b^2 Y \lambda}{4M_w} \qquad (15)$$

These results show that the specific modulus of the SWNT array is identical in form to that of the SWNT.

Weight Fraction Versus Volume Fraction

With the establishment of the density of the SWNT, it is now possible to develop a relationship between weight fraction and volume fraction of SWNT in a mixture with a second material such as a polymer. Consider a SWNT/polymer mixture of density, ρ_m with a SWNT volume fraction, V_n, a SWNT density of ρ_n and a polymer density of ρ_p. The SWNT volume fraction can be expressed as follows:

$$V_n = \frac{\rho_m - \rho_p}{\rho_n - \rho_p} \tag{16}$$

It is also possible to express the SWNT volume fraction in terms of its weight fraction, W_n:

$$V_n = \frac{\rho_m}{\rho_n} W_n = \frac{W_n \rho_p}{W_n \rho_p + (1 - W_n)\rho_n} \tag{17}$$

Combining Eqs. (20) and (8) we obtain:

$$V_n = \frac{W_n 3b \left(3b^2 \Lambda^2 + 2\sqrt{3}b\pi v\Lambda + \pi^2 v^2\right)\rho_p}{W_n \rho_p 3b \left(3b^2 \Lambda^2 + 2\sqrt{3}b\pi v\Lambda + \pi^2 v^2\right) + (1 - W_n)16\pi k\Lambda} \tag{18}$$

The relationship between volume fraction and weight fraction for SWNT-polymer mixtures is shown in Fig. 9 for (6,6), (12,12) and (18,18) SWNT for a polymer density of 1 g/cm^3. These results show that as the diameter of the SWNT is decreased the non-linearity in the relationship increases. Results in the dilute concentration regime are shown in Fig. 10 for convenience since many applications are in this range. The influence of variations in polymer density upon the relationship between volume and weight fraction is shown in Fig. 11 for the (10,10) SWNT. Here increases in polymer density reduce the non-linearity in the relationship.

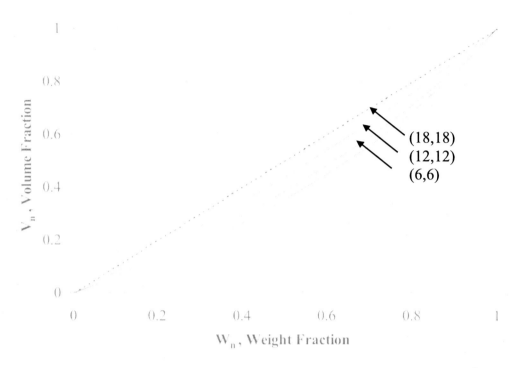

Fig. 9. SWNT volume fraction versus weight fraction (polymer density = 1 g/cm^3).

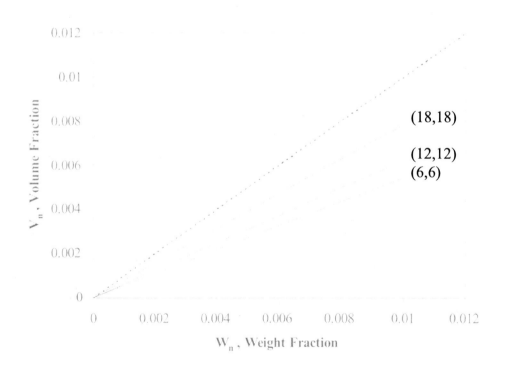

Fig. 10. SWNT results at dilute concentrations (polymer density = 1 g/cm^3) .

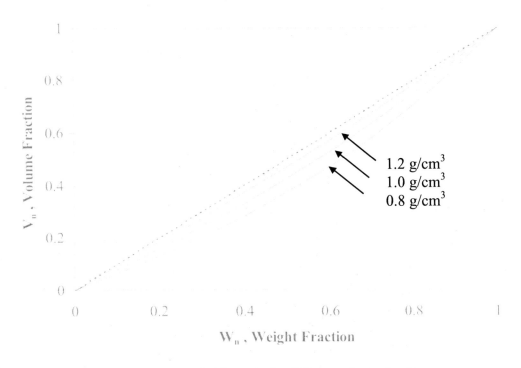

Fig. 11. SWNT (10,10) results for different polymer densities.

The relationship for SWNT arrays in a polymer mixture containing weight fraction, W_a and the volume fraction, V_a is:

$$V_a = \frac{W_a 3\sqrt{3}b\left(3b^2\Lambda^2 + 2\sqrt{3}b\pi v\Lambda + \pi^2\lambda^2\right)\rho_p}{W_a\rho_p 3\sqrt{3}b\left(3b^2\Lambda^2 + 2\sqrt{3}b\pi v\Lambda + \pi^2\lambda^2\right)+ \left(1 - W_a\right)8\pi^2 k\Lambda} \quad (19)$$

Fig. 12 illustrates a similar weight-volume fraction relationship for SWNT arrays as for individual SWNT suspended in a polymeric phase of density 1 g/cm^3. Arrays in dilute concentrations suspended in a polymeric phase are illustrated in Fig. 13 for (6,6), (12,12) and (18,18) SWNT. Finally, the influence of polymer density upon the relationship between polymer mixtures containing SWNT arrays is shown in Fig. 14.

These relationships, based on a consistent definition of volume and thereby, volume fraction, are necessary for micromechanics calculations to determine the effective properties of nanotube reinforced polymers.

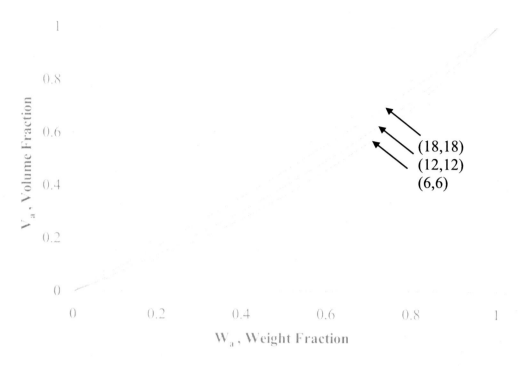

Fig. 12. Array volume fraction versus weight fraction (polymer density = 1 g/cm^3).

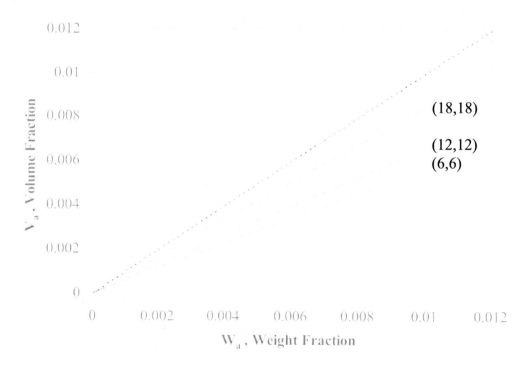

Fig. 13. Array results at dilute concentrations (polymer density = 1 g/cm^3).

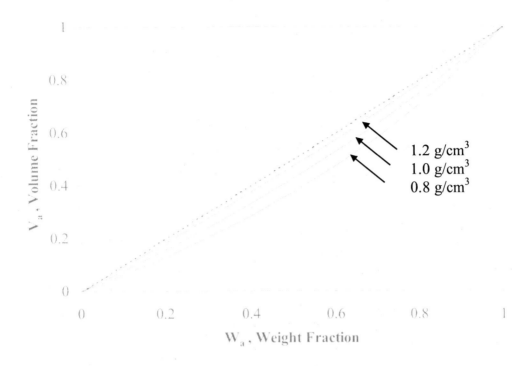

Fig. 14. SWNT (10,10) array results for different polymer densities.

Summary Table of Properties

Table 6 shows the set of physical properties of the SWNT and its hexagonal arrays as a function of the chiral vector, (n,m).

Observations for SWNT and Their Arrays

One of the primary objectives of this paper was to develop a consistent set of predictions for several of the physical properties of SWNT and their arrays to provide the foundation for developing an understanding of SWNT and arrays as reinforcements in polymers.

The density and moduli of the SWNT and the SWNT array are shown to differ only slightly since the van der Waals distances differ only modestly (0.318 versus 0.342 nm) and the array has a packing fraction of 0.906. The equilibrium separation distance for the SWNT hexagonal array, determined to be 0.318 nm, was seen to be in good agreement with the values of 0.313 nm and 0.315 published earlier in the literature.

When the simple model for the SWNT principal modulus, consisting of a hollow cylinder with properties of the cylinder wall equal to the stiffness of the graphene sheet was employed, the specific modulus of both the SWNT and its hexagonal arrays were found to be independent of SWNT diameter.

Table 6 Properties of SWNT and SWNT arrays.

Property	SWNT	SWNT Array
Radius (nm)	$R_{ne} = \dfrac{b}{2\pi}\sqrt{3}\Lambda + \dfrac{v}{2}$	$R_{na} = \dfrac{b}{2\pi}\sqrt{3}\Lambda + \dfrac{\lambda}{2}$
Density (g/cm³)	$\rho_n = \dfrac{16\pi M_w \Lambda}{3bN_a\left(3b^2\Lambda^2 + 2\sqrt{3}b\pi v\Lambda + \pi^2 v^2\right)}$	$\rho_a = \dfrac{\pi}{2\sqrt{3}N_a}\left[\dfrac{16\pi M_w \Lambda}{3b\left(3b^2\Lambda^2 + 2\sqrt{3}b\pi\lambda\Lambda + \pi^2\lambda^2\right)}\right]$
Modulus (GPa)	$E_n = \dfrac{4\sqrt{3}\pi b Y\Lambda v}{3b^2\Lambda^2 + 2\sqrt{3}b\pi\Lambda v + \pi^2 v^2}$	$E_{na} = V_a E_{na} = \dfrac{2\pi^2 b Y\Lambda\lambda}{3b^2\Lambda^2 + 2\sqrt{3}b\pi\Lambda\lambda + \pi^2\lambda^2}$
Specific Modulus (GPa)	$E_{\rho n} = \dfrac{3\sqrt{3}b^2 Y v}{4k}$	$E_{\rho a} = \dfrac{3\sqrt{3}b^2 Y\lambda}{4k}$
Volume-Weight Fraction	$V_n = \dfrac{W_n 3b\left(3b^2\Lambda^2 + 2\sqrt{3}b\pi v\Lambda + \pi^2 v^2\right)\rho_p}{W_n\rho_p 3b\left(3b^2\Lambda^2 + 2\sqrt{3}b\pi v\Lambda + \pi^2 v^2\right) + \left(1 - W_n\right)16\pi k\Lambda}$ $V_a = \dfrac{W_a 3\sqrt{3}b\left(3b^2\Lambda^2 + 2\sqrt{3}b\pi v\Lambda + \pi^2\lambda^2\right)\rho_p}{W_a\rho_p 3\sqrt{3}b\left(3b^2\Lambda^2 + 2\sqrt{3}b\pi v\Lambda + \pi^2\lambda^2\right) + \left(1 - W_a\right)8\pi^2 k\Lambda}$	
	$\Lambda = \sqrt{\left(n^2 + m^2 + mn\right)}$ b = 0.142 nm v = 0.342 nm Y = 1029 GPa	$M_w/N_a = 0.01995$ λ = 0.318 nm

For both the SWNT and array, the principal Young's modulus showed significant dependence upon SWNT radius. Predictions of array principal moduli were also shown to be in excellent agreement with those previously published in the literature for (3n,3n) "armchair" and (3n,0) "zigzag" SWNT. Maximum principal modulus was achieved at the smallest SWNT radius, while density decreased significantly for increased SWNT radius.

In dealing with SWNT/polymer mixtures, a useful equation was derived for the relationship between weight fraction and volume fraction for SWNT and hexagonal arrays. The relationship requires knowledge only of the components of the chiral vector of the SWNT and the weight fractions of the constituents to determine volume fraction of the SWNT or its hexagonal arrays in a mixture.

The equations developed in the present work may easily be adapted to a spreadsheet format and can be useful to researchers in the field of composite materials.

Multi Walled Carbon Nanotubes (MWNT)

MWNT Geometry

The geometry of the MWNT is illustrated in Fig. 15 where the external radius is defined in terms of the effective internal radius, R_{mi}, the number of layers, N and the van der Waals equilibrium distance, μ. Here the van der Waals distance is defined as the equilibrium separation between layers and is taken to be constant throughout the MWNT as illustrated in Fig. 15. The internal radius is clearly shown as the radius of the intermost graphene layer, minus on-half the van der Waals distance.

$$R_{me} = R_{mi} + N\mu \quad (nm) \tag{20}$$

The each layer may be described by a unique values of the interger pair (n,m) and the chiral vector need not be the same for all layers.

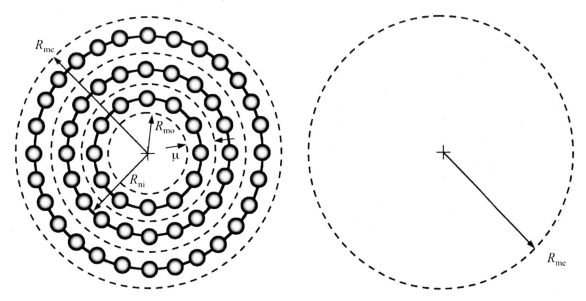

Fig. 15. MWNT geometry.

MWNT Density

The density of the MWNT is determined by summing the number of carbon atoms within all layers and dividing by the volume containing the atoms consistent with Eq. (20) for SWNT density.

$$\rho_m = \frac{\sum_{i=1}^{N} N_i k}{\pi R_{me}^2} \quad (g/cm^3)$$ (21)

Fig. 16 shows the increase in MWNT density with increase in outer diameter. Here the inside diameter of the MWNT is fixed by the choice of the innermost CNT layer and the diameter is increased by adding layers to achieve the increased diameter. Note that the MWNT density converges on a value of 2.2 gm/cm^3, the density of single crystal graphite.

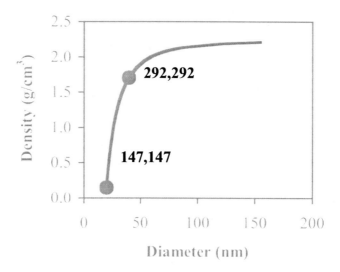

Fig. 16. MWNT density versus diameter.

MWNT Young's Modulus and Flexural Modulus

The Young's modulus of the MWNT can be expressed as the sum of the contributions of each of its layers:

$$E_m = \frac{\sum_{i=1}^{N} G\mu \left[(R_{ni} + \mu/2)^2 - (R_{ni} - \mu/2)^2 \right]}{R_m^2} \quad (GPa)$$ (22)

Eq. (22) is based on the relationship developed earlier by the authors and presented in Eq. (11). The value of G in this study was taken to be 238 N/m.

The increase in Young's modulus with diameter for the MWNT, shown in Fig. 17, mirrors its increase in density with increase in diameter. The innermost radius corresponds to the (147,147) CNT. It is noteworthy that the asymptotic value of Young's modulus of the MWNT exceeds that for high strength carbon fiber by more than a factor of 2.0.

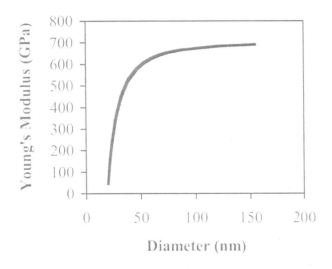

Fig. 17. MWNT Young's modulus versus diameter.

The flexural and extensional stiffnesses are not equal for the MWNT:

$$\frac{E_{mf}}{E_m} = \left[1 + \left(\frac{R_{mi}}{R_{me}}\right)^2\right] \qquad (23)$$

Note that the MWNT inner and outer radii, R_{mi} and R_{mo} are defined in Fig. 15, where $R_{mo} = R_{me}$. The ratio of flexural to extensional modulus is seen to approach unity as the outer radius is increased (Fig. 18). This is consistent with a uniformly filled cylindrical geometry.

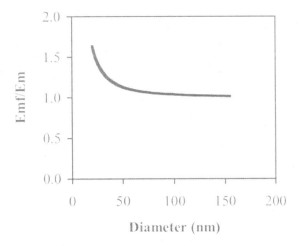

Fig. 18. Ratio of MWNT flexural to axial modulus.

Observations for MWNT

Relationships have been developed for the geometric and physical properties of MWNT that require only the specification of chiral vector properties and the van der Waals distances. A common volume element provides for density and modulus property predictions that are consistent. Results presented for the MWNT show that both density and principal Young's modulus increase with increase in diameter due to the addition of atomic layers that contribute both to density and modulus as the MWNT diameter is increased. The ratio of flexural to axial moduli of MWNT are presented and the results are shown to converge to unity with increase in diameter.

Van der Waals Forces and Transverse Properties of SWNT Arrays

Carbon nanotubes naturally tend to form crystals in the form of arrays with hexagonal symmetry that can be expected to exhibit a transversely isotropic constitutive behavior. Although the nanotube axial stiffness is on the order of 1 TPa due to a strong network of carbon-carbon bonds, the intertube interactions are controlled by weaker, nonbonding van der Waals forces which are orders of magnitude less. An accurate determination of the effective mechanical properties of nanotube bundles is important in order to assess potential structural applications such as reinforcement in future composite material systems. A direct method for calculating effective material constants is applied in the present study. The Lennard-Jones potential can be used to model the nonbonding cohesive forces. A complete set of transverse moduli are obtained and compared with existing predictions.

Future nanostructured composite materials are expected to incorporate carbon nanotube reinforcement either dispersed individually or as nanofilamentary bundles or ropes yielding unprecedented mechanical properties. Fig. 19 contains a cross-section of a bundle ensemble of individual nanotubes obtained through transmission electron microscopy (TEM) (Iijima, 1991).

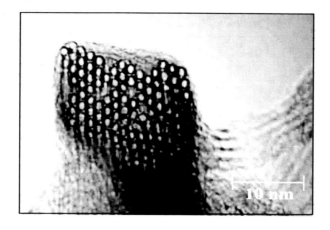

Fig. 19. Nanotube array or crystal. (Iijima, 1991).

Nanotube ensembles typically form hexagonally packed crystal configurations in which the intertube force interactions are due exclusively to non-bonding van der Waals effects which are much weaker than the valence forces and are highly nonlinear. Less consideration has been given to the transverse mechanical properties of nanotube bundles which depend on a good description of these non-bonding interactions. These intertube cohesive properties are of special interest for use in predicting the properties of carbon nanotube polymer composites (Frankland et al., 2002) and fibers of woven nanotubes (Pipes and Hubert, 2002). Selected moduli of nanotube bundles have been calculated with a continuum model based on the integrated average of the discrete Lennard-Jones potential (Girifalco et al., 2000), MD simulation using the Tersoff-Brenner potential (Tersoff and Ruoff, 1994) and lattice dynamic methods (Popov and Van Doren, 2000; Lu, 1997).

A direct summation of atom-pair potentials can be used to avoid any simplifications made to the nonlinear van der Waals interactions. Because the fundamental constituents of nanotube bundles are only resolvable at nanometer length scales, analyses to predict macroscopic properties must necessarily merge concepts and techniques from continuum elasticity theory and discrete molecular simulation. The basic approach of subjecting a molecular ensemble to applied strain modes and recovering effective moduli from energy measures has been used in molecular dynamic simulations (Theodorou and Suter, 1986; Fan and Hsu, 1992). The methodology developed herein combines a unit cell continuum model with molecular static calculations to determine effective moduli in aligned carbon nanotube bundles.

The Lennard-Jones potential is utilized to simulate the van der Waals interaction forces among carbon atom-pairs in aligned carbon nanotube arrays. An achiral "zig-zag" configuration is assumed for the carbon nanotubes with 12 graphene units around the circumference. Using the standard Hamada index notation (Hamada et al., 1992), this configuration is referred to as a (12,0) nanotube. The resulting tube radius is assumed small such that the cross-section can be considered rigid.

The objective of this work is to formulate a molecular mechanics model to predict the elastic properties of carbon nanotube arrays in the transverse plane. This method utilizes a unit cell approach for simulating a hexagonal crystal of aligned nanotubes. The calculated moduli are shown to exhibit a transverse isotropy which is anticipated for a material possessing hexagonal symmetry. The predicted moduli can be compared with available published data.

Material Constitutive Relationship

The coordinate system assumed for individual nanotubes and the form of the stress-strain relation is shown in Fig. 20. A transversely isotropic material is defined by five independent parameters, C_{11}, C_{12}, C_{44}, C_{22} and C_{23}. Only two independent elastic constants are required to define an isotropic material system. For the transverse planes in a bundle, these constants are given by the C_{22} and C_{23} stiffnesses.

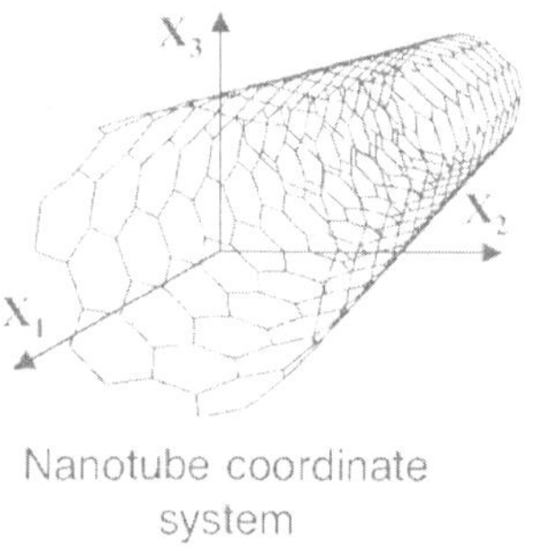

$$\begin{Bmatrix}\sigma_{11}\\ \sigma_{22}\\ \sigma_{33}\\ \sigma_{23}\\ \sigma_{31}\\ \sigma_{12}\end{Bmatrix} = \begin{bmatrix} C_{11} & C_{12} & C_{12} & & & \\ C_{12} & C_{22} & C_{23} & & & \\ C_{12} & C_{23} & C_{22} & & & \\ & & & C_{22}-C_{23} & & \\ & & & & 2C_{44} & \\ & & & & & 2C_{44}\end{bmatrix}\begin{Bmatrix}\varepsilon_{11}\\ \varepsilon_{22}\\ \varepsilon_{33}\\ \varepsilon_{23}\\ \varepsilon_{31}\\ \varepsilon_{12}\end{Bmatrix}$$

Fig. 20. Nanotube coordinate system and constitutive law for transverse isotropy

The plane strain bulk modulus and shearing modulus in the transverse plane for a hexagonal system are given in terms of the stiffness coefficients by:

$$\begin{aligned} K_{23} &= (C_{22} + C_{23})/2 \\ G_{23} &= (C_{22} - C_{23})/2 \end{aligned} \qquad (24)$$

where K_{23} and G_{23} are the transverse bulk and shear moduli, respectively.

Modified Unit Cell Formulation

In micromechanical analyses, the method of unit cells has been developed to determine the effective properties of heterogeneous materials by identifying and analyzing convenient domains of repeating microstructure. Applied to the determination of effective continuum elastic moduli of bundles of aligned carbon nanotubes, a repeating unit of nanotubes is defined and subjected to continuous field deformation modes. Because the potential energy of the system is due to atom pair interactions between adjacent nanotubes, the cell boundaries need to be defined as periodic. Under periodic boundary conditions (pbc's), cells of nanotubes in the transverse plane and nanotube segments in the axial dimension are treated as images of the constituents within the cell and used in the calculation of potential energy. This permits interactions between atom-pairs across the boundary to avoid introducing discontinuities in the force field. In general, these conditions ensure conservation of mass and energy, avoid surface or boundary effects, and mathematically give the primary unit cell a strict periodicity such that it can be considered to represent an infinite ensemble of molecules (Allen and Tildesley, 1987). An assemblage of a primary unit cell of nanotubes with surrounding image regions that are required in applying periodic boundary conditions.

The initial equilibrium configuration of the hexagonal unit cell is determined by minimizing the energy of the system as the nanotubes are moved radially outward from a fixed center. This establishes the equilibrium radius, R_{eq}, and the nanotube center-to-center separation distance, S, as shown in Fig. 21.

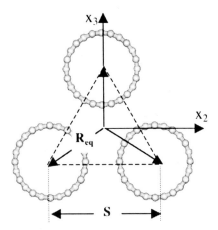

Fig. 21. Equilibrium radius definition for a hexagonal cell.

Potential Energy Calculations

The intertube forces are typically modeled by the Lennard-Jones potential to represent van der Waals interactions. The Lennard-Jones or '6-12' potential energy function is given by

$$\Phi = 4\varsigma \left[\left(\frac{\alpha}{r_{ij}} \right)^{12} - \left(\frac{\alpha}{r_{ij}} \right)^{6} \right] \qquad (25)$$

where ς is the depth of the energy well, α is the van der Waals radius, and r_{ij} is the separation distance between the ith and j^{th} atoms in a pair. (Traditionally, the parameters α and ς are referred to as σ and ε, which are used in this work for stress and strain, respectively.) The r_{ij}^{-6} term represents the attractive contribution to the van der Waals forces between neutral molecules. It includes permanent dipole-dipole interactions, the induction effect of permanent dipoles, and instantaneous dipole induced dipole interactions which are sometimes referred to as the London dispersion forces. The other component of the van der Waals interactions mimics the repulsion between overlapping electron clouds and is modeled by the r_{ij}^{-12} term which is short ranged (Allen and Tildesley, 1987; Moore, 1972). The combined effect of the attractive and repulsive interactions on potential energy is shown in Fig. 22.

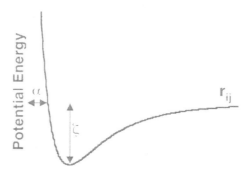

Fig. 22. Form of the Lennard-Jones Potential.

Analysis Methodology

The methodology used to determine selected nanotube crystal properties involves defining an appropriate pbc-unit cell, applying selected strain modes to the crystal, and computing the potential energy due to atom-pair interactions as a function of the deformation kinematics. A direct transformation to continuum properties is then made by assuming that the potential energy of discrete atom interactions is equal to the strain energy of a continuous substance occupying the volume of the unit cell: Φ_o (Potential energy density) $\rightarrow U_o$ (Strain energy density). Effective elastic constants are then determined from the variation in the system strain energy density as:

$$C_{ij} = \frac{\partial^2 U_o}{\partial \varepsilon_i \, \partial \varepsilon_j} \qquad (26)$$

where C_{ij} is the material stiffness, U_o is the strain energy density, and ε_k is an applied strain mode. Strain modes are applied to the nanotubes in the crystal by the imposition of specific deformation fields. The G_{23} shear modulus for a hexagonally packed nanotube array is calculated using a pbc-unit cell subjected to a pure shear strain mode as shown in Fig. 23.

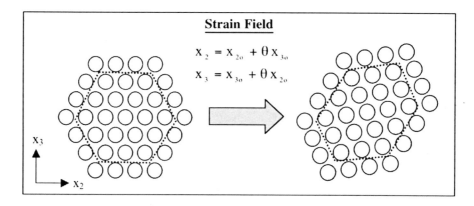

Fig. 23. Imposed shear deformation on hexagonally packed nanotube.

The magnitude of the shearing strain is given by twice the shearing angle or $\gamma_{23} = 2\theta$. During a progressive deformation with increasing θ, the potential energy is computed by summing all atom-pair interactions between adjacent nanotubes. The G_{23} shear modulus is then obtained from the elastic strain energy, U_o, using a finite difference approximation as

$$G_{23} = \frac{\partial^2 U_o}{\partial \gamma_{23}^2} = 2\frac{U_{o,i+1} - 2U_{o,i} + U_{o,i-1}}{\left(\gamma_{23,i+1} - \gamma_{23,i-1}\right)} \qquad (27)$$

The bulk modulus is computed by applying a dilatational strain as shown in Fig. 24. Because the strain in the axial dimension, ε_{11}, is assumed to be zero, the dilation is defined as $e = \varepsilon_{22} + \varepsilon_{33}$ with $\varepsilon_{22} = \varepsilon_{33} = \varepsilon$. The plane strain bulk modulus is obtained by applying Eq. (26) with equal biaxial strain components.

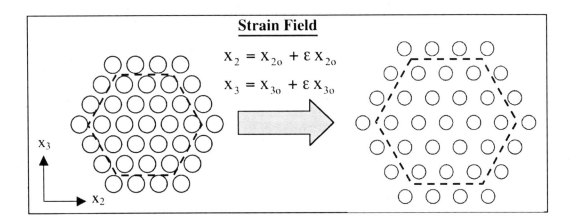

Fig. 24. Imposed dilatational strain on hexagonally packed nanotube.

Array Transverse Property Observations

A comparison between predicted elastic moduli using the current direct method and results obtained using alternate approaches is presented in Table 7. Popov and Van Doren (2000) and Lu (1997) utilize a lattice dynamics approach while Tersoff et al. (1994) is based on a molecular dynamic simulation. From the limited published results it is clear that there is a wide variation in predicted elastic moduli for nanotube bundles. All the results listed in Table 7 for the present analysis were computed independently, none were derived from a subset of other values.

Table 7 Comparison of predicted elastic constants.

Elastic Constant	Direct Method	Popov	Lu [**]	Tersoff
Bulk Modulus K_{23} (GPa)	45.8	42.0	18.0	33.6
Shear Modulus G_{23} (GPa)	22.5	5.3[*]	--	--
Young's Modulus E_{22} (GPa)	60.3	17.0	--	--
Normal Stiffness C_{22} (GPa)	68.3	42.0	78.0	--
Poisson Ratio υ_{23}	0.34	0.75	--	--

[*] Value derived using relationship in Eq. (1).

[**] Results generated using (7,7) chiral nanotubes with diameter = 0.94 nm.

In the direct summation and lattice dynamic methods, the physics of cohesion are identically represented by the same parameterization of the Lennard-Jones potential. Possible differences in predictions using lattice dynamics may be due to the inherent integral averaging of force constants used in the lattice dynamical matrix and the a priori selection of interacting nearest-neighbor atoms used in defining the primitive lattice cell, both of which are avoided in the direct method.

A potential source of inaccuracy affecting all methods is the form of the Lennard-Jones potential function itself. The Lennard-Jones potential was originally developed for noble gases and is known to produce poor results in other applications including graphite. It gives good results for the C_{33} modulus (interplanar separation), but the C_{44} parallel plane shear modulus is under predicted by an order of magnitude (Green and Bolland, 1974). Alternative potentials have been proposed (Pipes and Hubert, 2002) that yield accurate predictions for both the transverse normal and shear moduli in graphite. Additional study is warranted to assess the spatial interactions of delocalized bonds in carbon nanotubes that may be underestimated using a spherical Lennard-Jones model.

In an effort to assess the relative magnitude and character of the van der Waals cohesion exhibited in the nanotube crystal, it is instructive to examine a comparison between the transverse elastic properties of the nanotube crystal and a composite consisting of high strength carbon fibers surrounded by an epoxy matrix at a 60 percent fiber volume fraction. The results presented in Table 8 show that the effective elastic properties of the nanotube crystal in this work exceed those of the carbon fiber composite by a factor of four or more and are therefore indicative of an extremely well "bonded" assembly.

Table 8 Comparison of cohesion in a nanotube crystal and a carbon fiber composite.

Elastic Property, GPa	(12,0) Nanotube Crystal	Carbon Fiber Composite
C_{11}	651	158
C_{22}, C_{33}	68.3	15.3
ν_{23}	0.34	0.46
G_{23}	22.5	3.20
K_{23}	45.8	11.2

References

Allen, M.P., and Tildesley, D.J., 1987. *Computer Simulation of Liquids*. Claredon Press.

Arepalli, S., Nikolaev, P., and Holmes, W., 2001. Production and Measurements of Individual Single-Wall Nanotubes and Small Ropes of Carbon. *Applied Physics Letters,* **78**, (10).

Chesnokov, S.A., Nalimova, V.A., Rinzler, A.G., Smalley, R.E., and Fischer, J.E., 1999. Mechanical Energy Storage in Carbon Nanotube Springs. *Physical Review Letters*, **82**, 343-346.

Collins, P.G., and Avouris, P., 2000. Nanotubes for Electronics. *Scientific American*, 62-69.

Cornwell, C.F., and Wille, L.T., 1997. Elastic Properties of Single-Walled Nanotubes in Compression. *Solid State communications*, **101**, (8), 555-558.

Dresselhaus, M.S., Dresselhaus, G., and Saito, R., 1995. Physics of Carbon Nanotubes. *Carbon*, **33**, (7), 883-891.

Fan, C.F., and Hsu, S.L., 1992. Application of the Molecular Simulation Technique to Characterize the Structure and Properties of an Aromatic Polysulfone System. 2 Mechanical and Thermal Properties. *Macromolecules*, **25**, 265-270.

Frankland, S.J.V., Caglar, A., Brenner, D.W., and Griebel, M., 2002. Molecular Simulation of the Influence of Chemical Cross-Links on the Shear Strength of Carbon Nanotube-Polymer Interfaces. *J. Phys. Chem. B*, **106**, 3046-3048

Gao, G., Cagin, T., and Goddard, W.A., 1998. Energetics, Structure, Mechanical and Vibrational Properties of Single-Walled Carbon Nanotubes. *Nanotechnology*, **9**, 184-191

Girifalco, L.A., Hodak, M., and Lee, R.S., 2000. Carbon Nanotubes, Buckyballs, Ropes and a Universal Graphitic Potential. *Physical Review B*, **62**, (19), 13 104-13 110.

Green, J.F., Bolland, T.K., and Bolland, J.W., 1974. Lennard-Jones Interactions for Hexagonal Layered Crystals. *J. Chem. Phys.*, **61**, 1637-1646.

Hamada, N., Sawada, S.I., and Oshiyama, A., 1992. New One-Dimensional Conductors: Graphitic Microtubules. *Phys. Rev. Lett.*, **68**, 1579-1581.

Hernandez, E., Goze, C., Bernier, P., and Rubio, A., 1998. Elastic Properties of C and $B_xC_yN_z$ Composite Nanotubes. *Physics Review Letters*, **80**, 4502-4505.

Hyer, M.W., 1998. *Stress Analysis of Fiber Reinforced Composite Materials*. McGraw-Hill, 1998.

Iijima, S., 1991. Helical Microtubules of Graphitic Carbon. *Nature*. **354**, 56.

Krishnan, A., Dujardin, E., Ebbesen, T.W., Yianilos, P.N., and Treacy, M.M.J., 1998. Young's Modulus of Single Walled Nanotubes. *Physical Review B*, **58**, (20), 14 013- 14 019.

Lourie, O., and Wagner, H.D., 1998. Evaluation of Young's Modulus of Carbon nanotubes by Micro-Raman Spectroscopy. *Journal of Materials Science*, **13**, 2418-2422.

Lu, J.P., 1997. Elastic Properties of Single and Multilayered Nanotubes. *J. Phys. Chem. Solids,* **58**, 1649-1652.

Majumdar, A., 2001. Not Without Engineering. *Mechanical Engineering*, 46-49

Moore, W. J., 1972. *Physical Chemistry, 4th ed.* Prentice Hall Englewood Cliffs, New Jersey, 913-915.

Popov, V.N., Van Doren, V.E., and Balkanski, M., 2000. Elastic Properties of Crystals of Carbon Nanotubes. *Solid State Communications*, **114**, 395-399.

Popov, V.N., and Van Doren, V.E., 2000. Elastic Properties of Single-Walled Carbon Nanotubes. *Physical Review B*, **61**, (4), 3078-3084.

Pipes, R.B., and Hubert, P., 2002. Scale Effects in Carbon nanostructures: Self-Similar Analysis. Proceedings of the MESO 2002 Conference, Aalbourg, Denmark.

Pipes, R.B., and Hubert, P., 2002. Helical Carbon Nanotube Arrays: Mechanical Properties. *Comp. Science and Tech.*, **62**, 419-428.

Ruoff, R.S., and Lorents, 1995. Mechanical and Thermal Properties of Carbon Nanotubes. *Carbon.* **33**, (7), 925-930.

Saether, E., Frankland, S.J.V., and Pipes, R.B., 2002. Nanostructured Composites: Effective Mechanical Property Determination of Nanotube Bundles. Proceedings of the 43rd AIAA/ASME/ASCE/AHS/ASC Structures, *Structural Dynamics and Materials Conference*, Denver.

Salvetat, J.-P., Briggs, G.A.D., Bonard, J.-M., Bacsa, R.R., Kulik, A., Stockli, Burnham, N.A., and Forro, L., 1999. Elastic and Shear Moduli of Single-Walled Carbon Nanotube Ropes. *Physical Review Letters*, **82**, (5), 944-947.

Salvetat-Delmotte, J.-P., and Rubio, A., 2002. Mechanical Properties of Carbon Nanotubes: A Fiber Digest for Beginners. *Carbon*, in press.

Sanchez-Portal, D., Artacho, E.,Soler, J.M., Rubio, A., and Ordejon, P., 1999. Ab Initio Structural, Elastic and Vibrational Properties of Carbon Nanotubes. *Physical Review B*, **59**, 12678-12688.

Tersoff, J., and Ruoff, R.S., 1994. Structural Properties of a Carbon-Nanotube Crystal. *Phys. Rev. Lett.*, **73**, 676-679.

Theodorou, D.N., and Suter, U.W., 1986. Atomistic Modeling of Mechanical Properties of Polymeric Glasses. *Macromolecules*, **19**, 139-154.

Thess, A., Lee, R., Nikolaev, P., Dai, H., Petit, R.J., Robert, J., Xu, C., Lee, Y.H., Kim, S.G., Rinzler, A., Colbert, D.T., Scuseria, G.E., Toma'nek, Fisher, J., and Smalley, R.E., 1996. Crystalline Ropes of Metallic Carbon Nanotubes. *Science,* **273**, 483-87.

Yakobson, B.I., Brabec, C.J., and Bernholc, J., 1996. Nanomechanics of Carbon Tubes: Instabilities beyond Linear Response. *Physical Review Letters*, **76**, (14), 2511.

Contributors

LEON M. KEER is Walter P. Murphy Professor in the Department of Mechanical Engineering at Northwestern University. Dr. Keer, who earned his Ph.D. at the University of Minnesota, is a member of the National Academy of Engineering. Other honors and awards include Distinguished Speaker in the Prof. Chau Wai-yin Memorial Lectures on Science & Sciences Education at Hong Kong Polytechnic University, ASME Life Fellow, and Honorary Scientific Advisory Board Member of the journal *Mechanics of Materials*.

JAN D. ACHENBACH is Walter P. Murphy Professor in the Department of Civil and Environmental Engineering and Distinguished McCormick School Professor at Northwestern University. He is also Director of Northwestern's Center for Quality Engineering and Failure Prevention. Dr. Achenbach, who holds a Ph.D. from Stanford, is a member of the National Academy of Engineering and the National Academy of Sciences. He received the 2003 National Medal of Technology, the nation's highest honor for technology accomplishments. Dr. Achenbach is also a Fellow of the American Academy of Arts and Sciences, a Corresponding Member of the Royal Dutch Academy of Arts and Sciences, and recipient of the ASME Timoshenko Medal in 1992.

ALAN NEEDLEMAN is Florence Pirce Grant University Professor and Professor of Engineering at Brown University. Needleman, who holds a Ph.D. from Harvard University, is a Member of the National Academy of Engineering, a Fellow of the American Society of Mechanical Engineers, a Fellow of the American Academy of Mechanics, an Honorary Member of MECAMAT (Groupe Français de Mecanique des Matériaux), and a Foreign Member of the Danish Center for Applied Mathematics and Mechanics. He has been recognized by ISI (Science Citation Index) as a highly cited author in both Engineering and in Materials Science. In 1994, Dr. Needleman's work on 3D modeling of metallic fracture was a finalist in the Science Category for the Computerworld-Smithsonian Award.

ISAAC E. ELISHAKOFF is J. M. Rubin Foundation Distinguished Professor of Structural Reliability, Safety and Security in the Department of Mechanical Engineering at Florida Atlantic University. He earned his Ph.D. at the Moscow Power Engineering Institute. Dr. Elishakoff is a recipient of the Bathsheva de Rothschild prize, as well as Fellowships from the German Academic Exchange Office, and the National Technical Foundation of the Netherlands. In 1991, Dr. Elishakoff was elected a Fellow of American Academy of Mechanics, for outstanding contributions to random vibrations of structures and pioneering contributions to uncertainty modeling. He has published over 300 papers in leading national and international journals and conference proceedings.

ALLAN M. ZAREMBSKI is President of Zeta-Tech Associates in Cherry Hill, New Jersey. He earned his Ph.D. at Princeton University. Dr. Zarembski is a member of the National Academy of Sciences and the National Materials Advisory Board Committee on Nondestructive Testing of Longitudinal Force in Rails. He is the author of some 200 papers and articles on railroad track analysis and behavior and railway operations and maintenance, as well as a book, *Tracking R&D: Research and Development*.

JOHN P. DEMPSEY is Professor in the Department of Civil and Environmental Engineering at Clarkson University. He earned his Ph.D. at the University of Auckland. A professional engineer in the State of New York, Dempsey is a Fellow of the American Academy of Mechanics, the American Society of Civil Engineers, and American Society of Mechanical Engineers. In 1993, he received the ASCE Walter L. Huber Civil Engineering Research Prize. Dr. Dempsey's research focuses on the effects of scale on the strength and fracture properties of structural materials, the analytical and experimental mechanics of materials behavior, the fracture mechanics of quasi-brittle materials, contact mechanics, composites and new materials, and elasticity.

ZDENĚK P. BAŽANT is McCormick School Professor and Walter P. Murphy Professor in the Department of Civil and Environmental Engineering at Northwestern University. A member of the National Academy of Engineering and the National Academy of Sciences, he has won a number of awards, including the Prager Medal from the Society of Engineering Sciences, the von Karman Medal, the Newmark Medal, the Lifetime Achievement Award, the Croes Medal, the Huber Prize and the TY Lin Award from ASCE, the Warner Medal from ASME, and the Humboldt Prize. Dr. Bažant has also been designated a Highly Cited Scientist in Engineering by the Institute of Scientific Information.

ANASTASIOS M. IOANNIDES is Associate Professor in the Department of Civil and Environmental Engineering at the University of Cincinnati. He earned his Ph.D. at the University of Illinois at Urbana-Champaign. Dr. Ioannides teaches courses in geotechnical and transportation facilities engineering. His research interests include response modeling for rigid and flexible pavements; pavement behavior and performance data interpretation; applications of dimensional analysis; development of mechanistic design procedures for pavements; fracture mechanics; finite element analysis; supercomputing applications; and neural networks.

JACK R. VINSON is the H. Fletcher Brown Professor of Mechanical and Aerospace Engineering at the University of Delaware, with joint appointments to the Center for Composite Materials and the College of Marine Studies. He earned his Ph.D. at the University of Pennsylvania. In addition to being the Founding Director of UD's Center for Composite Materials, Dr. Vinson is a dedicated teacher, a prominent researcher in structural mechanics and composite materials, an author/co-author of seven popular textbooks, a highly active contributor to several professional societies, and an inspiring mentor to graduate and undergraduate students.

IVAR STAKGOLD is Emeritus Professor and former Chair of the Department of Mathematical Sciences at the University of Delaware. He earned his Ph.D. at Harvard University. His research interests include nonlinear partial differential equations, reaction-diffusion, and bifurcation theory. Dr. Stakgold is the former President of the Society of Industrial and Applied Mathematics and the author of a seminal text, *Green's Functions and Boundary Value Problems*. He is also the honoree of a book in the field—*Nonlinear Problems in Applied Mathematics: In Honor of Ivar Stakgold on His 70th Birthday*.

JOHN DUNDURS is Professor Emeritus in the Department of Civil and Environmental Engineering at Northwestern University. His work focuses on the theory of elasticity and its applications to stress analysis, materials, fracture, composites, and tribology. He has been the recipient of a number of awards and honors, including the Theodore Von Karman Award in 1990, the Nowinski Lectureship in 1994, and an American Academy of Mechanics Outstanding Service Award in 2003. A special issue of the *International Journal of Solids and Structures* was published in his honor in 1995. Dr. Dundurs holds a Ph.D. from Northwestern University.

R. BYRON PIPES joined the faculty at Purdue University as the John L. Bray Distinguished Professor of Engineering in 2004. Pipes was elected to the National Academy of Engineering in 1987 in recognition of his development of an exemplary model for relationships between corporate, academic and government sectors to foster research and education in the field of composite materials. As co-founder and director of the Center for Composite Materials at the University of Delaware, he developed an industrial consortium of over forty corporate sponsors from countries throughout the world. His most recent research programs focus on the application of nanotechnology to engineering disciplines including aerospace, composite materials and polymer science and engineering. Dr. Pipes also has active programs in the study of the advanced manufacturing science for composite materials.

Index